Introduction to the
Anisotropic
Geometrodynamics

K&E Series on Knots and Everything — Vol. 47

Introduction to the
Anisotropic
Geometrodynamics

Sergey Siparov
State University of Civil Aviation, Russia

 **World Scientific**

NEW JERSEY · LONDON · SINGAPORE · BEIJING · SHANGHAI · HONG KONG · TAIPEI · CHENNAI

Published by

World Scientific Publishing Co. Pte. Ltd.
5 Toh Tuck Link, Singapore 596224
USA office: 27 Warren Street, Suite 401-402, Hackensack, NJ 07601
UK office: 57 Shelton Street, Covent Garden, London WC2H 9HE

British Library Cataloguing-in-Publication Data
A catalogue record for this book is available from the British Library.

INTRODUCTION TO THE ANISOTROPIC GEOMETRODYNAMICS
Series on Knots and Everything — Vol. 47

Copyright © 2012 by World Scientific Publishing Co. Pte. Ltd.

ISBN-13 978-981-4340-83-0
ISBN-10 981-4340-83-9

Printed in Singapore.

Preface

Ptolemy was probably the first who designed a world machine that corresponded to what everyone saw and was based on some inherent ideas, but at the same time the mechanism was supplied with details that needed long and skillful observations. There were stars, Sun, Moon and planets that were obviously rotating around the Earth – along the circular orbits belonging to the heavenly spheres, but it turned out that from time to time the planets performed retrograde motions, and Ptolemy set inside various tilted epicycles.

For some time it worked smoothly, but then the bug started to manifest itself in such a way that the errors were accumulating with time. For Copernicus it became clear that the drawback was general. Still, the obvious must have been preserved, and its presentation should have been based on some inherent ideas, while the rest of the device should have been redesigned. The components were reshuffled: Sun took the central position, and the Earth, planets and stars started to rotate around it along circular orbits. It looked good but worked even worse than before, until Kepler noticed and revealed the new details – those planetary orbits were elliptical. This was implemented, and since then the machine seemed to work perfectly. But then the new engineers became curious, why it was so.

Galileo paid attention to the surrounding conditions and proclaimed that the working principles one might be looking for should be independent of place, time and straight uniform motion. These notions required a language with the help of which the strict definitions, demands and instructions could be formulated. And Newton performed this linguistic task and, in parallel, recognized that the independence on straight uniform motion could be achieved only if the reason for the motion is proportional not to the motion itself but to its change. He gave the short definition that became the foundation. Turning back to the world machine in possession of new knowledge, he discovered the law on the base of which it was functioning. The discouraging news was that there was a distant force acting between Sun and planets (stars were too far), but nothing else than to remain logically consistent could be done, and Newton had to attribute a universal character to this force.

The next magnificent linguistic achievement was made when they thought of the materials used for the device. One short phrase consisting of four parts pronounced by Maxwell not only perfectly displayed all the huge variety of electromagnetic phenomena, but, unexpectedly, revealed the deep relationship between electromagnetism and optics – that is explained the way we *see*. And then the thunderbolt came. The light did not obey the velocity addition rule following from the main principle of independence proclaimed by Galileo and known as "relativity principle".

Einstein accepted the blow and stood up with the realization that we see what we see – any time, any place and any straight uniform motion, and that the guesses should be dismissed however clever they are. When we see, it takes time for light to pass a path in space, therefore, the obtained images correspond not to the current positions of objects but to the abandoned positions, and if we want to know how it works, we should focus not on the objects themselves but on the processes involving their motions and light travel times. This approach preserved the Maxwell's achievement and produced the new one: the account for the light travel time added a new dimension to the world picture, and now the motions became not the trajectories on the flat plane governed by the distant forces, but the curved trajectories nested on the relief. The heavenly (and other) bodies produced rather potentials than forces, and these potentials produced the relief. This caused the problem with measuring details of the machine with the help of triangles: the triangle plane did not fit the curved surface. But already Newton was keen enough in regarding infinitely small parts of everything, since then they learned the ways, and Einstein made geometry indistinguishable from dynamics.

The residuary trouble is distances. Currently, we can neither penetrate into the regions of matter where electrodynamics is not enough nor reach the distant parts of the observable Universe. But the signals come from both regions, and we have some nice patterns for their interpretation. The golden dream is to unify the patterns, and the Einstein's rule "we see what we see" seems to be the most appropriate base for that.

This book is a single step in this direction. The need for the large amount of "dark notions" that appeared in astronomy in the last decades means that the machine seems to start coughing again. The introduction of those performed in order to make the mechanism fit the drawn specification makes one suspicious. It comes to mind to mimic Einstein's approach, change the point of view and try another model for the background of everything we could observe and register, i.e. for the space and time. A simple analogue of such possible model is known in general physics, it is the phase space. It cannot be perceived by our senses immediately, but it is not hard to get used to it. Such transfer resembles the pass from the Lagrange approach to the Hamilton one, and any physicist is aware of both. The results of the suggested description led to the simplification of the current world picture: there is no need for the dark matter and, possibly, no need for the dark energy in the explanation of the paradoxes in the known observations on the cosmological scale. At the same time all the achievements of the classical GRT on the planetary scale are preserved. In order to suggest a specific test for this modified theory, we need an effect on galactic scale, and it seems that the effect of optic-metrical parametric resonance will do. The effect is not widely known, the first promising observations were performed only recently, and the explanation of its theory requires some technical details. But the result is simple: we can notice the effect of the gravitational radiation of the periodical sources like short-period binary stars with the help of regular radio telescopes, because the spectra of some space masers obtain specific periodic components. A couple of the suitable radio sources has already been found and their signals studied. When the data statistics of such sources is sufficient, it would be possible to investigate the geometry of our galaxy and compare the results with the suggested approach.

Both special and general relativity theories are simple in their deep essence as it usually is with the fundamentals, they are really physical, despite the mathematical entourage that one has to overcome in order to understand them. Being a "land of possibilities" mathematics is very fascinating and meaningful especially when it comes to interpretations. That is why it could sometimes tempt one to go slightly too far and enjoy mirages - images that are surely related to certain physical properties and that are also "seen by everyone" - rather than other wonders of the real world that could appear none the less charming when it comes to the direct testing. Aren't these "other wonders" also mirages but on the deeper level, will be hardly ever known. But this book is a humble attempt to bring relativity closer to the general physics to which it belongs and to focus on the short and simple ideas discovered by Newton, Maxwell and Einstein.

It's an honor and pleasure to mention those whose efforts made it possible for this book to appear. The idea to write it was suggested by Peter Rowlands and supported by Louis Kauffman. Nicoleta Voicu and Vladimir Samodurov were my collaborators in obtaining some of the results. The essentials and general ideas were many times discussed with Sergey Kokarev, Michail Babich, Alexander Kazakov and Nickolay Razumovsky and also with colleagues from RI HSGP and its head (and soul) Dmitry Pavlov who does not share my views but supports my activity. The colleagues from the University of civil aviation (Department of Physics headed by Valery Arbuzov) took from me a part of my current job in spring semester. I am indebted to Edwin Taylor who long time ago asked me for the comments to the manuscript of his wonderful book 'Scouting the Black Holes' (written together with J.A.Wheeler), this book triggered the vague interest I had in this field before. Julia Hlynina helped me with the search in the electronic astronomical catalogues. And I would definitely be unable to cope with the editing problems in time without the help of Elena Ruchkova and especially of my elder kid George Laptev. I am really grateful to all these people.

Some of the results presented in the text were obtained during the work supported by the RFBR grants No. 07-01-91681-RA_a and No.08-02-01179-a, and also by the grant HC No.149.

April 2011
Sergey Siparov

Contents

Preface   v

1. Classical relativity: Scope and beyond   1

   1.1   Physics and mathematics: Long joint journey   1
       1.1.1   From Pythagoras to Kepler and Newton   1
       1.1.2   Curvature, forces and fields   3
   1.2   Inertial motion, relativity, special relativity   8
       1.2.1   Bradley experiment   10
       1.2.2   Michelson and Morley experiment   12
       1.2.3   Lorentz contraction   18
       1.2.4   Special relativity   21
   1.3   Space-time as a model of the physical world   29
   1.4   Generalized theory of relativity and gravitation   40
       1.4.1   Tensors: who and why   40
       1.4.2   Maxwell identities   43
       1.4.3   Least action principle   48
       1.4.4   Mass and energy   51
       1.4.5   Field equations   53
       1.4.6   Gravitational waves   56
   1.5   GRT - first approximation - predictions and tests   59
       1.5.1   Newton gravity   59
       1.5.2   Classical tests   62
       1.5.3   Gravitational lenses in GRT   66
   1.6   Exact solutions   68
       1.6.1   Star: static spherically symmetric case   68
       1.6.2   Universe: cosmological constant and expansion . .   71
   1.7   Observations on the cosmological scale   75

1.7.1	Rotation curves and their interpretation	77
1.7.2	Break of linearity in Hubble law	83

2. Phase space-time as a model of physical reality   **85**

2.1	Preliminary considerations		85
	2.1.1	Scales	85
	2.1.2	Boundaries	89
	2.1.3	Newton and Minkowski models for the intuitive space and time	95
2.2	Interpretation dilemma, variation principle, equivalence principle		101
	2.2.1	Dilemma: new entity or new equations	102
	2.2.2	Comparison of methods	106
	2.2.3	On the variation principle	109
	2.2.4	On the equivalence principle	113
2.3	Construction of the formalism		117
	2.3.1	Space and metric	117
	2.3.2	Generalized geodesics	120
	2.3.3	Anisotropic potential	122
	2.3.4	Field equations	124
	2.3.5	Back to Einstein method	130
2.4	Gravitation force in anisotropic geometrodynamics		132
2.5	Model of the gravitation source and its applications . . .		140
	2.5.1	Center plus current model	140
	2.5.2	Flat rotation curves of spiral galaxies	143
	2.5.3	Tully-Fisher and Faber-Jackson relations	145
	2.5.4	Logarithmic potential in spiral galaxies	147
	2.5.5	Classical tests on the galaxy scale	148
	2.5.6	Gravitational lenses in AGD	156
	2.5.7	Pioneer anomaly	159
2.6	Electrodynamics in anisotropic space		160
	2.6.1	Weak deformation of locally Minkowski metrics .	162
	2.6.2	Lorentz force	163
	2.6.3	New term - "electromagnetic" vs. "metric"	166
	2.6.4	Currents in anisotropic spaces	168
2.7	Approaching phase space-time		171
	2.7.1	Coordinate-free dynamics	171
	2.7.2	Generalized Lorentz transformations	172
	2.7.3	Geometry, groups and their contractions	175

2.8 Cosmological picture 179

3. Optic-metrical parametric resonance - to the testing of
the anisotropic geometrodynamics   185

3.1 Gravitation waves detection and the general idea of optic-
metrical parametric resonance 185
3.1.1 Space maser as a remote detector of gravitation
waves 193
3.1.2 Atomic levels 196
3.1.3 Eikonal 199
3.1.4 Motion of a particle 202
3.2 OMPR in space maser 205
3.3 Astrophysical systems 214
3.3.1 GW sources 215
3.3.2 Space masers 217
3.3.3 Distances 219
3.3.4 Examples 222
3.4 Observations and interpretations 225
3.4.1 Radio sources observation methods 225
3.4.2 Ultra-rapid variability and signal processing . . . 228
3.4.3 Search for the periodic components in space maser
signals 231
3.5 On the search for the space-time anisotropy in Milky Way
observations 237
3.5.1 Mathematical formalism and basic equations . . . 238
3.5.2 Weak anisotropic perturbation of the flat
Minkowski metric 242
3.5.3 Modification of the OMPR conditions 244
3.5.4 Investigations of the space-time properties 246

Appendix A Optic-mechanical parametric resonance   249

A.1 Brief review 249
A.1.1 Mono-chromatic excitation 249
A.1.2 Poly-chromatic excitation 250
A.1.3 Bichromatic fields 251
A.1.4 Mechanical action on atoms 252
A.2 Force acting on a two-level atom 255
A.2.1 Dynamics of an atom in the bichromatic field . . 256

	A.2.2	Stationary dynamics of atom	258
	A.2.3	Light action on an atom	262
	A.2.4	Groups of atoms in optical-mechanical parametric resonance	264
	A.2.5	Velocity change due to the force action	265
	A.2.6	Main result	266
A.3		Probe wave absorption	267
	A.3.1	Problem and its solution	268
	A.3.2	Assumptions and demands	278
A.4		Fluorescence	280
	A.4.1	Calculation of the fluorescence spectrum	284
	A.4.2	Driving the TLAs by the bichromatic radiation	288
	A.4.3	Assumptions, demands and possible applications	290
	A.4.4	Conclusion	292

Bibliography   295

Index   303

Chapter 1

Classical relativity: Scope and beyond

1.1 Physics and mathematics: Long joint journey

1.1.1 *From Pythagoras to Kepler and Newton*

Two basic concepts of modern science – quantum mechanics and relativity – are formalized very deeply. And this depth is not technical, that is, just mathematically complicated, but it is ideological, philosophical. Not paying attention to this circumstance, it is impossible to go beyond the specific objectives, however important they may be; it is impossible to suppose what could be a fundamental error if suddenly the whole class of paradoxical problems appears, or a whole series of observations contradicts the existing theory. And prominent scientists always paid attention to this circumstance. The strangeness of some mathematical ideas and notions broke through into the world of observations and measurements, and the results of the last forced the researchers to build hitherto unknown speculative constructions.

If we assume Pythagoras to be one of the forefathers of science, we can recall how having started with a simple "Egyptian" triangle, he proceeded to cut in half the square along its diagonal, and found out that the measurement of the diagonal has no numerical value in the usual sense of the word. This discovery shocked him so much that for the disclosure of this secret, for removing it outside the circle of associated "Pythagoreans" a person was subjected to the death penalty. "All is number" – Pythagoras believed, and this statement had profound meaning. And even though we live in a perceptible world, there is always some abstract mathematical object in the very basis of our understanding of the universe, for example, a number or a sphere – the most perfect geometrical figure.

The Heavenly sky above the Earth also appeared to have spheres – each planet orbiting the Earth had its own one, and the biggest, all-encompassing sphere (by the way, the seventh one – "seventh sky" – since then there were only four known planets and also the Sun and the Moon) was the sphere with the stars that didn't change their arrangement with time and comprised the constellations. This is the absolute, this is Harmony, and no observed effect should shake the ideal mathematical figure – the locus of points equidistant from one. But how is it possible to explain the observed retrograde motion of planets moving among the unchanging patterns of constellations? Those spheres cannot rotate non-uniformly! And in addition to the "differents" – the circles in which the planets move around the Earth, Ptolemy introduces "epicycles" – small circles in which the planets move around the points circling along the "differents". In order to fit more precise measurements, all the planes containing "differents" and "epicycles" had to be tilted at various angles. The calculation of the planets positions became a furious exercise, but, nevertheless, the most persistent researchers were able to find the result and reasonably predict the planets positions. Only in a few centuries there appeared a systematic error gradually increasing with time.

However, when Copernicus twisted inside out the geocentric Ptolemaic system, placed the Sun into the center, and let the Earth move in a circular orbit around it like other planets did (and then Galileo let it rotate around its own axis), the calculation of planetary positions – greatly simplified and not demanding epicycles – gave even worse results than a complex Ptolemy's model. Therefore, the Copernican Revolution at first step was an ingenious hypothesis, which caused an encouraging smile, but could be hardly considered relevant to reality.

In his turn, Kepler, who analyzed the results of the 20-year long measurements of the planets performed by Tycho Brahe and himself, was also confident of world harmony based on mathematics. He hoped that those circles in which the planets were likely to move around the Sun had the radii corresponding to the concentric spheres, into which the regular Platonic polyhedral were subsequently inscribed. And when he discovered that it was not so, moreover, the planets appeared to move not in circles but in ellipses, he probably survived the culture shock. But soon the ellipses measured by Kepler brilliantly came to calculations based on Copernicus hypothesis, eliminated the existing errors, and the world finally became heliocentric. The law relating the radii of the orbits and orbital period of planets discovered by Kepler as a result of observations and measurements, later led Newton to the law of gravitation in the form we know it now.

Newton gave a start to a new era in the science development. He refused to follow the *a priori* mathematical statements based on a naive desire for harmony, made his motto "Hypotheses non fingo" and firmly established induction as the main scientific method. So, when Kepler's measurements and law which Newton used when determining the ratios of planets accelerations, led him to the idea of the long-range attraction between the planets and the Sun (the force of attraction followed the inverse square law), Newton postponed the publication of his result for two decades, because such a "hypothesis" seemed inappropriate. In his letter to Berkeley, Newton wrote "The idea that the ability to excite gravity could be an inherent, intrinsic property of matter, and that one body can affect another through a vacuum at a distance and without something that could transfer the force from one to the other, this idea seems to me so preposterous that, I think, it could hardly come to mind to any man able to think philosophically". And when, reluctantly, he let the public become familiar with his achievements, he had no reason to assume that this "effect through the void" is not transmitted instantaneously, and the void itself which is just a "container" where it all happens, is not described by Euclidean geometry.

Thus, until the 18th century, mathematics and science, alternating in the lead, participated in the development of increasingly accurate world picture which made it possible to predict and calculate the future events. In this picture, time was, in a sense, the fourth coordinate whose various values could not be measured using only the senses. An instrument was needed, and this instrument was mathematics. Like any other tool, it distorted the picture by the rules that were not always known.

1.1.2 *Curvature, forces and fields*

At the end of the 18th century, Gauss was the first to constructively approach the question of the applicability of Euclidean geometry to describe the world. He directly measured in situ the sum of internal angles of a triangle. The vertices of the triangle were at the tops of nearby mountains. Gauss did not find any deviation of his calculations from the Euclidean geometry within the accuracy of his measurements.

At the beginning of the 19th century, Lobachevsky assessed the principal possibilities of astronomical observations and found the conditions in which it becomes impossible to assert anything definite about the geometry describing the phenomena. This inevitably led him to the creation of the first non-Euclidean geometry [Lobachevsky (1979)], whose applicability to astronomical scales could not be ruled out.

In the second half of the 19th century, Riemann developed the differential geometry, the theory of surfaces, in which such notions as curvature (proportional to the second derivative of the displacement of the point along the curve in the dimensionless parameter) and metric associated with the scalar product were used.

In physics at that time the Coulomb force, the Lorentz and Ampere forces were studied. These, like gravity, were considered long-range instantly acting forces also following the inverse square law. Maxwell used a new approach to describe the electromagnetic phenomena; it corresponded to the transition from the Newtonian theory of long-range acting forces and the relevant dynamics equations to the field theory, i.e. to the short-range links between such local characteristics as stresses and potentials of forces in the neighbor points. This formalism allowed Maxwell to theoretically predict, and then Hertz to experimentally detect the electromagnetic waves. Faraday's experiments identified the nature of these waves with that of light waves.

Already in the mid-19th century, Clifford consistently defended his idea that "all the manifestations of the physical world are experimentally indistinguishable from the corresponding changes in the geometrical curvature of the world geometry" [Clifford (1979)]. The simplest example is a smart worm that can move within a closed one-dimensional channel, in which it lives. If the channel has constant curvature (has the form of a circle), the worm's senses do not report the worm of its motion. If the curvature of the channel varies, then the form of the worm changes in its motion, and its senses will experience the impact that it could interpret as a signal from the external world. Thus, it was found that the relationship of mathematics (geometry) and physics, though not of a transcendental nature, as it was before Newton, who refused to invent hypotheses, was on the one hand, mutual and close (impossibility to distinguish between the curvature changes and "physical" phenomena), while on the other hand – quite arbitrary, depending on the level of our knowledge at the moment. Indeed, Euclidean geometry is quite convenient to describe what is happening on Earth, but, as it was shown by Lobachevsky, at a significantly larger scale, not accessible to direct observation and investigation, there is yet no convincing reason for choosing it.

Following these ideas, Einstein in the early 20th century brought the total, but the qualitative Clifford's statement to approve a narrower, but quantifiable one. Based on the requirement of general covariance of physical laws (and also postulating the speed of light constant and the equivalence principle for the inertial and gravitational masses true), he developed a theory according to which the gravity which is unavoidable in our world, is indistinguishable from the manifestations of the geometric properties of space-time. Assuming that the space-time geometry corresponds to that suggested by Riemann and Minkowski, Einstein in his seminal paper gave examples of the observable effects that should occur in this case and calculated their results. The observations showed good agreement with the calculations, and the new geometry rightfully entered the physical theory. At the macro level, it makes possible, for example, to obtain more accurate results than those of Newton gravitation theory with regard to the Solar system. At the micro level, Dirac's theory appeared in quantum mechanics. At the mega level, the modern theory of the expanding Universe (subsequently) reformed the cosmological picture.

Speaking of the observational data that should be consistent with the theory, one should always indicate the scale of consideration. In astrophysics, there are three such scales: the Solar system (the neighborhood of a point mass modeling a star), the Galaxy (about 10^{11} point masses far away from each other) and meta galaxy (i.e. the entire observable Universe, consisting of billions of galaxies). The achievements of the General Relativity Theory (GRT) based on the introduction of the new (Riemannian) geometry allowed the associating of the laboratory physics and the first of the scales. But the possibility of its application to larger ones was not that obvious however desirable. By the end of the 20th century, the new astrophysical data were accumulated. In order to explain them on the base of the classical GRT ideas, there appeared rather mysterious new concepts, such as dark matter and dark energy, supposedly embodying 96% of the total mass of the observable Universe. Some scientists feel enthusiastic about that, thinking of the vast field of the non-cognized awaiting them on the familiar way. But others feel anxiety, and think that such situation indicates at the possible need for some revision of the foundations of the theory. As always, both groups are right.

The fundamental possibility to use the speculative mathematical structures for describing the world as perceived by our senses is not obvious and is a postulate. Each of these alternatives – the construction of structures and the description of the world – is self-sufficient. So, the mentioned possibility implies certain arbitrariness. On the one hand, the "inconceivable effectiveness of mathematics in natural science" is well known. On the other hand, the systematized observational data reveal the non-obvious areas of nontrivial mathematical research. These circumstances are both the motivation and the regulator for the activity known as Theoretical Physics. Sometimes, the equivalent description of the same phenomena can be obtained by several mathematical approaches. As a rule, the scientific community recognizes only the first of them; all others are cut off by the "Occam's razor." The direction of the search for the new theory is naturally determined by the problems found in the old one. No new theory – no matter how beautiful it is and no matter how successful in coping with the old theory problems it is – can replace the previous physical theory until the observations corresponding to the predicted specific results take place.

There are hundreds, if not thousands, of books on Special and General Relativity Theories. Some of them are monographs for specialists in the field, others are textbook for students of physical faculties - either for those who are going to study relativity and gravitation or for those who study the other fields of physics, other books are intended for general public. Every time one is going to write a book because of some new facts discovered, because of some new ideas appeared, because of the new ways of teaching invented, because of one's personal attitude to the subject, one usually has to select a target group and adapt the contents of the book for this group of readership.

This book contains some ideas and approaches that are not widely known, and it also gives the results of new observations, therefore, it could be of interest to specialists both in theoretical physics and in astronomy. On the other hand, the new ideas and the new interpretations of the known observations that it contains, demand not as much knowledge of the frontiers of the relativity mainstream, but rather the understanding the meaning of its foundations. The impact of Einstein, Minkowski and others revolution in thinking was so profound and inspiring, produced so many beautiful possibilities, that the reason why and what for it all appeared has been sometimes abandoned, and the seeming breakthroughs led to the neighbor passes in the labyrinth. Having recognized in cold mind the reasons for the suggestions made by the founders, one could try to adapt them to the new situation that we have now with our new data. Therefore, the book could also be of interest to the students in Physics who look for a field to choose. It turns out that these target groups cannot be separated from the point of view of the contents, and the first chapter has to contain the brief outline of what took place up to now with the stress on why this or that point of view was taken and accepted. Of course, it is not a textbook, and there is no reason to advocate the relativity theory once again, but the mentioning of some simple enough and known results cannot be escaped. And, of course, it is not a kind of review monograph accounting for the latest developments, many of which present brilliant examples both in theory and experiment, but the explanations and physical interpretations are sometimes discouraging. The approach presented here is thought of as rather an attempt to follow the style of thinking that makes it possible to explain the observations effectively and quantitatively but with minimal complications, in the spirit and not in the letter of the laws, using Occam razor every time it is possible. That is why a survey of the way to the current state given in the first chapter is needed.

The second chapter contains the approach which is called here the anisotropic geometrodynamics, and it is the development of the initial relativistic ideas for the cosmological scales. The word 'anisotropic' should not be misleading, we are speaking about space-time, and generally, it is not supposed to have the properties of a crystal though such theories also exist. And in usual meaning the space-time in question here also does not have them. The different values of the light speed in various directions is not discussed here either. But anisotropy could mean not only the dependence of regarded variables on some fixed directions, but the dependence of variables on the derivatives of the coordinates alongside with the dependence on coordinates themselves. Actually, such anisotropy - which is the dependence on velocities - was always present in theories of gravitation though in the higher terms of the approximations series. Do we have reasons to re-estimate their meaning and importance, and modify the theory correspondingly, is a question that should be analyzed and answered. The answer found in the approach developed here appeared to be fruitful in explaining some important observations of nowadays. Simultaneously, it suggests new interpretations to some observations known earlier, and they could be used either instead of or in parallel to the conventional ones. This can affect the general cosmological picture in a way that seems promising. The approach is not a complete theory yet, that's why the book is only the introduction to the anisotropic geometrodynamics. The combination of these two words and also the notion of "phase space-time" introduced in the book were already used in literature on relativity and had slightly different meanings than they have here, the problem of terminology seems inevitable, and these terms were used not intentionally but in order to use names speaking for themselves.

In order to validate the anisotropic geometrodynamics approach, not only by the interpretation of the known observations but also in the direct way, it would be good to test and measure in more detail the geometrical properties of our galaxy. The third chapter of the book contains the theory and first observational results of the effect that is called optical-metrical parametric resonance. It can be observed in the radiation of space masers with the help of the usual radio telescopes. It is shown that when certain astrophysical conditions are fulfilled, the radiation of space masers can be affected by the periodic gravitational waves emitted by the short-period binary star systems. The signal appears to be specific and it can't be mixed with other signals, besides, it is large enough due to the parametric resonance character of the effect. Therefore, the presence of the gravitational waves and some of their characteristics can be registered without special instrumentation but with the use of special processing of the observed signal. As it is, this effect could be used for developing a kind of gravitational wave astronomy, because it makes possible to locate and use as beacons the sources of the periodic gravitational waves in our galaxy. In future, when the amount of observational data is large enough, it could also be used for measuring the geometrical properties of our galaxy, Milky Way.

1.2   Inertial motion, relativity, special relativity

Inertial motion is a more appropriate term than straight uniform motion that is frequently used in physics, though this last type of motion does not exist in the surrounding world but only in the simplified formal problems. But since we can observe and measure it with sufficient accuracy, we can easily imagine that it can take place. As soon as we see a body moving straightforward with constant speed, we are ready to call this motion inertial, but strictly speaking, the velocity of the body will be permanently and inevitably changing, because of the medium surrounding it, be it an interstellar gas with the density 1 atom per cubic centimeter, and because of the fields (or space-time curvature) produced by the other bodies, be them far away or have them tiniest masses or electric charges. An observer who is just observing the moving body also produces this effect by his very existence. Besides, and this is the main point we are going to discuss now, even neglecting all these tiny circumstances, one should be very distinct in the question: relative to what frame this motion is straight and this speed is constant? When Galileo or Newton or any other experimentalist studies the laws of mechanics in laboratory, this question hardly arises, and one should be Galileo to pose it right there and give an answer like: if we are doing science, it does not matter and it should not matter relative to what a body moves inertially, that is, the laws of physics are and should be the same in any inertial reference frame (now we know it as the principle of relativity). And an inertial reference frame is such a frame that preserves its state of rest or straight uniform motion.

Is it improper to ask again, relative to what? The space itself is an intuitive notion as philosophy underlines, the same is true for the state of a body. Therefore, as soon as we come to the concrete descriptions, the conventional character of the fundamental notions should be admitted and accounted for.

In a laboratory or near the railway we always understand what does this principle of relativity mean. We admire the genius of Galileo and agree that dribbling the ball inside a railway car and on the platform look absolutely the same when observing it from inside the car and from the nearby on the platform correspondingly. That is, the laws of its motion are the same. And the visible change of the ball motion when an observer on the platform looks at the player inside the car must be described by some simple technique like adding the car velocity to the ball velocity in every point of the trajectory of the ball. What is at rest seems clear, but besides, it does not matter. Newton's law of inertia according to which a body preserves its state of rest or straight uniform motion, unless there is a force acting on it, also has this idealistic shade. What is a force? It is something that changes the state of motion. This definition is given by the second Newton's law or principal law of mechanics. What nature does a particular force have, is a next question. Besides, there could be lots of various forces that can act on a body simultaneously and compensate each other in such a way that we would be unable to register them here and now, and do this only by speculation - recollecting similar experiments that took place earlier. We can guess that an uncompensated force exists only while observing acceleration in the motion of a body. Newton suggested to use the distant motionless stars as an inertial reference frame, and the accuracy of this assumption was far beyond the accuracy of measurements he used while deriving the gravity law. It can hardly be so when we pass to the discussion of the galactic scale events, but this we postpone for the sections to follow.

Everything changes when we leave the Earth surface or perform such an experiment that it is not so clear who moves and how - of course, we speak here of experiments with light signals propagating through the ether - specific matter whose properties drastically differ from the properties of the ordinary matter we know, and that we failed to observe in any other way than presumably in experiments with light.

1.2.1 *Bradley experiment*

The first quantitative picture of the Universe was constructed by Ptolemy (90-168 AD). The Earth was motionless and occupied the center. The Sun and planets took part in two motions: around the Earth - following the so-called differents, i.e. various geocentric circular orbits, and, simultaneously, performing circular "dance" following epicycles - smaller circular orbits drawn around points that actually moved along differents. All these circles had various tilts in order to suffice observations. This machinery was placed inside the sphere on whose surface the stars were located, they magnificently revolved around the Earth making one turn per day.

It is interesting to evaluate the total energy of such system. Transforming this picture quite a bit and distributing the stars uniformly not on the spherical surface but throughout the volume of a sphere according to the distance measurements performed later, we can evaluate the energy as $E_U = \frac{Jw^2}{2} = \frac{1}{2}\frac{2}{5}M_U R_U^2 \frac{4\pi^2}{T_d^2}$, where $M_U \sim 10^{80}$ masses of proton, i.e. $M_U \sim 10^{51}$kg is the now estimated mass of the visible Universe, $R_U \sim 10^9$pc $\sim 10^{25}$ m is the radius of the visible Universe, $T_d = 86400$ s is the period of Earth rotation. Then the calculation of energy for Ptolemy Universe gives $E_U \sim 10^{91}$ Jouls.

The complicated construction of tilted orbits worked for a while but, gradually, the errors between the calculations and observations increased. Galileo stopped the stars' rotation and made them motionless making Earth rotate instead, but even before that Copernicus turned inside out the internal mechanism, and made the Earth move around the Sun with velocity, V, while the planets moved smoothly with their own velocities along their "differents", no "epicycle" dancing presumed. The machine became much less energetic: the Sun rotation around its axis (discovered later) gave $T_{Sol} = \frac{J_{Sol}w_{Sol}^2}{2} \sim 6 \cdot 10^{16} J$; planets circling around the Sun added about 2% of this value; the energy of their interaction with the Sun was of the same order due to virial theorem; motionless stars, $N \sim \frac{10^{80}m_p}{M_{Sol}} \sim 10^{20}$ in number, were too far and could add only their rotation energy similar to that of the Sun. Therefore, the energy of the Newton Universe was about $E_U \sim 10^{36}$ Jouls.

This picture was directly testable, and the experiment was performed by

Bradley in 1728, a year after Newton's death, providing the direct physical evidence for the long discussion on the Universe design.

The stars are so far that we can consider the light beams coming from them to be parallel. If the Earth moves around the Sun, then the stellar aberration must be observed, that is a telescope must be directed not exactly at a light emitting star, but must be tilted at an angle defined by the Earth speed. Performing the observations from the opposite points of the Earth orbit, one can measure the doubled aberration angle as shown in Fig. 1.1.

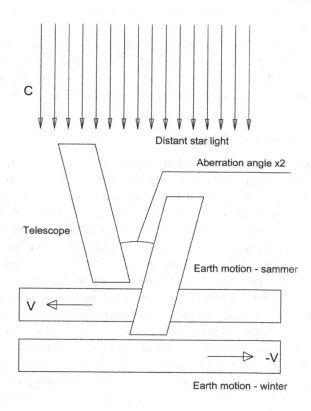

Fig. 1.1   Aberration of star light.

The measurements performed by Bradley gave a value of the aberration angle equal to

$$\beta = \arctan \frac{V}{c} \sim 20'',$$

his telescope being able to resolve $1''$ second of arc. What makes his observations even more valuable, is the fact that initially he could not understand why his readings for the same star in summer and in winter differ by 20 seconds of arc, and only after that found the theoretical explanation. It made him able to correct the value the speed of light estimated by RØmer in 1676. It should be underlined that, actually, the value of V in this formula is not a definite speed of the Earth measured relative to a star or to some other body, it appears only as a difference of the Earth velocities in summer and in winter.

The observation of stellar aberration did not demand any medium, and at that time the Universe was considered to be an empty space with celestial bodies in it, while light was a flow of particles traveling through space. Therefore, we can say that the Universe model was still Newtonian.

1.2.2   *Michelson and Morley experiment*

Ether became an important notion in physics when the light revealed its wave nature in experiments involving interference, diffraction and polarization. At first, they spoke of the luminiferous ether - an all-embracing medium in which light propagated with the speed of 300,000 km/s. And one could wonder what kind of medium it could be, since in an analogue to the acoustics phenomena, its density and rigidity providing such speed must be very high, while the Earth travels through such ether for billions of years with the speed of 30 km/s and does not tend to decelerate and fall on the Sun. The experiment performed earlier by Faraday proved that the light waves had the same nature as the electromagnetic waves, predicted by Maxwell, and ether became even more relevant. We have the waves in theory, and there are various wave effects observed in experiment, therefore, there undoubtedly must be a medium carrying these waves, and we call it ether and can try to study its properties however strange they are. At the same time, this ether could be used as an absolute reference frame - the one which would once and forever solve the problem of relative to what a body moves. The essential thing was to measure with high accuracy the speed of light propagating through ether. It was first done by Fizeau [Fizeau (1850)] as shown in Fig. 1.2. The cogwheel angular velocity was chosen such that the light that passed through the gap between the cogs was reflected by the mirror and then passed through the neighbor gap between the cogs. If the

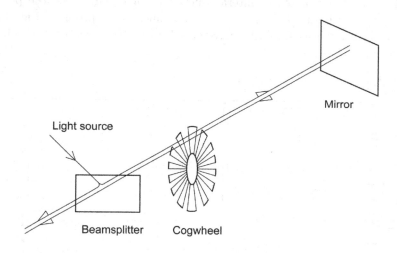

Fig. 1.2 Fizeau experiment.

setup does not move through ether, then we can use a clock to measure the time, t, needed for the wheel rotation by the angle corresponding to one cog and equate it with the time needed for the light to travel to and from the mirror. Then we get for the speed of light propagation through ether

$$t = \frac{2l}{c}$$

If the setup moves through ether in the direction of the light beam falling on the mirror, and its speed relative to ether is equal to V, then

$$t = \frac{l}{c+V} + \frac{c}{c-V} = \frac{2l}{c}\frac{1}{1-(V/c)^2} \approx \frac{2l}{c}[1 + (V/c)^2 + ...]$$

The speed of Earth orbital motion gives a difference proportional to $(V/c)^2 \sim 10^{-8}$. At least for summer and winter measurements, the difference would be observable, if there is a medium for the light to move through. But then the sensitivity of the instruments was too low. Fizeau also performed the experiments with the light traveling through the tubes filled with the water streaming in the tubes with a certain speed.

The development of the interferometric methods suggested the possibility that can be illustrated by the simple kinematical problem, Fig. 1.3. Two identical speed-boats that have speed c relative to water are racing.

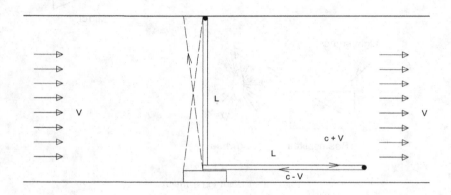

Fig. 1.3   Idea of Michelson and Morley experiment.

Both are supposed to make a round trip in the river with a speed of current equal to V. One boat goes downstream with the speed $c + V$ to the check point at the distance L equal to the width of the river, and then it goes upstream with the speed $c - V$. And the other boat goes the same distance across the river to the point opposite the berth and back. It is easy to show that the boat that goes across always wins.

In 1887 Michelson and Morley [Michelson and Morley (1887)] organized the optical version of this race with ether streaming by the Earth as the Earth moves along its orbit, and with light beams as speed-boats. In their famous experiment, the sensitivity was quite enough to measure the phase shift between the beams propagating in the orthogonal directions, Fig. 1.4.

Let the Earth with the set up on it move to the right. It will take t_1 for the parallel beam to travel from the beam splitter to the right mirror and back,

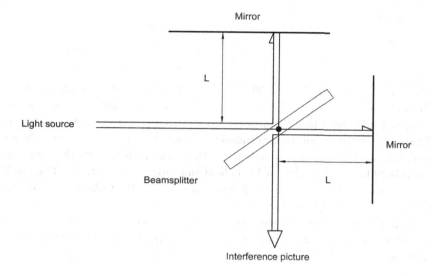

Fig. 1.4 Michelson and Morley experiment with interferometer.

$$t_1 = \frac{L}{c+V} + \frac{L}{c-V} = \frac{2L}{c}\frac{1}{1-(V/c)^2} \approx \frac{2L}{c}[1+(V/c)^2+...]$$

and it will take t_2 for the orthogonal beam to travel to the upward mirror and back with the speed $\sqrt{c^2-V^2}$, note that the upward mirror moves to the right at the same speed, V,

$$t_2 = \frac{2L}{\sqrt{c^2-V^2}} \approx \frac{2L}{c}[1+\frac{1}{2}(V/c)^2+...]$$

The path-length difference characterizing the interference maximum, $k\lambda$, will be equal to

$$\delta = c(t_1-t_2) = L\frac{V^2}{c^2}$$

Rotating the setup by 90 degrees, we get the path-length difference characterizing the interference maximum, $k'\lambda$, equal to

$$\delta' = c(t_2-t_1) = -L\frac{V^2}{c^2}$$

Therefore,

$$k' - k = \frac{2L}{\lambda} \frac{V^2}{c^2}$$

and for $L = 11$ m, $\lambda = 0.59 \cdot 10^{-6}$ m, $V = 3 \cdot 10^4$ m/s and $c = 3 \cdot 10^8$ m/s, Michelson and Morley expected to see the shift of $k' - k = 0.4$ fringes, while the setup could resolve 0.01 fringes. But no shift was found. This experiment is not an astronomical observation but a terrestrial based research, and it was performed many times since then with the accuracy essentially larger than that in Michelson and Morley experiment. The result remained the same, ether never revealed itself in a direct observation of any type.

When they thought of this result at the end of the 19th century, they regarded the following distinct possibilities in the description of the general situation:

(1) There is space filled with ether and there are celestial bodies in it. The ether resembles fluid piercing all the bodies and can be entrained when a body (celestial body) moves. In the experiment the ether is dragged by the moving Earth, that is why the light that propagates in ether and measured by the terrestrial experimental setup produces no phase shift originating from the Earth motion. Galileo relativity principle is valid (Hertz);

(2) There is space and there are celestial bodies in it. The speed of light is constant relative to its sources, i.e. to stars, Sun, lamp (or with some more details, to the surfaces and objects that reflect or re-emit it). Light demonstrates rather ballistic than wave behavior. That is why we have a null result in Michelson and Morley experiment. Galileo relativity principle is valid again (Ritz);

(3) There is space filled with ether which is absolutely motionless, and there are celestial bodies in it. The speed of light is constant relative to ether, but not a single material body in the Universe (including all the instruments we use for measurements) can be regarded as an absolute solid, it consists of particles that can have charges and interact electromagnetically. When a body moves through the motionless ether, the electric and magnetic fields corresponding to all the particles change, and the motion causes the deformation of a body. It could stretch or contract in a way that corresponds to the null result of Michelson and Morley experiment, and it is possible to calculate the exact deformation

of this kind [Lorentz (1892)]. It could also be shown that the motion leads to time dilation in the moving system. The principle of relativity is supplemented by the ether existence which still cannot be used as an absolute reference frame (Lorentz, FitzGerald, Larmor);

(4) As we see from Michelson and Morley experiment, the light has constant speed relative to the receiver (observer), and this is the only fact that can be deduced from their observations. The very existence of ether is not proved by any direct registration, therefore, there is a need of a thorough analysis of all the details of measurement procedure. If the principle of relativity is supposed to sustain, it means that our notions and models of space and time should be improved in such a way that they agree with observations (Einstein).

The first explanation can be dismissed on the ground that it means that we would be unable to observe any aberration in the visible positions of stars. The lack of this effect resembles the situation when the plane sound wave falls perpendicularly upon the wall of the moving train: whatever speed the train has, the direction of the sound waves inside it does not alter. But we do observe the aberration of light.

The second explanation suggested and investigated by Ritz in 1908 gave the aberration all right, but led to the specific prediction for the observations of the brightness change for the double stars orbiting around the common center of mass. The light emitted by one of the stars from the different parts of its orbit path would travel towards Earth at different speeds calculated according to a simple Galilean rule of velocity addition. The faster light emitted during the approach stage would be able to catch up with and even overtake the slower light emitted earlier during the moving off stage, and the star would present an image whose brightness would not be periodic at all. De Sitter presumed that the effect should have been seen if the model was correct. He made a statistical study of double stars in 1913 but found no cases [De Sitter (1913)] where the stars' images appeared scrambled. Later, the speeds of light reflected by the bodies like Moon and Jupiter that definitely had to have different values if the ballistic explanation was right were also measured [Majorana (1918); Tomaschek (1924); Brecher (1977)]. But the speed values were the same.

It is instructive to discuss the third explanation in more detail. It came to mind independently to Lorentz [Lorentz (1892)] and FitzGerald and is referred to as Lorentz-FitzGerald contraction or Lorentz contraction because FitzGerald had only a right guess while Lorentz gave a physical explanation of the obtained formulas.

1.2.3   *Lorentz contraction*

Lorentz used a reasonable idea that if the electric potential of a charged particle which is at rest is spherically symmetrical and is proportional to q/R, then the potential of the same particle which moves through ether with velocity V in x-direction is already not spherical but has the symmetry of the ellipsoid of rotation, the ellipsoid being flattened in the direction of motion. He calculated that the diameter, l in the direction of motion is a factor of $1/\sqrt{1-(V/c)^2}$ less than the diameter in the perpendicular direction, the latter being equal to the diameter l_0 of the spherical surface with the same potential when the particle is at rest

$$l = l_0 \frac{1}{\sqrt{1-(V/c)^2}}$$

Every macroscopic body - the arms of the interferometer as well - presents the superposition of such charges, and therefore, contracts as a whole in the direction of motion. Lorentz worked on the "theory of electron" for several years beginning from 90-s of the 19th century and up to 10th of the 20th century and improved it taking into consideration the change of size of the particles themselves.

We can do without much of electrodynamics in the speculations showing Lorentz reasoning [Silagadze (2008)]. An observer who is motionless in the reference frame, S, measures the velocity of a certain body and finds it equal to v. Another observer who is motionless in the reference frame S' that is in a straight and uniform motion with velocity V relative to S, measures the velocity of the same body and finds it equal to v'. According to the principle of relativity, we demand that v is a sum of v' and the relative velocity V. This gives the following rule to calculate the coordinates of the body, x and x', measured (with a solid meter stick) in the corresponding frames for the simplest case of relative motion along the Ox axis

$$x = Vt + x' \tag{1.1}$$

where t is time measured in S with the help of an additional instrument (clock) in which some periodic process takes place all the time. At first glance, the expression eq. (1.1) seems obvious, but since Galileo's times, we learned something more about solid sticks and can ask a question: does anything happen to it when the observer takes it from S to S' and back? Really, the ruler is material, there are charges inside, and at rest there is only the electrostatic interaction between charged particles. The charges moving with speed V have the deformed electric potentials and can produce magnetic field, and, generally, the length of the stick might change. If absolutely the same happens to the observer's body and other objects when they start to move, it would be really difficult to notice the changes in coordinates while measuring. Therefore, the observer must be very attentive when measuring the distance between points with the help of either moving or motionless meter stick.

Let us account for that and introduce a factor responsible for the possible change in the stick length. Equation (1.1) becomes

$$x = Vt + k(V^2)x' \tag{1.2}$$

Here we assume that the space is isotropic and the scale factor depends only on relative velocity, V of the frames. Then

$$x' = \frac{1}{k(V^2)}(x - Vt) \tag{1.3}$$

and the lengths measured in the moving frame and in frame at rest differ. Now let the observer be motionless in S'. It means that S (together with the clock) moves relative to him at velocity $-V$. Then, according to the relativity principle, all the primed values in eq. (1.3) could be marked with the unprimed symbols and vice versa. We obtain

$$x = \frac{1}{k(V^2)}(x' + Vt') = \frac{1}{k(V^2)}[\frac{1}{k(V^2)}(x - Vt) + Vt'] \tag{1.4}$$

and the last expression gives

$$t' = \frac{1}{k(V^2)}[t - \frac{1 - k^2(V^2)}{V}x] \tag{1.5}$$

This means that when the clock starts to move relative to the observer who performs the measurements, the clock readings that he is going to use in calculations change. This happens every time when factor $k(V^2)$ differs from unity.

Thus, the existence of electric charges severely threatens the very possibility to have definite values of lengths and also shows that the Newtonian idea of absolute time is at least intestable. Equations (1.3, A.2) give the velocity addition formula which accounts for the possible changes

$$v' = \frac{dx'}{dt'} = \frac{dx - V\,dt}{dt - \frac{1-k^2}{V}\,dx} = \frac{v - V}{1 - \frac{1-k^2}{V}v} \tag{1.6}$$

Again changing primed and unprimed symbols and substituting $-V$ instead of V, we get

$$v = \frac{v' + V}{1 + \frac{1-k^2}{V}v'} \tag{1.7}$$

Introducing new notation $\frac{1-k^2}{V^2} \equiv \frac{1}{c^2}; c = const$, we can rewrite the expressions in eqs. (1.3, A.2) and eq. (A.6) in the following way

$$x' = \frac{1}{\sqrt{1 - (V/c)^2}}(x - Vt) \tag{1.8}$$

$$t' = \frac{1}{\sqrt{1 - (V/c)^2}}[t - \frac{V}{c^2}x]$$

$$v = \frac{v' + V}{1 + \frac{V}{c^2}v'} \tag{1.9}$$

If the new constant, c, that has the dimensionality of speed is regarded as the speed of light in motionless ether, the last expressions present the final results obtained by Lorentz on the base of explicit electrodynamics approach.

Thus, the explanation of the result of Michelson and Morley experiment given by Lorentz seems consistent. It preserves the Newtonian space which is filled with the undetectable ether, the principle of relativity remains valid, and even in 1913 Lorentz argued that there is little difference between his theory and the etherless relativity theory of Einstein and Minkowski that appeared in 1905-07, and that it is a matter of taste which theory to prefer. His own choice was the one in which all the bodies in the Universe including instruments were permanently changing their forms and dimensions with regard to their velocity measured relative to omnipresent ether in such a way that the measured velocity of light propagation remained the same. Lorentz's speculations were complicated but fair and physically grounded. This point of view is rightful, because, as we will see below, the results obtained in both theories have the same form. It is the interpretation of these results that was completely different in these theories. Even now students and even scientists can mix these two interpretations, and this can cause misunderstanding. Ether is not bad and its undetectability makes it very adapted in the evolution of physical theories. Nowadays, it has changed its name and is usually known as physical vacuum. Still, it seems that nobody knows how to use it as an absolute reference frame.

1.2.4 *Special relativity*

Finally, the fourth explanation developed by Einstein in his famous 1905 paper [Einstein (1905a)] became known as Special Relativity Theory.

Let us regard a couple of formal mathematical expressions

$$x' = x - Vt \tag{1.10}$$
$$t' = t \tag{1.11}$$

If we have (intuitive) notions of 3D empty space and time as Newton had, we can regard x and t as space and time coordinates of the *event*, and consider these formulas as describing a certain motion. It is impossible to introduce a reference frame for an *empty* space, unless this frame is related not to one but to every point of space and, at the same time, it does not affect anything. Therefore, we have to consider a kind of omnipresent and non-interacting ether already in laboratory mechanics (which is not so obvious as in the case below). So, if we have an absolute reference frame - the motionless ether filling the space completely through - then a laboratory (an observer) can move in it. Let the measurements of mechanical events performed in the laboratory correspond to x_m and t_m. For brevity, we would not deal with all the three dimensions but mention only one of them. If V_m is a constant with the dimensionality of speed, $V_m = const$, then in the absolute (etherial) frame with absolute time, we will have

$$x'_m = x_m - V_m t_m \tag{1.12}$$
$$t'_m = t_m \tag{1.13}$$

and this motion will be called the straight uniform one. In physics, the principal law of mechanics (Newton dynamics law) describing the motion of a massive body with mass, m, is

$$m\frac{d^2x}{dt^2} = \sum_k F_k \qquad (1.14)$$

and if the sum of forces is equal to zero, the only physical characteristic left is the inertial mass, and we see that in this case, the solution of the (homogeneous) dynamics equation will be

$$x = x(0) + V_m t$$

with $V_m = const$. This is what is usually called inertial motion which is here straight and uniform. Here appears the notion of Inertial Reference Frame in mechanics (IRF_m), i.e. such that all the mechanical phenomena including all kind of forces are described by the same equation - that of the dynamics law. This is so, because the absence or presence of the straight uniform motion does not affect the forces. Equation (1.14) is actually the definition of force as of something that changes the observable state of motion and makes it not straight and uniform (and is proportional to acceleration). We see now that the transformation eq. (1.12) corresponds to the change of the reference frame from initial one whose position is fixed in space (by the condition $t = const = 0$) and does not depend on time, to the final one which is in the straight uniform motion. Both of them are IRF_ms. The demand that *all mechanical phenomena are described by the same equation in any inertial reference frame* is known as Galileo's principle of relativity. It is instructive to pay attention to the fact that this principle of relativity was given by Galileo in 1632, that is *before* the equations of dynamics like eq. (1.14) came into being. Actually, he did not speak about equations but stated that all the physical laws must be the same whether a body is at rest or moves with constant velocity. With Galileo, it had the meaning of philosophical, rational demand (now we would call it physical sense) and was based on the intuitive notions of space and time.

In order to regard velocities of objects, we take a pair of events in IRF_m, for example, let them correspond to the motion of a body along its trajectory, and consider this trajectory in both IRF_ms. Then

$$dx'_m = dx_m - V_m dt_m \qquad (1.15)$$
$$dt'_m = dt_m \qquad (1.16)$$

We see that if it takes no time to measure the length of a body (apply the ruler to its ends), i.e. the signals from its ends travel with the infinite speed, and if we take $dt_m = 0$, then $dx'_m = dx_m$ and $dt'_m = dt_m$. Then, if $dt_m \neq 0$, we divide the upper equation by the lower one and get the so-called velocity addition formula in mechanics

$$v'_m = v_m - V_m$$

In electrodynamics, we deal with specific physical phenomena described by Maxwell field equations

$$curl\mathbf{E} = -\frac{1}{c}\frac{\partial \mathbf{H}}{\partial t} \qquad (1.17)$$
$$div\mathbf{H} = 0$$
$$div\mathbf{E} = 4\pi\rho$$
$$curl\mathbf{H} = \frac{1}{c}\frac{\partial \mathbf{E}}{\partial t} + \frac{4\pi}{c}\mathbf{j}$$

From the point of view of dynamics, there could be written an expression for Lorentz force, and then it can be introduced into dynamics equation. But this expression follows from Maxwell equations together with many other physical laws, therefore, for the electromagnetic phenomena Maxwell equations give more general description. If there is no sources, i.e. charges or currents, the remaining equations still have sense, and the solution of the homogeneous equations has the form of the waves propagating with the constant speed, c. If we use the designation x_e and t_e and make a substitution similar to eq. (1.10) into Maxwell equations, we will see that the equations change and the speed of waves propagation changes its value. Lorentz found such transformations of coordinates x_e and t_e that preserve the constancy of c for the wave solutions of homogeneous Maxwell equations, that is for the case when there were no bodies but only fields. They have the form

$$x'_e = \frac{1}{\sqrt{1-\beta^2}}(x_e - V_e t_e) \qquad (1.18)$$

$$t'_e = \frac{1}{\sqrt{1-\beta^2}}\left(t_e - \frac{V_e}{c^2}x_e\right)$$

$$\beta \equiv \frac{V_e}{c} \qquad (1.19)$$

where $V_e = const$. Treating this, at first, strictly mathematically, we can say that now *these* expressions can be regarded as the relation between space and time coordinates of a certain body moving in the intuitive 3D-space and time measured in the laboratory reference frame, (x_e, t_e), and in the absolute (etherial) reference frame, (x'_e, t'_e). The motion that we can attribute to the frame which gives these new coordinates, preserves the constancy of the speed of light, i.e. $c = const$. And, besides, it corresponds to the constancy of a certain vector V_e. We can interpret this vector as a velocity of a laboratory which is a new reference frame that can also be called "inertial" one (IRF$_e$), though we don't expect the electromagnetic field to have mass in Newtonian sense and there is nothing to be inertial ("straight uniform" here is all right). Let us further assume that $V_e = V_m \equiv V = const$.

For a pair of events, that is for a couple of points on a body's trajectory described in both IRF$_e$s, we get

$$dx'_e = \frac{1}{\sqrt{1 - \beta^2}}(dx_e - V dt_e) \tag{1.20}$$

$$dt'_e = \frac{1}{\sqrt{1 - \beta^2}}(dt_e - \frac{V}{c^2} dx_e)$$

Now when we have thought about light and vision and its use in experiments, it turns out that it takes some time to measure the length of a body (using, for example, a radar technic based on the constant light speed), i.e. the signals from the ends of the body travel with finite speed which means that even performing measurements in the frame where the body is at rest, we get $dt_e \neq 0$. It means that now dx_e and dt_e measured in the frame which is motionless relative to the laboratory set up are not equal to dx'_e and dt'_e measured in the frame relative to which the laboratory frame moves, $dx'_e \neq dx_e$ and $dt'_e \neq dt_e$.

Dividing, we get the so-called velocity addition formula in IRF$_e$

$$v'_e = \frac{dx_e - V dt_e}{dt_e - \frac{V}{c^2} dx_e} = \frac{dx_e - V dt_e}{dt_e(1 - \frac{V}{c^2} v_e)} = \frac{v_e - V}{1 - \frac{V}{c^2} v_e} \tag{1.21}$$

A question arises: how to correlate the results obtained for IRF$_m$s and presented by eqs. (1.1,1.15) and the results obtained for IRF$_e$s and presented by eqs. (1.18,1.20)? They presumably correspond to the inertial frames moving with one and the same velocity V relative to the absolute frame (ether), but give different descriptions of the trajectory of one and the same body, which the hypothetical etherial observer can also see and measure. The etherial results must obviously be

$$x'_m = x'_e \qquad (1.22)$$
$$t'_m = t'_e$$

A human observer can visit a laboratory but is hardly able to survive in heavenly ether. Sticking to common sense, he should demand that both in IRF_m and in IRF_e, moving with the same speeds relative to ether, the same values must be registered (Galileo's relativity principle)

$$x_e = x_m \qquad (1.23)$$
$$t_e = t_m$$

Substituting eq. (1.23) into eq. (1.18), one obtains

$$x'_e = \frac{1}{\sqrt{1-\beta^2}}(x_m - Vt_m)$$
$$t'_e = \frac{1}{\sqrt{1-\beta^2}}\left(t_m - \frac{V}{c^2}x_m\right)$$

Substituting in the rhs of the last equation primed variables instead of unprimed with the help of eq. (1.1), we get for the space coordinate

$$x'_e = \frac{1}{\sqrt{1-\beta^2}}(x_m - Vt_m) = \frac{1}{\sqrt{1-\beta^2}}x'_m$$

and for the time coordinate

$$t'_e = \frac{1}{\sqrt{1-\beta^2}}\left(t_m - \frac{V}{c^2}x_m\right) = \frac{1}{\sqrt{1-\beta^2}}\left[(1-\beta^2)t'_m - \frac{V}{c^2}x'_m\right]$$

Finally, we obtain

$$x'_e = \frac{1}{\sqrt{1-\beta^2}}x'_m \qquad (1.24)$$
$$t'_e = t'_m\sqrt{1-\beta^2} - \frac{V}{c^2\sqrt{1-\beta^2}}x'_m$$

and there is no possibility to satisfy both the demand of a human observer visiting laboratories and seeing one and the same body with his own eyes and the demand of the etherial observer. Something must be wrong.

Both approaches started with mathematical structures that were used to interpret physical situations. IRF_m and IRF_e had a feature which made them different - the presumable infinite and finite speeds of signal propagation. Though we have never registered any signal with the speed higher than that of light, we, for a moment, cannot exclude that it will happen sometimes and look for another possible drawback in our reasoning. IRF$_m$ and IRF$_e$ have one feature in common - the non-perceivable ether filling the 3D-space. Therefore, staying on the experimental ground, we have to suggest that it is the model with ether (equivalent to absolute empty space) which is not good. As a kind of material substance, ether has very exotic properties, we can't test its existence by direct measurement, since it does not interact with anything, and it is needed only as a medium for electromagnetic waves propagation. If we do away with it, then the passage from an IRF to an IRF should be performed only between any two *real laboratories* that can move (inertially) relative to one another. And all that we will have to predict is the results of measurements when passing from one such frame to another. So, now we have to reject the absolute reference frame and underline the conditional character of rest and motion for two IRFs. Then in the last formulas, we could assume the primed variables with $_e$-subscript to correspond to the observer who is at rest relative to the object under investigation

$$x'_e \equiv x^{(r)}; t'_e \equiv t^{(r)}$$

and primed variables with $_m$-subscript would correspond to the observer who moves with $-V$ relative to the first one

$$x'_m \equiv x^{(m)}; t'_m \equiv t^{(m)}$$

Then the pair of events related to the one and the same object moving along its trajectory can be expressed as

$$dx^{(r)} = \frac{1}{\sqrt{1-\beta^2}} dx^{(m)} \tag{1.25}$$

$$dt^{(r)} = dt^{(m)}\sqrt{1-\beta^2} + \frac{V}{c^2\sqrt{1-\beta^2}} dx^{(m)}$$

The last expressions give the formulas for length contraction and time dilation

$$L^{(m)} = L^{(r)}\sqrt{1 - \beta^2}$$
$$\Delta t^{(m)} = \frac{\Delta t^{(r)}}{\sqrt{1 - \beta^2}}$$

that must be true for the measurements performed from the frame that moves relative to the frame that is at rest relative to the experimental set-up. So, when the rod parallel to the direction of flight is placed inside the rocket that flies relative to the Earth and measured there by an astronaut, the ruler readings appear to be larger than the readings of the instrument on the Earth which is used by an astronomer to measure the same rod distantly. It resembles the situation with the perspective: when we look at the segment, the closer it is to the observer, the larger it seems, the further it is, the smaller it seems, and the actual length of the segment remains the same. The same takes place with the clock: flying on the rocket and counting the clock ticks for a certain process taking place in the rocket, we get the smaller number than observing this very process from the Earth and counting the Earth clock ticks. The analogy with perspective or "angle of sight" is less usual now, but the meaning of it remains the same. It's not the process that changes with passing to another IRF, it's only the measurements of it performed from another IRF that differ. It is the measurement procedure that must be analyzed, and measurement is literally the geometry, and geometry is what we use to measure space, and it seems that now time is also involved.

In other words, if we, as well as Galileo, want to preserve the relativity principle and do away with the discrepancy that could appear when the electromagnetic phenomena are included into the world picture, we should demand that not only mechanical, but *all the physical phenomena are described by the same equations in all inertial reference frames*. Then we must admit that it's only the different circumstances of measurements that can lead to differences. The last is due to the fact of distant measurements that are performed by the instrumentation exploiting light waves propagating with finite speed, and not to the processes themselves. This statement is known as Einstein principle of relativity.

The velocity addition formula eq. (1.21) now also takes the familiar form

$$v^{(r)} = \frac{v^{(m)} + V}{1 + \frac{V}{c^2}v^{(m)}} \tag{1.26}$$

coinciding with eq. (1.9). And we see that the form of results obtained by Einstein is the same as those obtained by Lorentz, but their meanings and the ways they were obtained are entirely different. It is important for the future, thus, we underline it once again.

Lorentz trusts the most profound and well-established theory of his days - Maxwell electrodynamics - with all the implicit prejudices it contains including the non-perceivable ether as a medium for light waves propagation. In a sense, he considers all this the final truth, the dogma, the paradigm. When the observations show strange results - the speed of light remains the same in moving frames - he tries to find out such physical mechanism that would be inherent to the current concept of electrodynamics (including implicit prejudices), that would make it stay. He thinks of physical properties of matter that could be accounted for in order to explain the observations, and discovers that all the meter sticks should contract together with the measured objects when oriented in parallel to the direction of their common motion through ether. If so, the constancy of light speed can be explained. He also finds the sensible reasons for time dilation of the similar sort. One can say, that all the efforts that were successfully undertaken by Lorentz were aimed to prove that there is no way to register and measure ether experimentally. Though it exists.

Einstein also trusts electrodynamics and wants to preserve it, but he acts in the completely opposite way. It is the well-established result of the observation that he starts with. Light speed, paradoxically, appears constant, and one should, first of all, seek for a weak point in the existing theory and not for the theory extension. Physics is an experimental science, and there should be as few untestable notions as possible. Einstein dismisses ether as a motionless IRF and focuses upon the procedure of measurements. He uses the light speed constancy and the finite character of its value as starting point in the analysis of measurements. It turns out that we can't but register the contraction and dilation in an experiment, because this is the only way we can watch it - when performing measurements at large speeds comparable to that of light and using the signals with highest known propagation speed, i.e. that of light. We cannot measure the contraction itself (and Lorentz showed why), we cannot measure the properties of ether (and nobody can), but there is no need to do it, because all these are mere illusions following from the actual possibilities of observation. The relations between the readings of the instruments can be explained not by "why they are such" but by "why we cannot see them other". It appears that everything follows from the "angle of sight". In a sense, it is the law of geometrical perspective that must be considered when observing a 2D picture presenting the three dimensional objects, and not the law of physics that deforms them.

The existence of etherial observer who learns everything instantaneously and inhabits the absolute and inconceivable reference frame seems to be rather a theological than a physical question. Einstein obviously tries to remain on the humble scientific position, and the famous Newton's motto "Hypothesis non fingo" fits his reasoning even better than those of Newton himself. But strictly speaking, Einstein himself never fought against ether, on the contrary, he did much to bring to peace the complex of profound physical ideas, one of which was ether. He said [Einstein (1920)] that Lorentz in his theory already deprived ether of all its properties but motionlessness, and what he himself did was just showing that this property cannot be attributed to ether in a testable way.

Finally, of the three possibilities to treat the results of Michelson and Morley experiment, namely:

- ballistic theory of Ritz in which the speed of light is constant relative to the *emitter* (any emitting, reflecting, re-emitting object);
- electromagnetic theory of Lorentz in which the speed of light is constant relative to the *medium* (motionless ether that presents the absolute reference frame); and
- special relativity theory (SRT) of Einstein, in which the speed of light is constant relative to any inertial *receiver* (observer),

the last one appears to be the most consistent *physical* theory which became the cornerstone practically in all fundamental fields of physics. Since even textbooks are sometimes not very distinct about the difference between Lorentz theory and SRT and do not mention ballistic theory at all, there still exist scientists who prefer the alternatives to SRT. And of course, the SRT is not the final step, new observations and data appear and bring to life doubly special relativity theory, very special relativity theory and so on. The power of the special relativity essence and approach is the profound and balanced link between physical and mathematical ideas.

1.3   Space-time as a model of the physical world

As we saw, the entity of ether appeared to be redundant, but after that the very notion of absolute space - obviously three-dimensional and Euclidean and supplied by the absolute time, demanded additional attention. It is this complex object that was used for modeling the physical real world since Newton times. The speculations of mathematicians like Lobachevsky and his non-Euclidean geometry or Clifford and his world curvature were interesting but hardly relevant to the available observations until the results of Michelson and Morley's experiment became known and the subsequent discussion took place.

Let us return to the velocity addition formula eq. (1.26), remove $^{(r)}$ and $^{(m)}$ superscripts, and get the formula equivalent to eq. (1.9), that is $v = \frac{v'+V}{1+\frac{V}{c^2}v'}$, where $\frac{1}{c^2} = \frac{1-k^2}{V^2}$. It has certain specific features.

The obtained expression is a symmetric function of its arguments, v' and V, [Silagadze (2008)], and, thus, we can rewrite the term in the denominator

$$\frac{1-k^2}{V}v' = \frac{1-k^2}{v'}V \qquad (1.27)$$

which gives

$$\frac{1-k^2}{V^2} = \frac{1-k^2}{v'^2} \qquad (1.28)$$

If $\frac{1-k^2}{V^2} \equiv \frac{1}{c^2}$, where c has the dimensionality of speed, then $\frac{1}{c^2} = const$, and c is a constant too, because this is the only way to provide eq. (A.8) for arbitrary v' and V. Let $\beta = V/c$, as before, and let us use a graphical illustration of the mutual positions of two IRFs, Fig. 1.5, where O is the position of the observer who is motionless and close to the event O, and

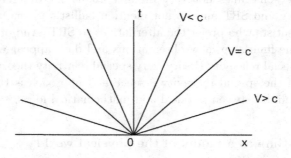

Fig. 1.5　Inertial motion in 2D space-time.

Ot is the time axis, Ox is the coordinate (space) axis and the fan-like set of straight lines correspond to IRFs moving with zero speed (Ot axis) and with speeds less than c, equal to c and larger than c relative to the observer in O.

Let us investigate the following three cases.

If $k^2 = 1$, then $\frac{1}{c^2} = 0$. It means that the following is true: there exists an absolute solid, as Lorentz remarked; the Galilean transformation of coordinates and time, eq. (1.1), are valid; velocity addition formula, $v = v' + V$ is applicable; parameter c with the dimension of speed has infinite value. In this case the $V = c$ line coincides with the Ox-

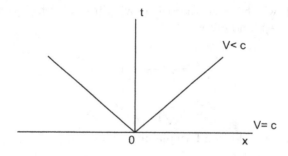

Fig. 1.6   Inertial motion in 2D space-time: absolute solid exists.

axis, the situation on Fig. 1.5 reduces to Fig. 1.6. One can say that the "IRF-fan" was opened wider and a little tightened because the sectors between $|V| = |c|$ and $|V| > |c|$ were "contracted" into the edges that were made into a flat angle coinciding with Ox-axis. Now any speed is possible for an IRF, and one can suggest that c could be interpreted as a speed of light or some other signal that supports the instantaneous correlation between all the events in the world. The last means that all the events registered with the help of this c-signal by the observers traveling in all the IRFs appear to be absolutely simultaneous. This corresponds to the Newton's idea of a model for space and time, i.e. to Newtonian mathematical model of real physical world in which the phenomena take place. However, the scientists are still unaware of signals that are faster than light, while the light speed itself is high but still finite. Therefore, this model is an idealization and does not correspond to what we know for certain. Newton *did* invent a hypothesis though, probably, did not re-alize it. This was because the light speed is so high that Newtonian model excellently fits for the description of almost all the processes that we usually deal with.

There are several problems with this model. We won't dwell upon them but only mention (see the detailed discussion in almost any textbook or, for example, in [Bohm (1965)]) that the model with infinite speed of a signal or with signal speeds higher than that of light allows time loops, that is the possibility of the event to affect its own past, thus, providing the possibility to affect the present and change it in various ways. Such intellectual play is hardly valid for a physical theory, and if we want the world to be cognizable, we should avoid such models.

If $k^2 > 1$, which means that the moving meter stick becomes longer, we take $\frac{1-k^2}{V^2} \equiv -\frac{1}{c^2}$ and obtain

$$x' = \frac{1}{\sqrt{1+\beta^2}}(x - Vt) \qquad (1.29)$$

$$t' = \frac{1}{\sqrt{1+\beta^2}}\left(t + \frac{V}{c^2}x\right)$$

with the velocity addition formula given by

$$v = \frac{v' + V}{1 - \frac{v'V}{c^2}} \qquad (1.30)$$

and all the xOt plane in Fig. 1.5 is accessible for the moving IRFs. In this case, the model also contains a logical defect. If both v' and V have the same direction, for example, they have positive values, and their product $v'V$ is larger than c^2, which is now finite, then v appears to be negative, which is nonsense. And if $v' = V = c$, then the measured value of the moving object velocity, v, is infinite which is hardly better.

The origin of this nonsense can be ascribed a profound meaning. If we designate $x_0 = ct$, and $\cos\theta = \frac{1}{\sqrt{1+\beta^2}}$, then eq. (A.9) can be rewritten as

$$x'_0 = x_0 \cos\theta + x \sin\theta \qquad (1.31)$$

$$x' = x \cos\theta - x_0 \sin\theta$$

Notice, that the coordinate x and moment of time t_0 measured in the rest frame correspond to a vector with length $x_0^2 + x^2$ on the Euclidean plane, (we don't speak about y and z coordinates for simplicity). When we pass to the moving frame and get x' and x'_0 described by the transformation eq. (A.11), we see that the initial vector has just rotated by the angle θ counter clockwise. Therefore, there is an invariant that is preserved by the coordinate transformation given by eq. (A.11), it is the sum $x_0^2 + x^2$, i.e. the length of the vector. The primed event whose coordinate is x' and moment of time is x'_0 remains on the same circumference. It means that if we measure a pair of events in the rest and in the moving frames and get

$$dx'_0 = dx_0 \cos\theta + dx \sin\theta \qquad (1.32)$$
$$dx' = dx \cos\theta - dx_0 \sin\theta$$

then not only time intervals between events dx_0 and dx'_0 are different, but it is impossible to specify uniquely the time order inside a pair. Both events are on the circle but which precedes which is impossible to say. This situation resembles an attempt to order the points in the usual 3D-space. Therefore, in case $k^2 > 1$, (that is a stick parallel to the direction of motion stretches in the direction of motion), it is impossible to uniquely specify and preserve causality which is associated with the time ordering.

From the classical point of view, the absence of time ordering and causality makes a theory non-physical. But the purely classical theories almost expired when the undeniably successful quantum mechanics took the scene. Now there are speculative theories that allow such Euclidean structure to exist, at least, for some very special regions. These regions have no time ordering, they are timeless, therefore, they cannot have a human observer inside and, thus, there is no reason to speak about reference frames there. The thing we can speak about in this case is not the causality but the concept of symmetry transformations that leave the space structure invariant, which is the concept of the theory of groups.

Finally, if $k^2 < 1$, which means that the moving meter stick becomes shorter, then the time and coordinate in the moving frame will be defined by

$$x' = \frac{1}{\sqrt{1-\beta^2}}(x - Vt) \qquad (1.33)$$
$$t' = \frac{1}{\sqrt{1-\beta^2}}\left(t - \frac{V}{c^2}x\right)$$

and the velocity addition formula eq. (1.9) remains valid

$$v = \frac{v' + V}{1 + \frac{v'V}{c^2}} \qquad (1.34)$$

If we keep to what we know for certain and do not speculate about the signals (or IRFs) traveling faster than light, then the region accessible for IRFs shrinks and occupies only the sector bounded by $V = c$ lines, Fig. 1.7.

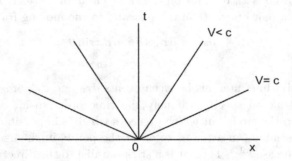

Fig. 1.7   Inertial motion in 2D space-time: observable world.

One can see that according to eq. (A.14), however close the values of v' and V might be to c, still remaining less than it, $v' < c$ and $V < c$, the measurement of v performed in S frame always gives a value that is less than c. When $v' = V = c$, the measurement of v performed in S will give $v = c$. It is important to remind that in these speculations c is just a constant with the dimensionality of speed. Strictly speaking, it is only the historical tradition that we always have in mind the light and Michelson and Morley experiment. If there is a demand that Maxwell equations of electrodynamics describing the charges behavior and the electromagnetic waves propagation remain invariant under Lorentz transformations eq. (1.33), then c must be naturally chosen equal to the speed of light. But the speculations given above don't need light, and we usually mean it only because the whole discussion of these problems stemmed from physics. One can see that we spoke about a parameter with the dimensionality of speed that could be finite or infinite and maximal or not [1]. Thus, some choices led to certain drawbacks in the model of space and time that can be used to describe the

---

[1]But why speed of light and electrodynamics? It comes to mind that in a fictitious acoustical civilization of bats who have no vision but use the ultrasonic signals to radar the surroundings, a similar role could be, probably, played by the speed of sound. This is an example of science fiction for scientists.

physical world, but there was one that is suited.

Einstein's principle of relativity accounts for electrodynamics alongside with mechanics and is the natural extension of Galileo's principle of relativity, but the way it was obtained by Einstein is different from that of Galileo. The problem of relative motion was realized and discussed throughout all the preceding three centuries but mainly in the philosophical qualitative sense. Einstein gave it an operational definition, that is, in order to construct a model of space and time, he started not with the concept but with the procedure that was expected to test it, and attributed the character of a postulate to the result of this procedure. Large variety of experiments with electricity and magnetism are described by one and the same set of Maxwell equations. The equations predicted the existence of electromagnetic waves that were subsequently discovered by Hertz, and Faraday's experiments proved that the light and electromagnetic waves had the same nature. As one can test in an experiment, the speed of light propagation is a) finite and b) appears to be the same when measured in any inertial frame. As it was underlined by Poincaré already in 1898, item a) ruins the absolute simultaneity of events, i.e. the absolute character of time. Item b) is equivalent to the following statement: with Galileo transformations of coordinates, eq. (1.10), that preserve the intuitive model of absolute space with or without ether, the results of the solution of Maxwell equations and even the form of these equations start to depend on the chosen IRF, that is on the state of an observer; with Lorentz transformations eq. (1.33) that preserve causality and the form of the testable Maxwell equations in every IRF, the intuitive model of space requires modification. Not to lose its identity, science has to stick to measurements and take the second choice, i.e. has to prefer testable to intuitive.

It was the former Einstein's calculus teacher Hermann Minkowski [Minkowski (1909)] who unequivocally recognized the problem with the modeling of space and time in a couple of years after Einstein's famous paper [Einstein (1905a)] appearance in 1905. He discussed the meaning of the diagrams like those presented on the figures above and stressed the meaning of the fact that we describe not the objects in the 3D space and track down their location with time, but we draw the processes themselves. The figures given above are the examples of the picture of the *world*, they are the combinations of the so-called *world lines* of the events, that is the pictures present the broach bits, the frozen histories of processes. Taking various projections of these processes that obviously have four dimensions, time being one of the coordinates, we can come back to the familiar descriptions, but all in all, it is something different. The difference is: when we recognize the specific demands originating from physics - like those of the impossibility of absolute simultaneity or the maximal character of the speed of light - it turns out that not all of the usual space is accessible for any given process. The deterministic part of this 4D space is so to say "squeezed" and limited by the conical surface. The axis of this *light cone* coincides with the time axis, and its top is the current position of the observer. Observer looks at the events and measures the events that are the points in this 4D space which should be now better called space-time. The lines connecting events - world lines - are the edges of the 4D figures (or their projections) that can be drawn. The upward cone is the cone of the future, all the events that can be found there will take place later than event O, all the events that can be found in the downward cone took place in the past, and the downward cone is the cone of the past. If we take one event inside the cone of the past and one event inside the cone of the future, they can be linked by a cause and effect relation. It means that for an arbitrary pair of events both located in the light cone, we can shift the origin (the top of the cone) to the location of any of them, and we will definitely find the other either in the new cone of the past or in the new cone of the future. It means that any two events that can be regarded as cause and effect would always be such whatever IRF we take. The time duration between them can change depending on the IRF chosen for observations, their space coordinates may also become different when measuring from different IRFs, but their cause and effect character will be preserved. Actually, we can sometimes connect them with the straight line that will correspond to the straight uniform motion. The observer in the IRF corresponding to this motion will be motionless relative to these events and observe them subsequently in the same place, i.e. in his own laboratory. That is why any interval between events in the interior of the light cone is called time-like. This is not so for the events (points of the 4D diagram) outside the light cone. These events do exist, we can observe them, but we are never able to say for certain if one was the cause of another, because there could be chosen such IRFs that in one of them the observer would see event b after event a, and in the other he will see event a after event b. It also means that there could be chosen such an IRF that they both take place simultaneously but in different places. The corresponding interval between such events (points on the 4D diagram) in the outside of the light cone is called space-like.

We can say that we see a good example of the space-like interval between events when looking at the photo of the night sky with stars. Every star on the photo is an event, but they are so far from us and from each other that it takes light hundreds or thousands of years to reach them from a star to us or to another star. When we look at them, we see them not as they are now, one of them is as it was 100 years ago, another one is as it was 200 years ago and so on. The images we see on the photo are distributed not only in space (in the heavenly sky) but in time also, and we see only a certain projection of the 4D events on the 2D sphere above us. Of course, we have the same situation just looking at the night sky.

In his lecture [Minkowski (1909)] in 1908 Minkowski underlined:

> "The views of space and time which I wish to lay before you have sprung from the soil of experimental physics, and therein lies their strength. They are radical. Henceforth space by itself, and time by itself, are doomed to fade away into mere shadows, and only a kind of union of the two will preserve an independent reality".

Actually, instead of the habitual geometry of point objects located in the three-dimensional Euclidean space, he came to the geometry of events in the four-dimensional space-time, described by non-Euclidean geometry with a different notion of "distance" now known as an interval

$$\Delta s^2 = c^2 \Delta t^2 - \Delta x^2 - \Delta y^2 - \Delta z^2 \tag{1.35}$$

His considerations were influenced by physics and physical discussions that took place for several years before that, but these considerations could be purely mathematical as well. The speculations given above can be supported by more strict and detailed mathematical statements dealing with geometry and group theory. We would only notice that in a space endowed with Euclidean geometry, a vector $(\Delta x, \Delta y, \Delta z)$, connecting two points, that is

$$\Delta r^2 = \Delta x^2 + \Delta y^2 + \Delta z^2$$

is invariant under rotations and translations. We can formally pass to four dimensions and consider the fourth coordinate t, then nothing principal will change, and the invariant will be

$$\Delta r^2 = \Delta t^2 + \Delta x^2 + \Delta y^2 + \Delta z^2$$

Suppose, we want to use this 4D space to model space and time of the physical world. Then the coordinate of this 4D-space that is appointed to be time, t, has to be measured in the same units as the other three, and, therefore, we have to introduce a conversion constant, c, that has the dimensionality of speed and rewrite

$$\Delta r^2 = c^2 \Delta t^2 + \Delta x^2 + \Delta y^2 + \Delta z^2$$

But we saw in our 2D example that such structure is not good for physical interpretations because of the causality failure. As we saw, this can be repaired by the usage of the following expression

$$\Delta s^2 = c^2 \Delta t^2 - \Delta x^2 - \Delta y^2 - \Delta z^2 \tag{1.36}$$

which is just the interval mentioned above that plays the same role in a space with pseudo Euclidean geometry as distance plays in a space with Euclidean geometry. It is this expression that will be invariant now with the pass from IFR to IRF accompanied by the coordinates transformation. The group of transformations that includes rotations and boosts defined by eq. (1.33) has six parameters and preserves the interval eq. (1.36) invariant is called the Lorentz group, $\mathcal{G}_c$. From a purely mathematical point of view, Lorentz group has a simpler structure than Galilei group, $\mathcal{G}_\infty$ under which the equations of Newtonian mechanics are invariant, because Lorentz group is a real noncompact form of the semi-simple Lie algebra $A_1 \times A_1$, whereas Galilei group is not semi-simple. Supplementing Lorentz group by another four parameters corresponding to spatial and time translations, that is forming the semi-direct product of $\mathcal{G}_c$ with $\mathcal{T}_4$, we obtain ten-parametric Poincaré group, $\mathcal{P}$ (or inhomogeneous Lorentz group) which is the basis of any theoretical approach now used. Strictly speaking, the invariance group of Newtonian mechanics is also not the six-parametric $\mathcal{G}_\infty$, but the ten-parametric Galilei group $\mathcal{G}$ which is a semidirect product of $\mathcal{G}_\infty$ with $\mathcal{T}_4$. Notice also that neither $\mathcal{P}$ nor $\mathcal{G}$ is a semi-simple group.

It should be mentioned that actually, Einstein and Minkowski dared to take the challenge and perform what Poincaré had already discovered independently [Poincaré (1904)] several months before Einstein's paper [Einstein (1905a)] appeared, but he put to doubt the profit-to-efforts ratio in the realization of such program. In his paper he mentioned "the postulate of relativity", he regarded the group characteristics of the "Lorentz group" (this name was given by Poincaré), he demonstrated the invariance of the interval, but finally concluded with doubts about the possible effect of translating all the physics into this new and unusual language.

Concluding this section, we should say that Einstein's generalization of Galileo's principle of relativity was not that direct as we used to think. Not only the principle of relativity was applied to electromagnetic phenomena alongside with mechanical ones, it was true in a different space - in the 4D Minkowski space-time which was the new model suggested to describe the physical world. The important feature of Minkowski space-time model is that it is not just possible to use it (the contemporary scientists were at first rather sceptical about it), but it appears inevitable, and the inevitability follows from the experiments and from Einstein's insight. The previous model, i.e. the Newtonian one, based on the infinite speed existence appears to be inconsistent both logically (time loops) and physically (unable to explain high speed motion together with electrodynamics).

Relativistic point of view developed by Einstein and Minkowski transformed the world view in physics. It turned out that the account of the "angle of sight" is sufficient in many cases where earlier the inconsistencies existed or complicated physical mechanisms were suggested. In a sense, it turned out that it is the law of geometrical perspective that must be considered when observing an $(N$-1$)$-dimensional picture presenting the N-dimensional objects, and not the law of physics that "deforms" them. Here the profound link appeared between mathematical and physical ideas that made this approach extremely powerful.

Let us also mention the direct experimental proof of the special relativity predictions. It deals with mesons - the unstable particles that can appear in the cosmic rays (or in laboratory when a substance is bombarded by the high energy particles from the accelerator). Mesons decay and transform into other particles. Their average lifetime is measured and known for every particular type of mesons. Let the lifetime corresponding to a meson at rest or to a slowly moving meson be equal to τ_0. Then the lifetime, τ, of a meson moving with speed V can be calculated according to Lorentz formula

$$\tau = \frac{\tau_0}{\sqrt{1 - \beta^2}}$$

and then compared to the observations. The relativity principle demands that meson lifetime in the IRF where the meson is at rest does not depend on the speed of this IRF. That is the period τ_0 plays the role of the clock on the rocket. This means that when we use the laboratory on Earth to measure the lifetime of the mesons falling on Earth with cosmic rays and moving at high speeds close to c, we are supposed to find the value of τ that is essentially larger than τ_0. Many such tests were performed, the most simple and clear description given in [Frisch and Smith (1993)]. They calculated the number of mesons with a certain speed at the top of Mt.Washington, 1918 m high, and compared it to the number of mesons with the same speed at sea level. The meson trip took less time than its lifetime, and it was possible to evaluate the average number of mesons that would still exist at the sea level. The expectation can be calculated for both cases - with and without lifetime dilation demanded by the special relativity which in this case predicted $\tau/\tau_0 = 8.4 \pm 2$. The measured number of these mesons gave $\tau/\tau_0 = 8.8 \pm 0.8$ which is in good accord with theory.

1.4 Generalized theory of relativity and gravitation

1.4.1 *Tensors: who and why*

After the "geometrization" of physical observations performed in space and time and after the appearance of a unified space-time, it was natural in physics to regard only such notions that are "stable" in themselves but may have different projections on axes of this or that reference frame. Such notions do exist in geometry, they present geometrical objects independent of the coordinates change and are called tensors. In general physics, the simplest examples of tensors are scalars and 3D vectors related to the physical fields like temperature and velocity correspondingly. With rotation or translation of the reference frame, they always remain the same though their components (projections) may become different. A scalar has no projections on axes, therefore, nothing to be changed. A vector has projections changing with rotations and translations of the reference frame. All such reference frame motions can be described by the group theory. More complicated tensors (of rank two) are known in the theory of elasticity, in the description of dielectrics and so on. Actually, we cannot observe or measure a tensor as it is (with the exclusion of a scalar), but we can relate its components to some measurable values and analyze in which way the results of measurements change with the change of the reference frame. If they change according to the specific rules known in geometry for tensor's components, then one can guess that a tensor "exists" and use this notion in speculations.

Tensor analysis is the basis for the GRT because there exist tensor identities that do not depend on the chosen reference frame. If there is a curved surface with a given curvature and there are local reference frames in each point, then these identities preserve their forms for every point. That is why it is so convenient to use tensors for the observations interpretations, i.e. for the formulation of the physical laws in the covariant form. Such theory is geometric because the notions of space and time which are intuitively close to the human perception can be endowed with reference frames, and the mathematical results can obtain the testable interpretations.

Tensor construction of the GRT is actually the following. As long as we use an instrument to measure the phenomena that are weakly interacting with it, everything is more or less simple. But if we deal with situations where the instrument itself can change, we demand that the relations between its readings and the readings of other instruments remain the same as when we use it for the weakly acting phenomena. We speak about such changes of space and time, that the measurable values that are related to the components of a tensor will be related to the same tensor. In fact, the tensor theory is an aspiration to preserve the measurement procedure that is applicable in the usual case also in the case when an instrument changes under experimental conditions in the same way as the elements of phenomena do. In other words, the desired covariance of the physical laws is the logical limit of the "Lorentz contraction", i.e. the effect which is unable to destroy the measurement procedure and provides the possibility to use the experimental data.

Einstein's idea was to translate all the physical laws into the 4D form corresponding to the geometry of space-time, i.e. to the geometry of the consistent model of the real measurable world, that has appeared with the analysis of high speed motions and of the compatibility of mechanics and electrodynamics. In this case the laws would be presented in the covariant, i.e. frame independent way. Therefore, they should be expressed in tensor form.

This idea is not only elegant but, in a sense, unavoidable, because sometimes it is not possible to learn in which way does a reference system move. Thus, when the situation with inertial motion became clearer and the principally new approach of the special theory of relativity appeared, the natural continuation was to regard the accelerated frames.

In physics, an accelerated reference frame always means the appearance of inertial forces, the forces that have the kinematical origin and that sometimes are called fictitious however efficient they are. According to Newton law eq. (1.14), any force corresponds to acceleration, the proportionality coefficient is called *inertial mass* and, in case when the inertial force is the only one acting on a body, the question arises, what is the nature of this co-efficient? It is usually considered inherent to a body, Newton thought about the quantity of matter and tacitly considered the corresponding value to be the same as in his gravity law. But Mach suggested that it was related to the instantaneous action of the distant stars that could be thought of as motionless. Now they speak of Higgs mechanism imparting a property of mass to a particle similarly to imparting a property of electric charge by attaching an electron. The problem is really profound. As we will see below, in special relativity it obtained a new side.

There are many excellent textbooks and monographs on special and general relativity, [Misner *et al.* (1973); Landau and Lifshitz (1967)] being the obvious examples, we should also mention [Bohm (1965); Born (1962); Fock (1964); Taylor and Wheeler (1992)] and, primarily, the works of Einstein himself. In this book we are not going to derive and discuss the equations of the general relativity and their meaning, but we cannot avoid mentioning the main notions that will be needed later.

The rules formulated in geometry deal with the way the tensor components change when the frame basis changes. There exist two different ways of these components change - covariant and contravariant - that are defined by their comparison with the change of the basis. The components of a tensor that change in the covariant way have lower indices. The components of a tensor that change in the contravariant way have upper indices. There is a twice covariant fundamental tensor g_{ik} that provides the possibility to pass from the ith contravariant component of the tensor to the kth covariant component (to lower the index). The fundamental tensor is built out of the pairs of the unity vectors of the canonical basis

$$\vec{e_i}\,\vec{e_k} = g_{ik}$$

This matrix can also be used to build twice contravariant fundamental tensor that performs the transformation of the ith covariant component of the tensor to the kth contravariant component (raises the index). It should be underlined that covariant and contravariant manner of the change of tensor components with the change of reference frame are just those fundamental properties that provide the possibility to interpret them as physically observed values. That is when the observable values are changing in a covariant or contravariant manner with the change of reference frame, we learn that there is a tensor behind them.

Scalars have the quantitative (measurable) value independent of coordinates. Tensors possess quantitative characteristics that could be obtained through convolutions. For a rank 1 tensor, i.e. for a vector, it is its length that coincides with the norm of the vector calculated with the use of special rules related to the definition of the metric of space, that is to the definition of scalar product, that is to the choice of geometry. In Riemannian geometry the norm is calculated as a quadratic form

$$||x||^2 = g_{ik}(x)x^i x^k \qquad (1.37)$$

The geometry suggested by Minkowski to be used for the space-time description was a simple particular case of Riemannian geometry, it corresponded to Minkowski metric tensor consisting of constants, mainly zeros, $g_{ik}(x)_{Mink} \equiv \eta_{ik} = diag\{1, -1, -1, -1\}$. The choice of the metric defines the choice of geometry.

Tensors can be constructed from one another according to the special rules. For example, if φ is a scalar, then $A_i = \frac{\partial \varphi}{\partial x^i}$ are the components of the covariant vector. In order to obtain a rank 2 tensor in the same way, one will need certain numerical arrays describing the effects of parallel transport on the curved manifolds, i.e. the coefficients of Levi-Civita connection, ∇_i, or Christoffel symbols defined as $\Gamma^i_{\ jk} = \frac{1}{2}g^{ih}(\frac{\partial g_{hj}}{\partial x^k} + \frac{\partial g_{hk}}{\partial x^j} - \frac{\partial g_{jk}}{\partial x^h})$ with the usual summation rule for repeating indices. That is if $e_i = \frac{\partial}{\partial x^i}$ is the basis of the tangent space at each point of the manifold, then $\nabla_i e_j = \Gamma^k_{\ ij}e_k$. From the physical point of view it could be said that the connection coefficients describe the "changes in the instruments" when an observer takes them from one point of the space-time to another, thus, they present a mathematical model of what Lorentz and Larmor thought about. If so, the components of rank 2 covariant tensor can be expressed as $A_{ik} = \frac{\partial A_i}{\partial x^k} - \Gamma^j_{\ ik}A_j$. Notice, that the connection coefficients do not possess the tensor properties, that is their components change neither covariantly nor contravariantly with the change of coordinates. This is important when constructing physical theory in terms of tensors. For example, $\Gamma^j_{\ ik}A_j$ is already a tensor.

1.4.2 *Maxwell identities*

Relativistic physics was brought to life by the results of the experiments with light, that is with the electromagnetic phenomena. Maxwell's field theory of electromagnetism was presented by the impressively simple, convenient, embracing all sorts of electromagnetic phenomena set of equations that possessed predictive power which was tested by Hertz who discovered electromagnetic waves. There is no exaggeration in saying that this was the peak of scientific thought. This can also explain why tensors appeared to be so convenient as an instrument of science.

Notice, that the following formal speculations *do not* refer to any physical contents. Let A_i be the components of an arbitrary covariant vector defined in the 4-dimensional space. If we construct the tensor

$$F_{ik} = \frac{\partial A_k}{\partial x^i} - \frac{\partial A_i}{\partial x^k} \qquad (1.38)$$

which is anti-symmetric and, therefore, has 6 independent components, then the following identity

$$\frac{\partial F_{ik}}{\partial x^j} + \frac{\partial F_{ji}}{\partial x^k} + \frac{\partial F_{kj}}{\partial x^i} = 0 \qquad (1.39)$$

is always true.

Let us particularize the space in which we regard this tensor and choose Minkowski space-time defined by the metric composed of constants

$$g_{ik}(x) = diag\{1, -1, -1, -1\} \equiv \eta_{ik} \qquad (1.40)$$

Then making formal designations

$$x^0 \equiv t; x^1 \equiv x; x^2 \equiv y; x^3 \equiv z \qquad (1.41)$$

and

$$F_{12} = E_x; F_{13} = E_y; F_{14} = E_z; F_{23} = -B_z; F_{24} = B_y; F_{34} = -B_x \qquad (1.42)$$

one can check that eq. (1.39) immediately gives the following pair of equations for the usual 3-dimensional vectors and usual vector operations

$$\frac{\partial \mathbf{B}}{\partial t} + curl\mathbf{E} = 0 \qquad (1.43)$$

$$div\mathbf{B} = 0$$

Following the geometrical receipt, we introduce the contravariant tensor $F^{ij} = \eta^{ik}\eta^{jm}F_{mk}$, and define the new 4-vector I^i according to

$$I^i = \frac{\partial F^{ij}}{\partial x^j} \qquad (1.44)$$

Then making additional formal designations

$$I^1 = \rho, I^2 = j_x; I^3 = j_y; I^4 = j_z \tag{1.45}$$

one can again use eq. (1.39) and definitions eq. (1.42) and obtain another pair of equations

$$curl\mathbf{B} - \frac{\partial \mathbf{E}}{\partial t} = \mathbf{j} \tag{1.46}$$
$$div\mathbf{E} = \rho$$

For a physicist, it is really hard not to recognize two pairs of Maxwell equations in eqs. (2.31,1.46), besides the immediate ideas of scalar potential, φ, vector potential, A_α ($\alpha = 1, 2, 3$), and electromagnetic tensor F_{ij} come to mind. But the equations (2.31,1.46) contain no charges, no experimental results and no electromagnetism whatsoever. They are pure geometry and formal definitions. As soon as we have eq. (1.38) and eq. (1.39), the notation present in eqs. (2.31,1.46) may mean anything we want.

Suppose, we examine the situation when there is no dependence on t. Then we see that the systems of equations (2.31,1.46) break into symmetrical E-part and B-part. For example, the E-part gives

$$div\mathbf{E} = \rho \tag{1.47}$$
$$curl\mathbf{E} = 0$$

where $\mathbf{E}$ is a 3D vector defined by eqs. (1.42) and (1.38). New designation $\mathbf{E} \equiv -\frac{1}{4\pi}\nabla\varphi$ together with the first equation of eq. (1.47) give Poisson equation

$$\Delta\varphi = -4\pi\rho$$

Now if we choose to express ρ with the help of δ-function, $\rho \equiv q \int \delta(\mathbf{r})dV$, where q is a constant, the application of Gauss theorem will give a formal expression

$$\mathbf{E} = \frac{q\mathbf{r}}{r^3} \tag{1.48}$$

For any physicist this is nothing but the stress of the electric field of a point charge, q, that results from the Coulomb law and has the potential $\varphi = -\frac{q}{r}$.

The B-part gives

$$div\mathbf{B} = 0 \qquad (1.49)$$
$$curl\mathbf{B} = \mathbf{j}$$

Here $\mathbf{B}$ is also a 3D vector defined by eqs. (1.42) and (1.38). In case of the t-independence, $\mathbf{B} = curl\mathbf{A}$, and the second equation in eq. (1.49) gives

$$\nabla(div\mathbf{A}) - \Delta\mathbf{A} = \mathbf{j}$$

As soon as we choose $div\mathbf{A} = 0$, we again obtain the Poisson equation

$$\Delta\mathbf{A} = -\mathbf{j}$$

which gives

$$\mathbf{B} = curl \int \frac{\mathbf{j}}{r} dV = \int \frac{[\mathbf{j}, \mathbf{r}]}{r^3} dV \qquad (1.50)$$

where the integration is performed in such a way that the radius-vector is directed from dV to the observation point. And any physicist would say that this is the Biot-Savart law for the magnetic field of a stationary current.

Continuing the play with the geometrical expressions, one can demand

$$\frac{\partial F^{ij}}{\partial x^j} = I^i = 0 \qquad (1.51)$$

and substitute $F^{ik} = \frac{\partial A^k}{\partial x_i} - \frac{\partial A^i}{\partial x_k}$ into the last expression. This will give

$$\frac{\partial^2 A^k}{\partial x_i \partial x^k} - \frac{\partial^2 A^i}{\partial x_k \partial x^k} = 0 \qquad (1.52)$$

Suppose now that the sum $\frac{\partial A^k}{\partial x^k}$ is equal to zero,

$$\frac{\partial A^k}{\partial x^k} = 0 \qquad (1.53)$$

Then the first term in eq. (A.47) vanishes, and there appears

$$\frac{\partial^2 A^i}{\partial x_k \partial x^k} = g^{kl} \frac{\partial^2 A^i}{\partial x^k \partial x^l} = 0 \qquad (1.54)$$

which is exactly the wave equation written in the 4-dimensional form.

The condition eq. (1.53) is known in physics as Lorentz gauge. If we use our notations eq. (1.41) and introduce $\varphi \equiv A^0$, then eq. (1.53) and eq. (1.54) give familiar expressions

$$\frac{\partial \varphi}{\partial t} + div\mathbf{A} = 0 \qquad (1.55)$$

for Lorentz gauge and

$$\Delta \mathbf{A} - \frac{\partial^2 \mathbf{A}}{\partial t^2} = 0 \qquad (1.56)$$

for the 3-dimensional "vector potential" with the D'Alembert operator $\Box = \Delta - \frac{\partial^2}{\partial t^2}$ instead of Laplace operator which we saw before. Obviously, eq. (1.56) suggests the existence of the wave solutions. (Notice, that our speculations correspond to the system of units in which $c = 1$.) One could wonder, what kind of limitations do the eqs. (1.53-1.55) (Lorentz gauge) impose? The answer is: the equations containing the vector field A^i must remain the same for any scale transformation. Historically, such scale transformation, first, corresponded to a multiplication by a scalar factor, later the factor became a complex value, thus, turning the scale transformation into the change of phase.

If one applies this formal scheme to electromagnetic phenomena, one can predict the existence of the electromagnetic waves. All the remaining rich and detailed mathematical apparatus developed for calculations in electrodynamics, naturally, remains the same but could mean anything else. But this apparent simplicity can be deceiving.

Already Maxwell himself thought of the similarity between Newton law for gravity and Coulomb law for electrostatics and of the application of the field theory to the gravitation. But he concluded that the similarity is only formal, because the speculations dealing with energy reasons which were so fruitful for electricity demanded that the same sign of (gravitation) charges must cause not the attraction but repulsion. He suggested that a massive body contributes negatively to the energy of the gravitational field in its vicinity, and, thus, the change in energy when the distance between the two bodies increase would be positive and correspond to an attraction force. But since energy was proportional to the square of the field strength, a negative energy meant an imaginary field strength. He tried to assume that the gravitation arose from the action of the surrounding medium with a huge positive energy at every point, so as to obtain the resultant attraction when the bodies are driven apart, but finally admitted that a medium can hardly possess such properties and dropped the idea.

Nevertheless, one cannot but be amazed how simply a mathematician who only knew about the almost trivial identity eq. (1.39) could hold in his hands the whole of electrodynamics - which is not only one of the major parts of physics but, actually, is still the basis of our electricity based civilization - without a slightest idea of it, and how hard and long was the opposite way - from experimental observations and measurements to their generalization into the laws of physics and then to the recognition of the deep interrelation between the mathematical structures of these laws.

As soon as one recognizes that this road has two ways, many new opportunities appear. Nowadays, there are other approaches that also start with a pure and simple mathematical statement, and after a certain sequence of steps, one may arrive at the equations well-known in physics and theoretical physics. For example, the discrete calculus approach developed by Kauffman [Kauffman (2004)] seems to be very effective. The problem is the obtaining of new and testable physical predictions and performing the corresponding experimental measurements. They must show that the new approach is better than the previous one, especially if the predictions of the new and old approaches differ. Einstein's advantage was that he was a physicist and knew where to look for testing, actually, he discovered in the toolkit an instrument suitable for his needs, while in this toolkit there is a lot of other refined and many-functional devices whose specific purposes are still unknown. Of course, mathematics is far from being a simple toolkit, it has its own identity and far going properties that can penetrate into physics. We are still rather far from being acquainted with all these abilities of mathematics. Relativity theory also known as geometrodynamics appeared to be one of the fortunate and tempting examples.

1.4.3 *Least action principle*

There is another geometrical idea that has found even more general physical consequences. In mathematics it is known as the variation principle which is the basis for variational calculus. In physics it presents a formal postulate - the least action principle - which appears to be extremely fruitful. The variation principle establishes the correspondence between the geometrical objects and their physical analogues according to the following scheme. The curve that is the extremal of the functional of length is called geodesics. In Euclidean space the length is given by the functional

$$l = \int\limits_{t_1}^{t_2} \sqrt{(\frac{\partial x}{\partial t})^2 + (\frac{\partial y}{\partial t})^2 + (\frac{\partial z}{\partial t})^2} dt \qquad (1.57)$$

The extremal of the functional, S

$$S = b \int\limits_{t_1}^{t_2} [(\frac{\partial x}{\partial t})^2 + (\frac{\partial y}{\partial t})^2 + (\frac{\partial z}{\partial t})^2] dt \qquad (1.58)$$

coincides with the extremal of the functional of length, l. The choice $b = \frac{m}{2}$, where m is mass, attributes the physical sense of kinetic energy to the expression $\frac{m}{2}[(\frac{\partial x}{\partial t})^2 + (\frac{\partial y}{\partial t})^2 + (\frac{\partial z}{\partial t})^2]$, kinetic energy being a scalar characterizing the physical system. The coincidence mentioned above leads to suggestion that the extremal of the path and the extremal of the total energy give one and the same curve which is called geodesics.

Notice, that the measurable physical values should be related to one and the same type of the geometrical objects, for example, to the contravariant components of vectors. This means that when they (or maybe some tensors of rank two and higher), are used to construct a scalar, the fundamental tensor, i.e. the metric, has to be used in order to obtain (here) the covariant components needed to construct a scalar. Thus, the metric always enters the expression for action, though sometimes it is not written there explicitly. The reason for the last is that in the 3D Euclidean space co- and contravariant vectors coincide.

The Hamilton principle of least action is a postulate according to which

• the motion of a physical system corresponds to the extremal of the functional of action defined as

$$S = \int\limits_{t_1}^{t_2} L(q, \frac{\partial q}{\partial t}, t) dt \qquad (1.59)$$

where q is the generalized coordinate and L is the Lagrange function, that is the difference between kinetic and potential energies of the system. For example, in a classical Newton problem, a body moving in a gravitation field has $L = \frac{mv^2}{2} - m\varphi$, where $\varphi = G\frac{M}{r}$ is Newtonian potential. For a free body that suffers no force, eq. (1.59) coincides with eq. (1.58). The variation of the functional of action leads to the Euler-Lagrange equations

$$\frac{\partial L}{\partial q^i} - \frac{d}{dt}\left(\frac{\partial L}{\partial(\frac{\partial q^i}{\partial t})}\right) = 0 \tag{1.60}$$

that present the equations of dynamics.

When applied to the relativistic mechanics with the phenomena taking place in the 4D space-time with non-Euclidean geometry, these ideas suggest that the curve which is now the world line can be obtained when eqs. (1.57, 1.58) are substituted by [Landau and Lifshitz (1967)]

$$S = -\alpha \int_a^b ds$$

where ds is the differential of the interval defined by eq. (1.35), a and b are the initial and the final events, integral is taken along the world line and α is a constant. Introducing the proper time, i.e. time measured by the watch moving together with the moving body, one gets

$$S = -\int_a^b \alpha c\sqrt{1 - \frac{v^2}{c^2}}dt$$

and the comparison with eq. (1.59) gives the relativistic expression for the Lagrangian of a free body

$$L = -\alpha c\sqrt{1 - \frac{v^2}{c^2}} \simeq -\alpha c + \frac{\alpha v^2}{2c}$$

If we want to preserve the form of the classical dynamics equations, the limit $c \to \infty$ for a free particle with mass m must give the usual expression for the kinetic energy which means that $\alpha = mc$, while the additive constant in the expression for L is not essential for calculations, because they deal with derivatives. Then finally, the scalar corresponding to the motion of the free particle is presented by the Lagrange function

$$L = -mc^2 + \frac{mv^2}{2} \tag{1.61}$$

If we turn to electrodynamics and regard interacting particles, the action must consist of three terms, S_f, S_m, S_{mf}, corresponding to electromagnetic field, charged particles and the interaction between particles and field. Particularly,

$$S_f = a \iint F_{ik} F^{ik} dV dt$$

where the field is produced by the charges and is described by a pair (φ, A_α), F_{ik} is the electromagnetic tensor and a is a dimensionless constant depending on the system of units. If we choose Gauss system of units $(a = -1/16\pi)$, introduce 4D volume according to $d\Omega = cdtdV$ and use the metric explicitly, then the action for the electromagnetic field takes the form

$$S_f = -\frac{1}{16\pi c} \iint g_{ij} g_{lk} F^{jl} F^{ik} d\Omega \qquad (1.62)$$

The full expression for the action that provides the possibility to obtain the equations of electrodynamics is [Landau and Lifshitz (1967)]

$$S = -\frac{1}{16\pi c} \iint g_{ij} g_{lk} F^{jl} F^{ik} d\Omega - \sum \int mc \, ds - \frac{1}{c^2} \sum \int A_k j^k d\Omega \quad (1.63)$$

where $j^k = (c\rho, \mathbf{j}), \mathbf{j} = \rho \mathbf{v}$ is the electric current density, and ρ is the density of electric charges.

The choice of the scalar that is supposed to become the starting point of the variational procedure may be the topic of discussion. Einstein thought that it should be as simple as possible, but the criterion of simplicity - such that the chosen scalar will necessarily lead to the essential results, is hard to give. So, the construction of a scalar to be varied in order to get the sensible physical equations demands certain skill.

1.4.4 *Mass and energy*

The expression for the Lagrangian of a free body eq. (1.61) that usually corresponds to kinetic energy, now contains two specific novelties that deserve special discussion. First, there is a constant term $-mc^2$ which depends neither on the velocity of a body nor on some other bodies or fields but only on the mass of the body itself. It is proportional to the constant which characterizes the geometrical properties of the chosen model of space and which is identified with the speed of light. It is now known as the rest energy of a body, and it is a huge value in comparison with the second term. It makes one think that mass characterizes not only the inertia of a body but also its energy, i.e. its ability to perform work, even if the body is static. The appearance of such a term is suitable for testing the new theory. Obviously, as long as there are no changes with the bodies that can interact, for example, in an elastic way, and their initial and the final states are the same, nothing can point at this entity. But suppose the changes did take place. For example [Kokarev (2010)], one of the two colliding bodies with masses m_1, m_2 broke into two equal pieces with mass $\frac{m'}{2}$. The energy conservation gives

$$-m_1c^2 + \frac{m_1v_1^2}{2} - m_2c^2 + \frac{m_2v_2^2}{2}$$
$$= -\frac{m'}{2}c^2 + \frac{m'u_1'^2}{4} - \frac{m'}{2}c^2 + \frac{m'u_2'^2}{4} - m_2c^2 + \frac{m_2v_2'^2}{2} + A$$

where A is the binding energy of the first body or the work of splitting. The rest energy of the second body cancels out, while the rest energy of the first body redistributes between all the interacting particles. If $v_2 = 0$ and $v_1 \to 0$, the process is an illustration of a spontaneous decay of the first body. Then we get

$$(m' - m_1)c^2 - A \geqslant 0$$

where the equality corresponds to the case of the motionless fragments. We see that the spontaneous decay is possible only if the total mass of fragments m' is not equal to the mass of the compound body m_1. Since we do observe the spontaneous decay and can even measure and compare the values of all the terms in the last equation, we can make sure that the theoretical result is in good accord with measurements. The difference of the rest energies corresponding to the mass defect, $E = \Delta mc^2$, is a notion that is well known and is now used in nuclear plants. This makes one think that either there is something more than formalism in this space-time modeling, and that Einstein found a deep physical relation between the basic notions of science, or the whole picture and the measurements procedures are so self-consistent that there could hardly be any reason to doubt it.

For another example similar to that given in [Einstein (1905b)], let a body emit two photons with energies $h\nu$ in the opposite directions. Let us calculate the differences between its initial energy and final energy in two IRFs: in the first of them the body is at rest and has energy E_i, in the second it moves with speed v which is less than the speed of light, and has the energy W_i. Then

$$E_i = E_f + 2h\nu$$

$$W_i = W_f + \frac{2h\nu}{\sqrt{1 - v^2/c^2}}$$

The square root in the second term reflects the frequency changes in the moving IRF. Introducing the notation $K = W - E$ and subtracting, one can find

$$K_i - K_f = 2h\nu\left(\frac{1}{\sqrt{1 - v^2/c^2}} - 1\right) = \frac{2h\nu}{c^2}\frac{v^2}{2}$$

and see that the emission of massless photons causes the change of the body mass. Therefore, the mass and energy seem to be the same, heating of a body should cause the increase of its inertia, the spring stretching would do the same, and as soon as we mention the equivalence principle, we would have to admit that all forms of energy interact gravitationally.

The second important feature that is present in eq. (1.61), namely, in the second term, is the effective dependence of the body's mass on its velocity. Taking the constant factor out, we obtain

$$L = -c^2 m\left(1 - \frac{v^2}{2c^2}\right) = -c^2 m(v)$$

that is the kinetic energy of a free body is related to the mass dependence on the velocity of this body. Recollecting the origin of the term in brackets, we see that

$$m(v) = \frac{m}{\sqrt{1 - v^2/c^2}}$$

where m is the rest mass, i.e. mass measured in the IRF where the body is at rest. The last expression is sometimes called relativistic mass.

1.4.5 *Field equations*

The geometrization of physics that was begun in SRT by the introduction of the postulate of the finite and constant speed of light was to be continued with account for the accelerated reference frames. When the motion was straight and uniform, Minkowski diagrams presented pseudo-Euclidean flat space with straight world lines on it. The accelerations made the world lines curved, the flat space transformed into curved one, Minkowski metric was now not enough, the metric started to depend on coordinates that is became Riemannian. Now the motion took place on the relief, and the famous thought experiment with the accelerating lift and lift with the cabin standing on the Earth surface led to the equivalence principle which postulated that

- there is no local physical experiment with the help of which it is possible to distinguish inertial and gravitational forces.

This meant that the generalization of the special relativity theory to the accelerated motions was going to be the theory of gravitation. At the same time, it meant that a body traveling freely in the gravitation field, i.e. on the curved relief, goes along the geodesics, that is along the extremal between two events on the relief defined by the equation of geodesics

$$\frac{d^2x^i}{ds^2} + \Gamma^i_{jk}\frac{dx^j}{ds}\frac{dx^k}{ds} = 0 \qquad (1.64)$$

Since $\frac{d^2x^i}{ds^2}$ is the 4-acceleration, the expression $-m\Gamma^i_{jk}\frac{dx^j}{ds}\frac{dx^k}{ds}$ can be called the "4-force" acting on a probe particle with mass m in the gravitation field. We see that the "stress", Γ^i_{jk} of this force is calculated by taking the derivative of a "potential" coinciding now with the metric tensor, g_{ik}. Thus, the equation of motion of a probe particle in the field of gravitation forces now corresponds to a curve originating from geometry.

In the field theory language, Newton gravity law is equivalent to Poisson equation

$$\Delta\varphi = 4\pi\rho \qquad (1.65)$$

where φ is the gravity potential and ρ is the density of matter. This equation is essentially non-relativistic, because it corresponds to the infinite speed of a signal which instantly "informs" the surroundings that the distant matter (for example, a star) has redistributed (for example, exploded). In other words, Laplace operator is not scalar with regard to boosts. The equation of motion of a test particle with mass, m in this field is given by the Newton law of dynamics

$$m\frac{d^2x}{dt^2} = -\nabla\varphi \qquad (1.66)$$

The first metric theory of gravitation based on a vector gravitation potential was suggested by Minkowski already in 1908, and the first metric theory of gravitation that maintained stable orbits for planets and was based on scalar gravitation potential was presented in 1913. It had a character of a series of ingenious guesses, one of which was the straightforward substitution of D'Alembert operator instead of Laplace one into eq. (1.65), but neither could both of them be derived from the "first principles" (with the help of variation approach), nor did they agree with observations. For example, the direction of the Mercury orbit precession calculated in Nordström theory appeared to be opposite to the known one. Besides, it had an inherent fault due to its linearity. According to special relativity, mass is equivalent to energy and energy to mass, therefore, the energy must produce additional gravitation which means that the field equations must be non-linear.

Einstein came to deeper understanding of the role of geometry and of the properties of tensors through the discussions and joint work with Marcel Grossmann [Einstein and Grossmann (1914)] who was a mathematician and Einstein's friend and classmate. Finally, Einstein suggested [Einstein (1916a)] the following system of 10 non-linear differential equations to describe the gravitation field

$$G_{ik} = \frac{8\pi G}{c^4} T_{ik}, \tag{1.67}$$

where $G_{ik} = \frac{\partial \Gamma_{ik}^j}{\partial x^j} + \Gamma_{im}^j \Gamma_{kj}^m$ is the so-called Einstein tensor, $T_{ik} = \frac{1}{c} \int \Lambda \sqrt{-g} d\Omega$ is the energy-momentum tensor in which Λ is the Lagrangian density, $g = \det\{g_{ik}\}$ and G is the gravitation constant.

A scalar suitable for the variation procedure leading to the physically sensible results, that is to Einstein field equations, was suggested by Hilbert [Hilbert (1915)] and by Einstein [Einstein (1916b)], it is the so-called scalar curvature, R, obtained from the Ricci tensor, R_{ik} by the convolution

$$R = g^{ik} R_{ik} \tag{1.68}$$

Ricci tensor is symmetric and is defined as

$$R_{ik} = \frac{\partial \Gamma_{ik}^l}{\partial x^l} - \frac{\partial \Gamma_{il}^l}{\partial x^k} + \Gamma_{ik}^l \Gamma_{lm}^m - \Gamma_{il}^m \Gamma_{km}^l \tag{1.69}$$

Ricci tensor in its turn is obtained also by convolution

$$R_{ik} = g^{jm} R_{jimk} = R^j_{ijk}$$

where R^i_{klm} is the curvature tensor defined in differential geometry as

$$R^i_{klm} = \frac{\partial \Gamma^i_{km}}{\partial x^l} - \frac{\partial \Gamma^i_{kl}}{\partial x^m} + \Gamma^i_{jl}\Gamma^j_{km} - \Gamma^i_{jm}\Gamma^j_{kl} \tag{1.70}$$

The Hilbert-Einstein action is defined as

$$S_g = \int \sqrt{-g} R d\Omega \tag{1.71}$$

The variation is performed with regard to the condition that there is no variation on the boundary, and it gives

$$\delta S_g = -\frac{c^3}{16\pi G} \int (R_{ik} - \frac{1}{2}g_{ik}R)\delta g^{ik} \sqrt{-g} d\Omega$$

Taking into consideration the part of action, S_m that deals with matter, one has

$$\delta S_m = \frac{1}{2c} \int T_{ik}\delta g^{ik} \sqrt{-g} d\Omega$$

Now there is no need to account for S_{gm} because in Riemann geometry all the derivatives in Lagrangian were substituted by the covariant derivatives that is with regard to the connection coefficients. Thus, the varying leads to Einstein field equations in the form

$$R_{ik} - \frac{1}{2}g_{ik}R = \frac{8\pi G}{c^4} T_{ik} \tag{1.72}$$

which is equivalent to eq. (2.27). These equations can be used to find the explicit form of the metric which also enters the equations of geodesics eq. (1.64) and finally gives a curve corresponding to the equation of motion that can be related to the observations.

1.4.6 *Gravitational waves*

There is a case when the field equations can be integrated [Einstein (1916c)]. Let us regard a region far from masses, that is the right-hand side of eq. (1.72) is equal to zero, the field is weak and the space is almost flat.

This means that the metric can be approximated by the linearly perturbed Minkowski metric η_{ik}

$$g_{ik}(x) = \eta_{ik} + h_{ik}(x)$$

The perturbation $h_{ik}(x)$ is so small that all the terms of orders higher than one that appear in Einstein equations eq. (1.72) when calculating the derivatives are negligible. In order to simplify the field equations, let us use the notation $h = \eta^{jl}h_{jl}$ and introduce new variables

$$\widetilde{h}_{ik} = h_{ik} - \frac{1}{2}\eta_{ik}h$$

Then we choose the conditions similar to Lorentz gauge eq. (1.53) which are used in electrodynamics

$$\frac{\partial \widetilde{h}^{ik}}{\partial x^k} = 0 \tag{1.73}$$

This brings the field equations in empty space to the final form

$$\Box \widetilde{h}_{ik} = \left(\frac{\partial^2}{\partial(x^i)^2} - \frac{1}{c^2}\frac{\partial^2}{\partial t^2}\right)\widetilde{h}_{ik} = 0 \tag{1.74}$$

and we see that the deformation of the metric tensor can be presented in the form of waves. In eq. (1.74) we introduced the factor $1/c^2$ explicitly in order to remind how it appeared. If the theory is constructed in such a way that it must include a constant with the dimensionality of speed, and if there are reasons to choose this constant equal to the physically meaningful speed of light, c, then the passage to the description of the non-inertial motion that involves gravitation shows that the equation for the metric in a slightly curved space could have the wave-like solutions, and the speed of these waves has to be the same with that of light.

The simplest solutions of eq. (1.74) are the monochromatic plane waves

$$\widetilde{h}_{ik} = Re[a_{ik}\exp(iK_jx^j)] \tag{1.75}$$

that suffice eq. (1.74) only if the wave vector, K_j of the gravitational waves suffices the condition

$$K_iK^i = 0$$

that is K^i is the light-like vector, and the gravitation waves propagate in the direction $\frac{1}{K_0}(K_x, K_y, K_z)$ with the speed of light. Lorentz conditions eq. (1.73) will, therefore, take the form

$$a_{ik} K^i = 0$$

which means that the gravitational waves are transversal. The appropriate choice of the coordinates transformation makes it possible to pass to the convenient form of the metric in which $\tilde{h}_{ik} = h_{ik}$, $h_{i0} = 0$, which means that only space components of h_{ik} differ from zero, and $h_{kk} = 0$, that is the trace of the tensor is equal to zero. Such choice is called transversal traceless gauge (TT-gauge) which is convenient for the investigations of the gravitational waves in vacuum. For example, if the wave propagates along Ox axis, then the perturbation part of the metric tensor in the TT-gauge obtains the form

$$h_{ik}^{TT} = \begin{pmatrix} 0 & 0 & 0 & 0 \\ 0 & 0 & 0 & 0 \\ 0 & 0 & h_{yy} & h_{yz} \\ 0 & 0 & h_{yz} & -h_{yy} \end{pmatrix} \tag{1.76}$$

and the plane wave can be presented as a combination of two polarizations. Thus, for the polarized gravitation wave propagating in the Ox-direction the metric tensor is

$$g_{ik} = \begin{pmatrix} 1 & 0 & 0 & 0 \\ 0 & -1 & 0 & 0 \\ 0 & 0 & -1+h & 0 \\ 0 & 0 & 0 & -1-h \end{pmatrix} \tag{1.77}$$

where $h = aRe \exp[-iD(t - \frac{x}{c})]$ and D is the frequency of the gravitational wave.

The research aimed at the direct experimental detection of the gravitational waves was begun by Weber in 60-s [Weber (1960)] and is still one of the main challenges in fundamental physics. The problem is that the amplitude of the gravitational wave is still far below the instruments' sensitivity. Two main directions of the experimental research are focused around the interferometric method (measurement of the distance between two free falling bodies) and resonance antenna method (measuring the resonance when the gravitational wave frequency coincides with the eigenfrequency of the system of two bodies interacting by non-gravitational forces). Both methods are based on the use of the geodesic deviation equation, the first exploits its homogeneous version, the second deals with the non-homogeneous version. The geodesic deviation equation describes the change of the distance between two bodies when the gravitation wave falls upon them. Some more details are given in Chapter 3.

Nowadays, there is solid but indirect evidence of the existence of the gravitational waves. It is possible to calculate the loss of energy emitted by the object in the form of gravitational radiation. And the appropriate object was found and investigated by Hulse and Taylor [Hulse and Taylor (1975)] during many years. It was a binary neutron star system PSR1913+16, one of the components of which is a pulsar, and the measured orbital period decay [Taylor and McCulloch (1980)] appeared to be equal to the calculated value of such decay presumably due to gravitational radiation within 10^{-3} accuracy.

1.5 GRT - first approximation - predictions and tests

Let us remind how the mathematical structure that Einstein obtained generalizing the relativity theory to the accelerated systems has transformed into the physically testable predictions for the gravitation phenomena. Of course, first of all, there must be shown the way to the simple Newton formula for the gravity law which works well enough for the Solar system scale. After that the other qualitative and quantitative predictions would be discussed.

1.5.1 *Newton gravity*

There are several ways of presenting the derivation of the Newton formula. Here we follow the original Einstein paper [Einstein (1916a)]. Let us regard the weak gravitational field, i.e. such that the metric is almost flat and can be presented as $g_{ik}(x) = \eta_{ik} + h_{ik}(x)$, $h_{ik}(x) = \epsilon \widetilde{h}_{ik}(x)$; $\epsilon << 1$ and $g_{ik}(x) \rightarrow \eta_{ik}$ at infinity where the space is considered to be empty. Let us use the following assumptions:

(1) The approximation is linear, that is all the terms of the order higher than h_{ik} are neglected. Then the motion is described by the geodesics $\frac{d^2x^i}{ds^2} + \Gamma^i_{jk}\frac{dx^j}{ds}\frac{dx^k}{ds} = 0$, where $\Gamma^i_{jk} = \frac{1}{2}\eta^{ih}(\frac{\partial h_{hj}}{\partial x^k} + \frac{\partial h_{hk}}{\partial x^j} - \frac{\partial h_{jk}}{\partial x^h})$.

(2) All the speeds are much slower than the speed of light, c, and we define $c = 1$, that is $\frac{dx^1}{ds}, \frac{dx^2}{ds}, \frac{dx^3}{ds} << 1$; $\frac{dx^0}{ds} \sim 1$, and the field is quasi-static.

The first assumption presumes that $\Gamma^i{}_{jk} << 1$, and the geodesics equation contains only one term which corresponds to $k = j = 0$. Therefore,

$$\frac{d^2 x^i}{dt^2} = -\Gamma^i_{00}$$

where $ds = dx^0 = dt$ was used. The second assumption suggests to neglect the time derivatives in the expression for Christoffel symbol and get

$$\frac{d^2 x^\alpha}{dt^2} = -\frac{1}{2}\frac{\partial h_{00}}{\partial x^\alpha}; \alpha = 1, 2, 3 \tag{1.78}$$

Thus, we see that $\frac{h_{00}}{2}$ can be related to a gravitational potential whose gradient is proportional to acceleration, that is to a force according to Newton dynamics law. The components of the energy-momentum tensor for the dust-like matter are: $T^{00} = \rho c^2$ - energy density, $T^{0\alpha} = \rho c v^\alpha$ - components of momentum density, $T^{\alpha\beta} = \rho v^\alpha v^\beta$ - the flow of the momentum α-component in β-direction. Then the field equation

$$\frac{\partial \Gamma^j_{ik}}{\partial x^j} + \Gamma^j_{im}\Gamma^m_{kj} = -\kappa(T_{ik} - \frac{1}{2}g_{ik}T) \tag{1.79}$$

in our approximation contains only $T^{00} = \rho$. In the left-hand side of eq. (1.79), only the first term remains because of the chosen approximation, that is

$$\frac{\partial \Gamma^j_{ik}}{\partial x^j} + \Gamma^j_{im}\Gamma^m_{kj} \simeq \frac{\partial}{\partial x^0}\Gamma^0_{ik} - \frac{\partial}{\partial x^1}\Gamma^1_{ik} - \frac{\partial}{\partial x^2}\Gamma^2_{ik} - \frac{\partial}{\partial x^3}\Gamma^3_{ik}$$

Besides, we can omit the time derivatives and let $i = k = 0$ as before and notice that

$$-\frac{1}{2}(\frac{\partial^2 h_{00}}{\partial(x^1)^2} + \frac{\partial^2 h_{00}}{\partial(x^2)^2} + \frac{\partial^2 h_{00}}{\partial(x^3)^2}) = -\frac{1}{2}\Delta h_{00}$$

The last equation together with eq. (1.79) gives the Poisson equation

$$\Delta h_{00} = \kappa\rho \tag{1.80}$$

Finally, we see that eq. (1.80) and eq. (1.78) are equivalent to the Newton gravitation law expressed by eqs. (1.65-1.66). Substituting the solution of Poisson equation into the metric, one gets

$$g_{00} = 1 + \frac{2\varphi}{c^2}; \varphi = G\frac{M}{r} \tag{1.81}$$

One can see that the acceptance of the equivalence principle in case of the weak gravitation field results in the practical identity of the picture obtained by the geometrical approach and of the picture obtained by the approach dealing with forces. The result given by eq. (1.81) is called the Newtonian limit, and in general relativity the components of metric other than g_{00} could be accounted for only in the next - post-Newtonian - approximation in which the energy of a particle-field interaction depends on the particle velocity.

At first glance, it seems that the meaning of the obtained result is the possibility to compensate any gravitational field by means of using an accelerated, freely falling frame. But this is not the general case. The complete compensation is possible only in a single point of a free falling cabin - in its center of inertia. Performing measurements inside the cabin that has a finite volume, one could discover forces, and the further is the point of measurement from the center of inertia, the larger are the forces. That is several bodies suspended inside the freely falling cabin can start to move relative to one another. This effect results from the possible inhomogeneity of the gravitational field. The corresponding forces are called tidal, they act inside any body of finite volume, and they cannot be compensated by an acceleration. Thus, we see that the homogeneous gravitational field or the field of inertial forces appearing in the uniformly rotating frame can be eliminated by the passage to the suitable reference frame. But in other cases there is no such possibility and certain "true" gravitation forces are intermixed with the forces of inertia and can't be distinguished from one another.

Still, the existence of the Newtonian limit as the first approximation to the geometrical general relativity theory is encouraging. The point-like probe particles taking part in the gravitational interaction move along the geodesics corresponding to the metric of space. There is a frame in which this motion is a free fall. If the metric is flat, i.e. the curvature tensor is identically zero, the force field has a purely kinematical inertial nature. If the curvature tensor does not vanish identically, there is a true gravitational field related to the curvature, and it cannot be separated from the field of inertial forces. World lines visualizing the histories of processes are now drawn on the curved surface and, therefore, they are not obligatory straight. If the metric is Riemannian, the local correspondence between special relativity and general relativity takes place, and GRT inherits all the specific effects genuine for SRT.

1.5.2   *Classical tests*

The presence of the force potential presumes that the velocities of local IRFs in different points of space-time differ. Therefore, the measurements performed in the different points of the curved space-time could show that the lengths of meter sticks and the time durations between the clock tickings also differ for these points. If the gravitational field is directed parallel to Ox axis, then the measurement of the segment of unit length, dx, which is placed into the field and oriented in parallel to it will show contraction. In the Newtonian limit (weak field) the segment length becomes equal to $dx' = (1 - \frac{\varphi}{c^2})dx$. The measurement of the segment perpendicular to the direction of the field does not show any change in its length. Therefore, the geometry of space in the gravitational field is not Euclidean, but the difference is very small, and it is hard to measure.

1.5.2.1   *Gravitational redshift*

Gravitational redshift of light demonstrates the change in the time interval duration. The measurement of the proper time in the gravitational field will read $d\tau = \sqrt{g_{00}}dt > dt$ which means that the higher is the gravitational potential, the slower are the process there when observed from outside of this region. In order to come to the possibility of the direct experimental testing, it is convenient to rewrite this relation for frequencies and get $\nu_1\sqrt{g_{00}(1)} = \nu_2\sqrt{g_{00}(2)}$, that is

$$\frac{\Delta\nu}{\nu} = \sqrt{\frac{g_{00}(1)}{g_{00}(2)}} - 1 \qquad (1.82)$$

In the Newtonian limit the potential close to the planet surface gives $\Delta\varphi = g\Delta h$, where g is the free fall acceleration and Δh is the difference in the altitude. Then the last expression leads to the formula

$$\frac{\Delta\nu}{\nu} = -\frac{g}{c^2}\Delta h$$

which can be used in measurements.

The first attempt to measure the gravitational redshift in the spectra obtained from Sirius atmosphere was made by W.Adams in 1925. But the decisive test took place only in 1959 when Pound and Rebka [Pound and Rebka (1959)] the two sources placed at the top and at the bottom of Harvard University's Jefferson tower and measured the relative redshift with the help of Mossbauer effect. Since then such measurements were also repeated many times, and the latest test was reported in 2010 by H.Muller [Muller (2010)] who found the effect to be in exact correspondence with GRT within an accuracy 10,000 times better than before.

1.5.2.2 *Orbit precession*

In the history of Solar system astronomy of the 19th century there are two dramatic events related to the Newton theory of gravitation. The first of them is the success in discovering a then unknown planet, Neptune, that belongs to Le Verrier and Adams who did it "with the end of the pen" trusting Newton theory. The second is a failure in explaining the rate of Mercury's orbit perihelion precession around the Sun. As it is known from classical mechanics and mathematics, the only situations when the closed (elliptical) orbits are possible are the motions of a body in the Newtonian potential and in the potential of the harmonic oscillator. Any possible perturbation will make the orbit unclosed, and when the perturbation is small, the effective picture is the precession of the orbit's pericenter around the force center. For Mercury, the measured rate of the perihelion precession was measured to be 5644 angular seconds per century. In order to find the cause of this effect, the precession of the Earth's orbit axis was taken into account and gave 5026 angular seconds per century. Some 575 angular seconds more were found to result from the gravitational action of other planets, but 43 seconds per century were missing, while the accuracy of measurements could account for fractions of angular seconds. And from the point of view of Newton gravitation theory this was really puzzling.

In 1915, Einstein attempted to find a solution of this problem taking into account the second approximation for his theory. The result was a shock for him as he wrote, because it gave precisely the needed 43 seconds [Einstein (1915)]. In 1916, Schwarzschild [Schwarzschild (1916)] found the exact solution of Einstein field equations for the stationary spherically symmetric case which gave the same formula and, besides, reproduced the close approximation to the 3-d Kepler law.

1.5.2.3 *Deflection of a light beam*

The effect of deflection of a light beam in the gravitation field of a spherically symmetric body presents a special interest for several reasons. First, this problem can be solved in classical mechanics and, thus, give a reference pattern; second, the vulgar treatment of the relativistic approach gives a prediction that is worse than that of the classical theory from the point of view of comparison with measurements; third, the calculated angle presents not a small correction to the classical result and, therefore, a measurement can be a profound evidence pro or contra general relativity; and fourth, the effect has a widely known picturesque application - the gravitational lenses that are now used for the further research in astrophysics and cosmology.

The classical mechanics equation describing the motion of a particle in a Newtonian potential has the form

$$(\frac{du}{d\varphi})^2 = -u^2 + \frac{2GM}{L^2}u + \frac{2E}{L^2} \qquad (1.83)$$

where $u = 1/r$, φ here is an independent variable, i.e. an angle, M is the mass of the gravitation center, E and L are the specific energy and specific angular momentum. The analogous equation for the isotropic geodesics provided by the general relativity approach is

$$(\frac{du}{d\varphi})^2 = r_S u^3 - u^2 + \frac{r_S}{l^2}u - \frac{1-\varepsilon^2}{l^2} \qquad (1.84)$$

where $r_S = \frac{2GM}{c^2}$ is Schwarzschild radius (or gravitational radius or radius of the events horizon), $\varepsilon \equiv \frac{E}{mc^2} \approx 1 + \frac{v^2}{2c^2} + \frac{\varphi_N}{c^2}$ and $l \approx \frac{L}{c^2}$ are the relativistic analogues of specific energy and specific angular momentum. It can be shown [Kokarev (2010)] that the first term in eq. (1.84) which is the relativistic correction to eq. (1.83) can also be interpreted in a couple of other ways. At large distances ($r >> r_S$) the correction may be regarded as the result of the Newton gravitation law modification $F = G\frac{Mm}{r^2}(1 + \frac{3L^2}{c^2r^2})$. Similarly, the same situation may be treated as the relativistic increase of the inertial mass due to the orbital motion, $m = m_0(1 + \frac{3L^2}{c^2r^2})$. These ideas could be discussed in connection with MOND theory (mentioned below).

For a hyperbolic orbit that can be obtained for Newton potential in classical mechanics [Landau and Lifshitz (1967)], the deflection angle, see Fig. 1.8, appears to be

$$\delta = 2 \arccos[-(1 + \frac{v_\infty^4 d^2}{G^2 M^2}] - \pi$$

Now one can set the particle speed at infinity equal to light speed, $v_\infty = c$, and pass to the limit $d/r_S >> 1$ where d is the impact parameter.

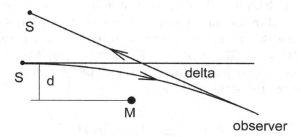

Fig. 1.8   Light bending in the gravitation field of point mass.

The classical deflection angle will then be

$$\delta_N \simeq \frac{d}{r_S} \qquad (1.85)$$

The value predicted by the general relativity calculation gives twice as large value

$$\delta_E \simeq \frac{2d}{r_S} \qquad (1.86)$$

which cannot be regarded as a simple correction. Strictly speaking, the derivation of eq. (1.85) is doubtful, because it presumed $E = \frac{mv^2}{2}$ for $v = c$, which is incorrect even in special relativity. But a direct attempt to account for that in the classical derivation and setting $m = E/c^2$ and $L = Ed/c$ to suffice special relativity brings to $\delta'_N \simeq \frac{\delta_N}{\sqrt{2}}$.

The corresponding measurements were first undertaken by Eddington who headed the 1919 European expedition that worked in two towns in Brazil and one in South Africa to perform the observations of stars that became visible near the Sun ridge during the total Solar eclipse. He found out that the measurements supported the calculations based on Einstein's theory that gave 1.75 seconds of arc. The accuracy accessible at that time and the method of data processing caused some doubts, and the observations were repeated several times since then until in 60-s the radio frequency measurements took the floor, and the question was settled to support the GRT within 1% accuracy.

1.5.3   *Gravitational lenses in GRT*

The deflection of light is a phenomenon which is well known in electrodynamics of continuous media and even in geometrical optics. And it can be shown that the deflection of light produced by a stationary gravitation field can be described in the same terms as in optics. Let us presume that eq. (1.39) contains the meaningful electromagnetic tensor F_{ik} and designate $E_\alpha = F_{0\alpha}; D^\alpha = -\sqrt{g_{00}}F^{0\alpha}; B_{\alpha\beta} = F_{\alpha\beta}; H^{\alpha\beta} = \sqrt{g_{00}}F^{\alpha\beta}; \alpha, \beta = 1, 2, 3$. Then the calculations [Landau and Lifshitz (1967)] that account for the stationary character of the field lead to

$$\mathbf{D} = \frac{1}{\sqrt{g_{00}}}\mathbf{E}; \mathbf{B} = \sqrt{g_{00}}\mathbf{H} \tag{1.87}$$

And one can see that the stationary gravitational field acts on the electromagnetic field as the medium with the electric and magnetic permittivities $\varepsilon = \mu = 1/\sqrt{g_{00}}$, that is the gravitation field appears to be equivalent to the medium with the refraction index equal to $n = 1/g_{00}$.

Therefore, there could be gravitational lenses. This idea first came to Chwolson [Chwolson (1924)] in 1924 and then was analyzed in detail by Einstein [Einstein (1936)] in 1936. If a point light source is located behind the spherical source of the gravitational field, then the observer could see the light beams deflected by the gravitation center in the form of a ring of finite radius, the so-called Einstein ring, instead of or together with the point source of light. When the eye, the light source and the gravitation source do not belong to the same straight line but are still not far from it and the condition $d/r_S >> 1$ is still fulfilled, the calculations show that a point source splits into two images. Unlike an optical lens, the closer to the gravitation center is the path of light, the larger is its deflection and vice versa. Consequently, a gravitational lens has a focal line instead of single focal point. The theory of gravitational lenses (e.g., [Blioh and Minakov (1989)]) gives the following formula

$$\eta = \frac{D_s}{D_d}\xi - D_{ds}\frac{2r_S}{\xi} \tag{1.88}$$

where ξ is the minimal distance between the light beam and the gravitation center, i.e. the visible "vertical" coordinate of the light source in the plane of the lens orthogonal to the axis of sight; η is the "vertical" coordinate of the source in the plane of the source orthogonal to the axis of sight; D_s is the distance between the light source and the observer; D_d is the distance between the lens and the observer; D_{ds} is the distance between the source and the lens. The situation when the source, the lens and the observer are

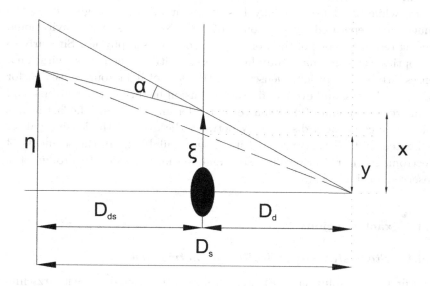

Fig. 1.9   Gravitational lens.

on the same straight line corresponds to $\eta = 0$. Then

$$\xi = \xi_0 = \sqrt{2r_S \frac{D_d D_{ds}}{D_s}} \qquad (1.89)$$

is Chwolson-Einstein radius. The corresponding angle is

$$\theta_0 = \frac{\xi_0}{D_d} \qquad (1.90)$$

and if the light source is a quasar and a gravitational lens is one of the nearest galaxies with mass 10^{12} Solar masses which is about $D_d \sim 100 \ kpc$ away, then $\theta_0 \sim 200''$. The first observation of a gravitational lens effect was registered in 1979 when the radiation characteristics of two close quasars SBS 0957+561 A and B appeared to be identical. Since the identity of the sources seemed hardly probable, the object was considered one and the same while the observed images were considered to be due to gravitational lensing. The distance to this quasar was estimated to be 8.7 billion light years, while the lensing galaxy lies at about 3.7 billion light years, the images are separated by 6 seconds of arc. Nowadays, the gravitational lensing research is one of the working methods in astrophysics. Similarly to the optics case, one may study light sources with the help of gravitational lenses, study gravitational lenses with the help of light sources, search for exosolar planets, and even study the expansion history of the Universe and estimate the amount of dark matter in the lensing galaxies. The last means that there could be a discrepancy between the mass of the lensing galaxy estimated by the observations of its image available in all the segments of electromagnetic radiation diapason, and the mass needed to produce the observed lensing effect.

1.6   Exact solutions

1.6.1 *Star: static spherically symmetric case*

The first exact solution of Einstein equation was found by Schwarzschild [Schwarzschild (1916)] soon after the publication of Einstein paper [Einstein (1916a)]. The last contained the final result of Einstein's long efforts in establishing the properties of the relativistic approach to gravitation. In this paper he gave the approximate solution of field equations which though being meaningful, was lacking the mathematical beauty which in some cases produces a strong effect. It happened so with the solution obtained by Schwarzschild, and it is hard to overestimate the role which it played in physics. Here we are not going to reproduce the solution, it can be found in any textbook, but it is important for the following to underline its characteristic features.

Schwarzschild regarded the line element

$$ds = \sqrt{g_{ik}(x)dx^i dx^k} \tag{1.91}$$

whose variation was the geodesics

$$\frac{d^2 x^i}{ds^2} + \Gamma^i_{jk} \frac{dx^j}{ds} \frac{dx^k}{ds} = 0 \tag{1.92}$$

treated as the equations of motion of a probe particle in the gravitation field of a certain mass, M, located at the space coordinates origin. The equations for the components $\Gamma^i{}_{jk}$ of this field, $\Gamma^i{}_{jk} = \frac{1}{2}g^{ih}(\frac{\partial g_{hj}}{\partial x^k} + \frac{\partial g_{hk}}{\partial x^j} - \frac{\partial g_{jk}}{\partial x^h})$, in the region where the mass was absent (empty space), were given by a system of 10 non-linear differential equations

$$\frac{\partial \Gamma^j_{ik}}{\partial x^j} + \Gamma^j_{im}\Gamma^m_{kj} = 0 \qquad (1.93)$$

Besides, the condition for the determinant

$$Det g_{ik} = -1 \qquad (1.94)$$

was considered to be fulfilled. Equation (1.94) provides the invariance of the form of eq. (1.93) with such change of variables that the Jacobian of the coordinates transformation is equal to unity. If the point mass, M, in the origin does not change with time and the motion at infinity is *straight and uniform*, the following restrictions are assumed:

(1) All the metric components do not depend on x^0, that is on time (stationarity).
(2) The demand for the components $g_{i0} = g_{0i} = 0$ is fulfilled identically.
(3) The solution possesses the spherical symmetry.
(4) At infinity, the metric becomes the Minkowski one, $g_{ik}(x) = \eta_{ik}$, i.e. the space-time at infinity is flat.

Thus, the problem was to find the line element coefficients sufficing eqs. (1.93-1.94) and conditions 1-4. Schwarzschild did the following. He wrote down the expression for the interval, ds, explicitly, passed from the Cartesian coordinates to the spherical coordinates and noticed that the Jacobian is not equal to unity. Then he changed the variables once again in such a way that the total Jacobian became equal to unity, while the final coordinates preserved the symmetry properties of spherical coordinates. Then the expressions for the components of the line element eq. (1.91), that is the solutions of the field equation eq. (1.93), was possible to find, and it appeared to be

$$ds^2 = (1 - \frac{r_S}{r})c^2 dt^2 - (1 - \frac{r_S}{r})^{-1}dr^2 - r^2(\sin^2\theta d\varphi^2 + d\theta^2) \qquad (1.95)$$

where $r_S = const$ is the so-called Schwarzschild radius that appeared as an integration constant whose value must provide the continuity of the solution.

Substituting the components of this metric into the geodesics equations and solving them, one gets the quasi-elliptical trajectories for a probe body with mass, m, total energy, $\mathcal{E}$, and angular momentum, $\mathcal{M}$, similarly to the classical case

$$\varphi = \int \frac{\mathcal{M} dr}{r^2 \sqrt{\frac{\mathcal{E}^2}{c^2} - (m^2 c^2 + \frac{\mathcal{M}^2}{r^2})(1 - \frac{r_S}{r})}} \qquad (1.96)$$

The integral can be transformed into the elliptical one, and one can check that the obtained trajectory describes the orbit pericenter precession. Schwarzschild also proved explicitly that taking a circular orbit instead of elliptical, one obtains the third Kepler law, $\frac{R_{orb}^3}{T_{orb}^2} = const$, with high accuracy. Besides, the light deflection formula eq. (1.86) also follows from his solution given by eq. (1.95). These impressive results were immediately acknowledged by the physical community and since then played an important role in all the subsequent development of relativity theory. Schwarzschild himself wrote that "it is always pleasant to have an exact solution of simple form" and expressed the hope that his formulas "would provide more purity to the brilliant result obtained by Einstein". Both feelings expressed by these citations were reinforced by the fact that those exact solutions not only formally gave the same predictions as the approximate solution of the new theory, but these predictions were experimentally testable, and the available tests were passed successfully.

Besides, the value of r_S expressed in the physically meaningful units appeared to be

$$r_S = \frac{2GM}{c^2} \qquad (1.97)$$

the expression which was no news in physics. Already Laplace in the 18th century pointed at the same relation between mass and radius of a spherical body for which the parabolic orbital velocity of a probe calculated on the base of Newton theory is equal to the speed of light. If c is the maximal accessible speed as the relativity theory presumes, then a spherically symmetric mass distribution sufficing eq. (1.97) will be able to absorb a signal - be it a particle or light, but will be unable to reflect it or send such a signal to the outside, i.e. it will behave as a hole which is completely black. But there was a problem with this solution. It is impossible to construct a stable mass distribution under the gravitational radius such that it will produce the field found by Schwarzschild. All the bodies under r_S and even light beams irresistibly fall to the center, there is a collapse. The picture seemed strange but acceptable, and it gave birth to the whole branch of science known as physics of black holes. Now there are not less than a dozen objects that are candidates nominated in astronomy for a black hole title. The galaxies' nuclei are also frequently considered to be giant black holes.

The introduction of this notion splitted all the singularities that could be present in the theory into two classes: dressed ones - hidden under the events horizon (another name for r_S), and naked ones. The presence of singularities is often interpreted as something pointing at the drawbacks of the theory, but the elegance of the exact solution pays for all. At the same time, with the demands 1-4 fulfilled and for $r > r_S$, Schwarzschild exact solution gives the same as Einstein approximate solution, that is the tails of the functions coincide, it has the same Newtonian limit, and it predicts the relativistic features that were found in observations in the Solar system.

1.6.2 *Universe: cosmological constant and expansion*

The general geometrical character of the theory appealed for the application of general relativity to the Universe as a whole. After it was done, it eventually completely changed our notion of the world. The smoothly rotating mechanism observed on a planetary scale whose precise details were discovered with the help of the new theory appeared to be a fragment of the remnants of the giant explosion. In order to see how it was possible to come to such picture, one should recollect that in the beginning of the 20th century when the new cosmology was coming into being, nobody including Einstein was aware of the scale of the Universe as we know it now. What they thought then to be the Universe, today would be rather called something like an immense single globular cluster - full of more or less uniformly distributed stars and nebulae. The discussion known as the Great Debate between astronomers Shapley and Curtis that concerned the size of the Universe and the very existence of other galaxies similar to our Milky Way and located at the distances that were orders of magnitude larger than the distances to the observable stars took place in 1920, its materials were published in 1921, and it did not produce a final decision on this point. Both sides had solid arguments in the interpretation of the observations in this or that way. Only the calculation of the distance to the Andromeda nebulae performed first by Epik in 1922 and then by Hubble in 1924 (both results were essentially less than those measured later but still ten times larger than the diameter of Milky Way) solved this problem in favor of the other island Universes existence. But the remains of these dismissed doubts were present in astronomy for more than a decade. It was this stationary "globular star cluster" idea of the Universe that was the starting point for the application of relativity to cosmology, and this must be taken into account when the problems in the obtained model appear. In his paper [Einstein (1917)] Einstein was very cautious and underlined that all he wrote was only considerations. Later Fock [Fock (1939)] in Russia also warned about the problems that general relativity might have in application to cosmology.

The stability of the spherically symmetric Universe consisting of stars seemed strange because of the analogy with gas consisting of atoms. The stars are so far from each other that the collisions are hardly possible, therefore, there is no equilibrium in the system. Gas would inevitably expand, and if the density at infinity is equal to zero, sooner or later, the density in the center would be zero too. This example comes to mind when one thinks of the boundary conditions for the gravitational field equations. The natural choice of flat Minkowski metric at infinity is good for a planetary system, but for the Universe it is not that obvious: empty space at infinity means no stationary Universe as we see it. Since the choice of the constant that can be added to the gravitational potential is arbitrary, it could be set infinite at infinity and used to save the situation. But this is also strange, because the gravitational potential is supposed to be produced by massive bodies, and the space at infinity is supposed to be flat, i.e. free from masses. This was the reason for the introduction of a certain *cosmological constant*, Λ, not related to any masses but improving the situation. This could be easily done by adding an extra term to the field equations. In Newton theory it looks like

$$\Delta\varphi - \Lambda\varphi = 4\pi\rho \tag{1.98}$$

If ρ_0 is a certain mass density corresponding to the uniform distribution of motionless stars in the infinite space, then

$$\varphi = -\frac{4\pi}{\Lambda}\rho_0$$

is the solution of the Poisson equation eq. (1.98). The value of ρ_0 could correspond to the real average density of the Universe, while the local inhomogeneities corresponding to stars could be described by the Newton law related to the term with the Laplacian. At constant and very low density, the matter is in equilibrium and there is no need for the existence of an internal pressure forces to preserve this equilibrium. In this case Λ should also enter the relativistic field equations that take the form

$$G_{ik} - \Lambda g_{ik} = -\kappa(T_{ik} - \frac{1}{2}g_{ik}T) \tag{1.99}$$

These equations have the solution

$$\rho = \rho_0 + \rho_1; \Lambda = \frac{\kappa\rho_0}{2}; \Lambda = \frac{1}{R^2}$$

the world is cylindrical, i.e. the 3D space has the constant curvature corresponding to the world radius, R, and a "world mass" related to ρ_0. The line element in this theory is

$$ds^2 = c^2 dt^2 - dr^2 - R^2 \sin^2(\frac{r}{R})[d\theta^2 + \sin^2\theta d\varphi^2] \tag{1.100}$$

De Sitter [De Sitter (1917)] has shown that the homogeneous (mass free) gravitation field equations have the solutions

$$\rho_0 = 0; \Lambda = \frac{3}{R^2}$$

the world is spherical, i.e. the 4D-space-time has constant curvature, R, and the line element in this case is given by

$$ds^2 = c^2 \cos^2(\frac{r}{R})dt^2 - dr^2 - R^2 \sin^2(\frac{r}{R})[d\theta^2 + \sin^2\theta d\varphi^2] \tag{1.101}$$

In 1922, Friedmann [Friedmann (1922)] rejected the limitation of the Universe stationarity and suggested that with or without the cosmological term, the space curvature radius could vary with time, that is $R = a(t)$

$$ds^2 = c^2 dt^2 - a^2(t)d\Sigma^2 \tag{1.102}$$

where Σ corresponds to a 3-dimensional elliptical, Euclidean, or hyperbolic space of uniform curvature written as a function of three spatial coordinates. The function $a(t)$ is known as the "scale factor". If the energy-momentum tensor is homogeneous and isotropic, the use of this metric leads to the differential equation for the scale factor

$$\frac{a''}{a} = -\frac{4\pi G\rho}{3} + \frac{\Lambda}{3} \tag{1.103}$$

This equation can be solved and used for interpretations. Then the stationary worlds described by Einstein and De Sitter become the particular cases of the solution obtained by Friedmann.

In 1929, Hubble [Hubble (1929)] reported his famous results of the observations of the redshift in the radiation of galaxies. He also discovered that the further they were, the larger was the redshift. This is known as linear Hubble law

$$v = Hr \qquad\qquad (1.104)$$

Hubble wrote:

"The outstanding feature, however, is the possibility that the velocity-distance relation may represent the de Sitter effect, and hence that numerical data may be introduced into discussions of the general curvature of space. In the de Sitter cosmology, displacements of the spectra arise from two sources, an apparent slowing down of atomic vibrations and a general tendency of material particles to scatter. The latter involves an acceleration and hence introduces the element of time. The relative importance of these two effects should determine the form of the relation between distances and observed velocities; and in this connection it may be emphasized that the linear relation found in the present discussion is a first approximation representing a restricted range in distance".

One can see that Hubble was keen enough not to exclude the relativistic time dilatation. But subsequently, the community came to the belief in the interpretation that involved Doppler effect: all the galaxies seemed to run away from the observer. Maybe, this had not only scientific, but psychological reasons as well. The picture with a balloon on the surface of which there are ink dots that run away from each other with the inflating of the balloon is easier to imagine than slowing down the time. Reinforced by Friedmann's solution of general relativistic equations, this interpretation turned upside-down the world view. The picture of expanding Universe which was purely theoretical obtained a powerful observational support, and the idea of a stationary Universe was abandoned. Instead, there appeared a Big Bang theory in which the Universe existed in a state of singularity and then lost stability as a result of quantum fluctuation. The evaluation of the redshifts and of the distances to the galaxies made it possible to evaluate the Universe age - the time that passed since Big Bang. The idea of this event led to a series of ingenious theories about the stages and character of the Universe evolution. Some of them, like relic radiation predicted by Gamow found the observational evidence.

Subsequently, various expressions for the time dependent metric of space-time were obtained. For example, Lemaitre managed to exclude the singularity at $r = r_S$ corresponding to the stationary solution obtained by Schwarzschild. He introduced new coordinates defined by

$$c d\tau = c dt + \sqrt{\frac{r_S}{r_L}} \frac{1}{1 - \frac{r_S}{r_L}} dr$$

$$d\rho = dt + \sqrt{\frac{r_L}{r_S}} \frac{1}{1 - \frac{r_S}{r_L}} dr$$

and obtained the Lemaitre metric

$$ds^2 = c^2 d\tau^2 - \frac{r_S}{r_L} d\rho^2 - r_L^2 (d\theta^2 + \sin^2\theta d\varphi^2) \qquad (1.105)$$

where $r_L = [\frac{3}{2}(\rho - c\tau)]^{2/3} r_S^{1/3}$ depends on τ. This metric had no singularity at $r_S = \frac{3}{2}(\rho - c\tau)$, but had the singularity at the center, corresponding to $\rho - c\tau = 0$. The expression eq. (1.105) was the first non-stationary metric that had this property, now similar metrics are known as Friedmann-Lemaitre-Robertson-Walker metrics, and they are widely used in cosmology.

1.7 Observations on the cosmological scale

The current situation in cosmology - that is in our general view of the Universe machine construction and functioning - is defined by the two notions that rather symbolically bear the name of "dark", they are *dark matter* and *dark energy*. The situation could be called confusing, because nowadays we have a strong observational support showing that the best theory we know for the Universe description - and, probably, the best physical theory we ever knew from the point of view of consistency and elegance, that is the general relativity based on special relativity - describes, strictly speaking, only 4% of what we can see and measure independently. The mainstream of the scientific research demonstrates understandable, so to say, humble obedience and even enthusiasm concerning the existence of such a vast field of still uncognized lying in front. Both reasons for that are already mentioned: the strong observational support, and the fundamental and persuasive character of general relativity.

The story of the "dark age" began with the observations of the Coma cluster of galaxies (then they were usually called nebulae) performed by Zwicky [Zwicky (1937)]. He evaluated the velocities of galaxies in the cluster and realized that if the interaction between them was just Newtonian attraction, then according to the virial theorem, the velocities were too high for this cluster to exist in stationary state. He then suggested that there was a lot of invisible "hidden mass" inside. In 1937, Zwicky was unaware of standard model, axions and the concept of nucleosynthesis, therefore, he thought not of *dark matter* but of some hidden mass like gas or some non-luminous bodies that were not the stars, now they are called MACHO (Massive Astrophysical Compact Halo Object), i.e. any astronomical body that could be responsible for additional gravitation inside the galaxy and emit little or no radiation. With time, more and more data that pointed at certain discrepancies in what we seemed to know about galaxies became available. The contradictions became flagrant when the rotation curves in spiral galaxies were measured (see below).

As it was already mentioned, the predicted effect of the gravitational lensing on the galactic clusters and on the field galaxies has been found. Since the results are sometimes several times larger than the predictions, they are also treated as pointing at the existence of some dark matter and used to estimate the amount of it in the lensing galaxy. Recently, the observations of the space laboratory "Chandra" [Clowe (2006)] delivered new data that provided the far going interpretation of the properties of dark matter. They obtained the x-ray image of the merging galaxy cluster 1E0657-558 and compared it to the optical image of the same cluster obtained by "Magellan" mission. The pictures present the collision of two clusters, and the equipotential lines for the gravitational potential are built with the help of optical distortions due to the gravitational lensing. The centers of potentials coincide with the centers of clusters. The x-ray image was used to find the gravitational centers of baryonic plasma that showed the shock-waves in the collision region. The location of these centers differ from those related to the galaxies distributions in clusters. Since the calculated masses needed for the observed lensing effect correspond to 2% for galaxies, 10% for intergalactic baryonic plasma and 88% percent for dark matter, they concluded that the dark matter does not take part in collisions and is nested in the galaxies.

1.7.1 *Rotation curves and their interpretation*

If the stars present the majority of a galaxy mass, then knowing the mass-to-luminosity ratio and assuming that it does not vary too much with radius, one can estimate the density of galactic matter with the help of star population brightness. Near the edge, the galaxy becomes dimmer, the average density of stars diminishes and the stars orbital velocity has to go to zero with distance from the center. The dependence of orbital velocity (or orbital velocity squared) on the distance from the center can be measured and is called rotation curve. It turned out that the observable rotation curves radically differ from expectations, and the measured velocities [Roberts and Whitehurst (1975); Begeman *et al.* (1991); Sofue and Rubin (2001)] are anomalously high for the densities obtained with the help of brightness measurements, Fig. 1.10. They suggested that at large distances from the galaxy center the principal role is played by the mass that reveals itself only

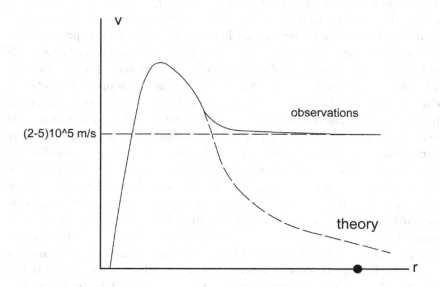

Fig. 1.10   Rotation curve: expectation (dashed line) and observation (soild line).

through the gravitation interaction. Independently, this hidden mass can be disclosed from the considerations dealing with the stability conditions of the star disk in spiral galaxy.

The experimental points obtained when measuring the dependence of orbital velocities, v, of stars and gas in spiral galaxies on the distances from the measurement points to the centers of those galaxies, R, can be described by the empirical formula [Mannheim (1997)]

$$v^2 = \frac{\beta^* c^2 N^*}{R} + \frac{\gamma^* c^2 N^* R}{2} + \frac{\gamma_0 c^2 R}{2} \qquad (1.106)$$

where c is the light speed, N^* is the number of stars in the galaxy (usually about 10^{11}), ß^* for the Sun is $\text{ß}^* = \frac{M_S G}{c^2}$ cm (M_S is the Solar mass, G is the gravitation constant), γ^* and γ_0 are universal parameters $\gamma^* = 5.42 \cdot 10^{-41}$ cm^{-1}, $\gamma_0 = 3.06 \cdot 10^{-30}$ cm^{-1}. All the three parameters become of the same order at the border of a galaxy, while the result of the Newton theory as well as the Schwarzschild's solution of the GRT equations predict only the decrease of the velocity corresponding to the first term in eq. (1.106). The calculations were performed with regard to the distribution of stars in a galaxy that can be taken as exponential with distance from the center. One cannotice that not only the second linear in R term which is proportional to the galaxy mass is present in eq. (1.106), but the third one too, and it is independent of the number of stars and also changes linearly with distance. The idea of the possible interpretation is the same. The mass of additional dark matter providing the observed motion of the visible matter appears to be thrice as much as the mass of this visible matter, and it neither emits, nor absorbs the electromagnetic radiation. One cannotice that, qualitatively, it is the same effect as in the clusters of galaxies. Obviously, if the existence of such specific dark matter is admitted, there is a need for a theory of the hypothetical elementary particles that form it. From the point of view of Standard Model that describes the strong, weak, and electromagnetic fundamental forces and uses mediating gauge bosons such possibility does exist. The species of gauge bosons are the gluons, W-, W+, and Z bosons, and the photons. The Standard Model contains 24 fundamental particles and predicts the existence of a boson known as Higgs boson, which is not discovered yet. The project aimed at the discovery of Higgs boson has been launched in CERN at Large Hadron Collider. Higgs field is related to the axions - the hypothetical particles constituting the dark matter. These particles presumably originate from the stars' nuclei, and there are several laboratories in which they are looking for these particles coming to Earth from the Sun nucleus.

But the dark matter influencing the cosmological scale observations, though generally acknowledged by the scientific community is not the only possibility. Something might be wrong with the theory. If no one wants to doubt Newton, the general relativity might be more vulnerable. The review of the theoretical results obtained on the way to modify the gravitation theory with regard to the problems stemming from the cosmological scale observations is given in [Mannheim (2006)]. The efforts of the theoreticians are aimed at modifying the existing theory in such a way that there is either no need for the extra quantity and type of matter or the corresponding terms are accounted for in the geometrical way characteristic for the relativity theory. Any change of the theory must preserve the existing phenomenology, the natural test is the existence of Newton gravity law for the Solar system scale and classical tests following from the Schwarzschild's solution and already observed and measured.

The most straightforward approach is the successive complication of the expression for the Hilbert-Einstein action

$$S_{HE} = -\frac{c^3}{16\pi G} \int d^4x (-g)^{1/2} R \qquad (1.107)$$

with account to the metric terms of the higher orders. DeWitt [DeWitt (1964)] was the first to suggest the terms like

$$S_{W_1} = -\frac{c^3}{16\pi G} \int d^4x (-g)^{1/2} R^2 \qquad (1.108)$$

or

$$S_{W_2} = -\frac{c^3}{16\pi G} \int d^4x (-g)^{1/2} R_{ik}^{ik} \qquad (1.109)$$

The corrections due to eq. (1.108) or (1.109) must give a negligibly small contribution to the Schwarzschild's solution. Besides, already this approach makes it possible to regard the cosmological constant in a way Einstein tried to do it himself. Accounting for the similar but higher order terms in the integrals, one comes to the so-called $f(R)$-theories of various sorts.

Another natural approach is the introduction of an additional macroscopic gravitational field, S, usually the scalar one. For example, Brans and Dicke [Brans and Dicke (1961)] suggested the following expression for the action

$$S_{BD} = -\frac{c^3}{16\pi G} \int d^4x (-g)^{1/2} (SR - w\frac{S_{;i}S^{;i}}{S}) \qquad (1.110)$$

where w is a constant.

One could also think of the increase of the number of space-time dimensions with the subsequent transfer to the Planck's scale of lengths. The corresponding works began with [Kaluza (1921)] and then lead to the mathematically developed modern theories of strings [Marshakov (2002)] and then of branes [Randall and Sundrum (1999)].

The geometry used to describe the space-time can also be modified. Already in 1918, Weyl [Weyl (1918)] stepped aside from the Riemannian geometry in order to unify gravitation and electromagnetism. He suggested to regard the transformations of the following form

$$g_{ik}(x) \rightarrow e^{2\alpha(x)} g_{ik}(x) \qquad (1.111)$$

$$A_i(x) \rightarrow A_i(x) - e\partial_i\alpha(x) \qquad (1.112)$$

in which the gravitation and electromagnetism were united by the common function $\alpha(x)$. The equations that can be obtained in this approach do not give the usual field equations obtained by Einstein; nevertheless, they contain the Schwarzschild's solution for the Solar system scale. Weyl called eq. (1.111) the gauge transformation, i.e. dependent on scale, but later this term was adopted by the other fields of physics mainly for the cases when the exponent was imaginary. In gravitation theory such transformations are now called conformal.

The evolution of these ideas leads to the theories of conformal gravitation where the metric has an additional symmetry, corresponding to eq. (1.111), the electromagnetic variables are not involved and this means that the geometry remains Riemannian. Formally such approach is analogous to that of DeWitt, but the choice of coefficients in eqs. (1.108) and (1.109) is specific. The field equations that appear in this approach are [Mannheim and Kazanas (1994)]

$$4\alpha_g W^{ik} = 4\alpha_g(2C^{ilkj}_{;l;j} - C^{ilkj}R_{lj}) = T^{ik} \qquad (1.113)$$

where α_g is dimensionless constant and C^{ilkj} is the so-called Weyl tensor which doesn't change with transformations given by eq. (1.111). Then in [Mannheim and Kazanas (1994)] they change

$$W^{ik}(x) \to e^{-6\alpha(x)} W^{ik}(x)$$
$$T^{ik}(x) \to e^{-6\alpha(x)} T^{ik}(x)$$

and transform the coordinates with the use of a certain function $B(r)$. After that, the source function $f(r)$ is introduced. As a result, the stationary version of eq. (1.113) gives the Poisson equation but not of the second order as usual, but of the fourth order

$$\nabla^4 B(r) = f(r) \tag{1.114}$$

If a physical problem has a spherical symmetry, eq. (1.114) has an exact solution. And this solution not only contains the Newton-Schwarzschild term but also the terms corresponding to eq. (1.106)

$$B(r > R) = -g_{00} = 1 - \frac{2\beta}{r} + \gamma r \tag{1.115}$$

$$2\beta = \frac{1}{6} \int_0^R dr' r'^4 f(r'); \gamma = \frac{1}{2} \int_0^R dr' r'^2 f(r') \tag{1.116}$$

The described approach does not require the introduction of additional (dark) matter, i.e. of the additional scalar field. Instead, it uses another choice of the scalar function in the variation principle. This preserves the Riemannian geometry of space-time but leads to the field equation in the form of eq. (1.113) which, by the way, does not have the structure of the wave equation for the empty space. This means that there are no solutions in the form of gravitation waves, and the effect of the orbital period decrease reported in [Taylor and McCulloch (1980)] which coincides with the former theory prediction is, thus, to be explained in some other way.

The rejection of symmetry in metric [Kursunoglu (1991); Moffat (1995)] can also lead to the suitable description of the rotation curves, while dark matter is not needed. Moffat [Moffat (1995, 2006)] postulated that the antisymmetric field may be massive and produced the theory of MOdified Gravity (MOG) with the addition of a vector field and also promoted the constants of the theory to scalar fields. This approach successfully accounted for cosmological observations. But the essence of it can be described as follows: far from a source the gravitation is stronger than Newtonian, but at shorter distances, it is counteracted by a repulsive fifth force (with its specific fifth force charge) corresponding to the introduced vector field.

Moreover, the classical foundation can also be revised. In the MOND phenomenological approach (MOdified Newton Dynamics) developed by Milgrom [Milgrom (1983)], there was introduced the new world constant with the dimension of acceleration. His suggestion was to change the dynamics law or, equivalently, to change the gravitation law in such a way that they take the forms

$$\mu(\frac{a}{a_0})\overrightarrow{a} = \overrightarrow{f} \text{ or } \overrightarrow{a} = \nu(\frac{f}{a_0})\overrightarrow{f} \tag{1.117}$$

It was suggested to find such fitting functions $\mu(x)$ or $\nu(x)$ and such value of a_0 that they match the classical result for the Solar system scale and give eq. (1.106) for the galaxy scale. In [Bekenstein and Milgrom (1984)] they performed the relativistic generalization of MOND (and produced the so-called TeVeS theory, Te for tensor, Ve for vector and S for scalar), where the scalar field ψ was introduced to give an additional term to the expression of Hilbert-Einstein action in the form

$$S(\psi) = -\frac{1}{8\pi GL^2} \int d^4x (-g)^{1/2} f(L^2 g^{\alpha\beta} \psi_{;\alpha} \psi_{;\beta}) \tag{1.118}$$

(f is a scalar function, L is constant). After that the MOND theory cannot be regarded as a pure phenomenology. Being adapted for it, this approach gives a good fit for the observed rotation curves described by eq. (1.106). It turned out that it could also be applied to some other cosmological observations though not to all. MOND gives the wrong prediction of the temperature profiles in clusters [Aguirre *et al.* (2001)].

The last reference, i.e. [Aguirre *et al.* (2001)], contains a whole list of the unavoidably necessary observational demands for any modification of the gravitation theory. It includes rotation curves and Tully-Fisher law for spiral galaxies, the globular clusters problem and some others. Globular star clusters with the orbits that do not belong to the spiral galaxy plane seem to obey Kepler law, while the stars orbiting around the spiral galaxy nucleus in the periphery of the galactic disk, do not obey it and fall into the flat rotation curves. Besides, relatively too many globular clusters are found in the vicinity of the galactic nucleus, while the Keplerian expectation suggests that they spend most of their orbital periods in the far away regions. The analysis given in [Aguirre *et al.* (2001)] shows that none of the known modifications copes with all the list. Even MOND approach which was empirically adjusted to fit observations appears inconsistent in some cases.

1.7.2 *Break of linearity in Hubble law*

Measuring the distance to the other galaxies is now performed by measuring the corresponding redshifts and then by applying the Hubble law which was found linear. The proportionality coefficient is called Hubble constant or more precisely, Hubble parameter. The value of this parameter has to be estimated before such measurements, i.e. the redshifts should be measured for the galaxies the distances to which are found by some other method. In astronomy, they use the so-called "standard candles" - the objects whose luminosities are known. The best standard candle is the type 1a supernova. They have very high brightness and burst only when the mass of the "old" white dwarf reaches the Chandrasekhar limit the value of which is known with high accuracy. That is, all the SN1a that are located at equal distances must have the same observable brightness. Comparing the brightness of supernovae in various galaxies, it is possible to estimate the distance to these galaxies.

In the end of 90-s, it was found [Reiss (2001, 2004)] that in the far away galaxies with the distances defined with the help of Hubble law, the SN1a have the brightness lower than expected. In other words, the distances to these galaxies calculated by the standard candle method were larger than the distances calculated with the help of the estimated Hubble parameter. The Hubble law became non-linear, and this was interpreted as the acceleration of the Universe expansion.

The earlier cosmological models suggested that the Universe expansion decelerates. This was based on the assumption that the major part of Universe mass is presented by matter - both visible (baryonic) and invisible (dark). On the base of observations interpreted as the acceleration of expansion, the existence of the unknown form of energy with the negative pressure (repulsion) was postulated. It was called *dark energy*. Estimating the value of energy needed to accelerate the given Universe, one can see that the relativistic mass-to-energy ratio gives the amount of mass corresponding to the dark energy to be about three times larger than the total mass of the visible and dark matter in the Universe. Lots of ingenious theoretical inventions, particular cases and situations can be and were regarded. The main problem is the origin of this energy (of repulsion). The geometrical foundation of general relativity and Einstein's attempts to introduce the cosmological constant make it possible to relate the intrinsic origin of the dark energy to the cosmological constant.

In the quantum field theory, they try to involve gravitation into the general scheme of interactions and would prefer to deal with some new field known as quintessence. The problem is the catastrophic discrepancy - hundred of orders - between the almost flat character of the observable Universe that corresponds to the very small value of cosmological constant, and the huge value of it demanded by the calculations based on the apparent break of the Hubble linear law. The problem of the dark energy is one of the most severe problems of the contemporary theory.

Chapter 2

Phase space-time as a model of physical reality

2.1 Preliminary considerations

2.1.1 *Scales*

First of all, let us mention once again the scales that are usually related to gravitation phenomena. The minimal scale is the planetary system one, i.e. the phenomena take place in the vicinity of a star (with planets orbiting around) or of a planet (with satellites orbiting around), and we have such systems at hand, and there is a well-developed theory - general relativity - which works there very well, with little nags being the flyby or Pioneer anomalies. The maximal scale is that of the Universe as a whole, we have the ways to establish distances to the far away galaxies and to measure some characteristics of these objects, amazingly, all these correspond to the deep past, and we can only calculate with certain accuracy and probability what could go on now and what could happen in future. The theory we have to describe the Universe is based on the theory that proved to be good for a planetary system, therefore, it is not for certain that the results obtained for the smallest scale would work for the largest one too. The intermediate scale is the galactic one, and we also have a galaxy at hand, this is our Milky Way, and several others are quite observable. There is the same problem with distances - all of them are so large that everything we see on the other side of our own Milky Way took place some 100,000 years ago. Far away and long ago seem to be very much alike on this scale too. As to the theory, we also try to apply the theory that proved good for a planetary system, but this time we have more materially-minded reasons to suspect that at least for galaxies other than ours, something goes wrong without, say, the introduction of a new entity called dark matter. The amount of the dark matter needed to repair any observable spiral galaxy and to make it work according to our theory is thrice larger than the amount of visible matter in this galaxy including stars and interstellar gas, the total mass of the latter is, by the way, 2-3 times larger than that of all stars in this galaxy.

So, the scale problem is definitely underestimated. It is also worth mentioning that the characters of the objects we describe with the help of our theories are very different. A planetary system is a system with one central body governing the situation around it. A classical model of Rutherford atom with heavy nucleus and tiny electrons around was built with regard to the construction of the Solar system. But a galaxy consists of stars that are so far away from each other that a galaxy resembles rather a rarefied gas with convection and other properties inherent to gas dynamics. The size of a galaxy is comparable to the distance between galaxies, therefore, it would be appropriate to speak of the Universe as of liquid fluid consisting of various large "molecules". We have enough observable examples of galaxies interacting with each other, distorting each other's forms and so on. It could be doubtful that one and the same theory could equivalently well describe such different systems. The hope is that the theory is very profound, it deals with the abstract theory of measurements, which here means with geometry. And the general principles of geometry, being mathematical in their nature, have to remain the same everywhere. But as we already saw, the appropriate geometries themselves could be different for different scales and conditions.

For a long time there was no need for other geometry than Euclidean one, until Lobachevsky in 1826 noticed and proved that there are regions in space for which we will never be able to test the feasibility of axioms and the applicability of this or that geometry directly. For those regions we should discuss only the observations performed remotely with the help of telescopes. And beginning from certain distances, the consequences of this or that system of axioms would be indistinguishable in observations. Therefore, there appears some interpretation freedom. Eighty years later, another mathematician, Minkowski accepted the reasoning that came from physics and suggested a fundamentally different mathematical model to describe the world. It was 4-dimentional, that is, it included time as a specific fourth coordinate, and it was based on another geometry, another way to calculate distances. This model made all the observations fit and inspired Einstein to complicate it and to suggest a generalized version - Riemannian geometry - to describe the world full of gravitating bodies that made each other move along the geodesics. But all we saw and all we could measure were these motions, therefore, the gravitation had to be involved into the testable mathematical description and vice versa. This idea of new geometry and its use in observations had a tremendous success and made it possible not only to quantitatively explain the still existing then observational problem (Mercury orbit precession) but also to give the correct quantitative predictions of the observable effects that were later discovered (light bending and gravitational redshift).

Then again came the stage when there was a need to improve the accuracy of measurements. The only small parameter of the theory was the v/c ratio (c is the characteristic velocity), and when the corrections became experimentally available, the velocities appeared in the next orders of the asymptotic expansion, thus, providing the evidence to test if the corresponding small corrections fit the theory. Do we really account for any velocities in the related problems? Even in the cases accounting for the rotation of the central body (e.g. using Kerr metric [Kerr (1963)]), we regard the motion of this central body as the only source of the field in which the probe body moves. It is clear that for a motion of a planet in, say, a double star system the dynamic problem essentially complicates, unless we regard the orbit that is far away from both stars, so that it is their common center of mass that matters, and the relative motion of stars can be considered a small perturbation. In this case, another small parameter appears, and it is r_0/r where r_0 is the characteristic space scale of the gravitation source thought of as a single one, and r is the distance to the planet. As long as this second small parameter has an order less than the first one, everything works. When it grows and starts to approach the first one, it is time to sit down and think. When it surpasses the first one, it is clear that it is not only the boundary conditions themselves, but the role they play in the theory that needs a revision. There appears a certain interference of the boundary conditions formulated for each component of the system of gravitating sources.

For a planetary scale theory, the choice $r_0 = r_S$ where r_S is Schwarzschild radius, fits excellently. Moreover, we can even pass to the description of an ordinary star interior, because in this usual case r_S is severely less than any reasonable (i.e. testable) distance that we could be interested in. But this cannot be the case already for galaxy scale. The planetary scale theory with its exterior problem solutions cannot be applied for the description of the galaxy interior and even for the close vicinity of a galaxy. There is a distribution of gravitation sources in the galaxy, and these sources can move in a more or less coherent manner. Therefore, the motion of a single observable star resembles rather a motion of a gas atom taking part in convection, it is not only the two-particle interaction that matters. What we see in this case observing from the far away, does not tell us much about the interaction between "atoms", but rather informs us of convection flows. The velocity of such convection has little to do with the velocities of separate atoms and cannot be related to the small parameter of the theory mentioned above in a way the velocities of separate gravitating bodies do for the planetary system scale.

How can we account for the corresponding correction of the theory sitting on one of such atoms and observing the others in our own galaxy? And if we observe such motions from aside, i.e. in the other galaxies, to what theory should we compare the results of observations? There appear doubts that the solutions related to the static spherically symmetric situation can fit. We cannot reconstruct the convection in a gas flow knowing only the microscopic parameters like the interaction potential of the atoms, we need something else. As long as we speak of gas, the introduction of the flow velocity into the atoms' interaction potential seems inappropriate and artificial. But as soon as we recollect that we have no other way to look at the "surrounding atoms" and to measure the corresponding characteristics than from the specially equipped site on one of them, we come to understanding that it is just the same as the recognition that the light speed is the maximal speed in the Universe. We have no means to observe with the help of anything which moves faster. We have no means to observe other than "sitting on an atom". And since our best notion of the "atomic interaction potential" is geometrical in origin, that is, it is based and founded on the measurements of the distances to the objects and between them, this convection velocity *is already present* in the interaction potential, that is, it must enter the metric.

Strictly speaking, the situation now resembles one that arose a hundred years ago, when out best notion of electromagnetism was Maxwell theory, and everyone wanted to preserve it in this or that way. Now our best notion of gravitation is Einstein theory, and there are good reasons to preserve it. A hundred years ago Maxwell theory was saved by the change of the geometrical model of the world and not by the introduction of an additional essence, i.e. ether. Maybe now it could and should also be done in the similar way?

It can also be noticed that there is a certain methodological dissimilarity between the treatments in quantum mechanics and in cosmological aspect of general relativity. They have similar starting points, meaning that in both of them there is an unavoidable difference between the scale of a measured object and the scale of an instrument. In quantum mechanics the measurement affects the result, that is the macroscopic instrument affects the microscopic object. In gravitation, the situation is reverse, the object unavoidably affects the instruments (meter sticks and watches). That is, the "megascopic" object affects the macroscopic instrument.

Because of this, the interpretation problems are similar but the ways to solve them are somehow different. In quantum mechanics they speak of wave functions, of averages and of uncertainty principle, and the role of measurement procedure is not usually discussed, that is the larger scale is disregarded. While in gravitation, they speak of geometry and properties of empty space, of cosmological constant, of dark matter and energy, the role of measurement procedure is again usually not discussed, but now it means that the smaller scale is disregarded.

In both cases, there exists an intermediate scale that provides the testing arena for our ideas of directly unmeasurable worlds. In the first case, it is substance, the properties of which must be explained by the general theories for atoms and give the observable results in larger scale. In gravitation, it is galaxies, whose properties must be explained by the general theory for planetary systems and give the observable results in larger scale. But the substances themselves must be described by the theories that include parameters that are absent on the atomic scale, for example, temperature, and include transfer phenomena. The same is true for galaxies. The intermediate scale phenomena are, fortunately, not the only ones that we can deal with, there are also such things as radiation - electromagnetic or gravitational, but the importance and the specific role of the intermediate scale cannot be argued. Between quantum theory of atom and classical mechanics of bodies there should be thermodynamics that could explain the properties of substance. The possibility to skip this scale and construct a consistent physical theory seems doubtful. The equation of state that is sometimes discussed in cosmology is usually applied to mathematical models, on the one hand, and also misses the intermediate scale, on the other hand.

2.1.2 *Boundaries*

As soon as we start investigating the roles of the small parameters, the boundary problem appears. The meaning of it is: which of the two problems should be regarded every time we turn to the comparison with observations, the inner one or the exterior one? One could easily see the consequences from the following simple example of a gravitation problem.

Let there be a spherical body with radius r_b, constant density through-out its volume and a given fixed pressure $p = p_0$ at the center. The pressure at $r = r_b$ is equal to zero, and we want to find the gravitational potential. The solution can be found for any r and then compared to the result of Schwarzschild solution obtained for the mass m appearing to be inside the sphere of radius r. If we demand that at the surface $r = r_b$ the result matches the usual Schwarzschild solution for the total mass M of the body, then the following relation will be found for the proper time τ and for the coordinate time t measured by the infinitely far watch

$$\frac{d\tau}{dt} = \frac{3}{2}\sqrt{1 - \frac{r_S}{r_b}} - \frac{1}{2}\sqrt{1 - \frac{r_S r^2}{r_b^3}}$$

Thus, the inner Schwarzschild problem for a constant-density sphere gives the solution which at $r = 0$ corresponds to the following expression

$$d\tau = dt \left(\frac{3}{2}\sqrt{1 - \frac{r_S}{r_b}} - \frac{1}{2} \right)$$

Therefore, if from the very beginning the radius of the body, r_b, is less than $\frac{9}{8}r_S$, or if the body passes this value during the collapse, then the inner solution above says that the proper time in the center changes its sign, and this will cause gravitational repulsion instead of the gravitational attraction. Thus, from the point of view of the inner observer, the collapse through the Schwarzschild horizon and the transformation into a black hole is hardly possible. Notice also that for the similar reasons the famous Newton's apple would travel inside the Earth (if it had such an ability) following Hooke's law and not Newton's law, that is the center of the Earth would coincide with the center of the elliptical orbit of the apple and not with the focus of it.

This becomes especially intriguing, if one evaluates r_b for such an object as our Universe. At first glance, the Schwarzschild radius associates with the value corresponding to a body with super-high density (for example, this was the case presumed in [Hobson *et al.* (2009)] where they suggested to apply this solution for the modelling of the interior of the neutron star), but actually, the density could be arbitrary if we choose the corresponding scale of its distribution. The problem of the choice of p_0 other than zero would appear, but let the imagination work. If our Universe happens to be inside its gravitational radius, then already classical approach will bring us to the inner problem for a homogeneous ball which then gives the Hooke's law for gravitation in the Universe. If the number of protons with mass 10^{-29}kg each is about 10^{80} (and this is a good place to recollect where this number well known from the literature has appeared from), then $r_S \sim 10^8$ pc which is close to the estimated radius of the observable (!) Universe. Notice, that in this case, from the point of view of the inner Schwarzschild problem, one might not need any additional dark energy to explain the gravitational repulsion causing the acceleration of expansion.

Of course, all this is just a speculation. Schwarzschild solution was not supposed to be used other than for the description of the field outside the isolated single static spherically symmetric source, and the only observable parameter is the mass of this source. Sometimes they speak of the solution in the vicinity of an isle system, but it seems that this isle usually belongs to an archipelago, and however important is this solution for GRT and astronomy, it is applicable to the less number of cases than it sometimes seems. The vicinities of stars could be those non-homogeneities that Einstein and others spoke about when dealing with the Universe as a whole, and usual classical GRT would work there all right. Maybe for such source of gravitation as a galaxy it would have also worked - sufficiently far away from it, but we would hardly be able to check it, because it is hard to find a suitable astrophysical system: isolated galaxy and a detectable probe body at a distance large in comparison with galaxy diameter orbiting in the empty space. As we know, already for globular clusters - objects belonging to galaxies, there are problems [Aguirre *et al.* (2001)] with their motion, sometimes it is Keplerian, sometimes it is not. And a problem of a permanent need for a correct model choice indicates that something is missing in the theory, for example, a parameter that would directly show when the Schwarzschild problem is applicable.

Then, what could be said of the redshift effect that we observe and consider related to the Universe expansion? Which solution to choose for the interpretation of it *now*? Does our "chosen" reference frame freely flies "outwards" together with others or freely falls "inwards" to the center of the Universe wherever it is? Doppler effect concerned, we just "see" how the observed horizon turns more red with distance. In the inner case, the Hooke's gravitational potential in the periphery is higher than in the center, and when we observe the linear redshift discovered by Hubble, it can now correspond rather to the gravitational one than to a Doppler shift which becomes not that simple. The equivalence principle gives the same, because the inward acceleration of the bodies at the periphery will be larger than in the inner regions, and this larger acceleration gives larger inertial forces and corresponds to the larger gravitation in the same formal (linear) way as Hubble law. Thus, the cosmological picture might start to differ from what we have now. But it doesn't seem that we should again hastily change it with regard to another kind of exact solution corresponding to observations as it was already done before, when Friedmann's solution appeared to fit Hubble's observations so well.

The model of a homogeneous rarefied ball does not seem inconsistent, say, for a galaxy and can provide the criterion for the applicability of the usual Schwarzschild solution for the description of the gravitation problem on the galactic scale. Usually, one takes the total mass of a system and finds its gravitation radius r_S, but with regard to the above said, it is clear that one should take the density of a finite system and find its gravitation radius as the radius of a homogeneous ball outside which the usual Schwarzschild solution could work. Let us find the critical density, ρ_c from the demand that $r_b = r_S$:

$$r_S = \frac{2GM}{c^2} = \frac{2G}{c^2}\frac{4\pi}{3}\rho r_b^3$$
$$r_b = r_S \Rightarrow \rho = \rho_c$$
$$\rho_c = \frac{3c^2}{8\pi G}\frac{1}{r_b^2} \approx 2\cdot 10^{26}\frac{1}{r_b^2}$$

Thus, if there is a finite region characterized by a certain radius, r_b and a certain average density of matter, $\rho_a = \frac{3M}{4\pi r_b^3}$, and $\rho_a > \rho_c$, then we can hope that at r_b and further away Newton law is valid. But if $\rho_a < \rho_c$, then there is no wonder if at r_b something like Hooke's law is preferable.

Example: Let a galaxy with radius 100kpc (i.e. 10^{21} m) contain 10^{11} stars of mass 10^{30}kg each. Hence, its average density is $\rho_a = \frac{3M}{4\pi R^3} \sim 10^{-44}$kg/m^3, while ρ_c here is proportional to 10^{-16}kg/m^3, and one has no reasons to expect to observe the $1/r$ decrease of the potential at the border of the galaxy.

For a spiral galaxy some specialization with regard to matter distribution is needed (see, for example, the calculations of density distribution in [Mannheim (2006)]), but obviously it cannot help much. Since the potential is proportional to the orbital velocity squared, the observed rotation curves should seem paradoxical and contradictory rather if they followed the $1/r$ law than if they radically differ from it.

The distances between galaxies are not more than a couple of orders larger that the galaxy radius. Therefore, it is hard to see how the Schwarzschild solution could be applied in cosmology. From the point of view of classical mechanics, the situation should be described not by the solution of Kepler problem but by the solution of many-body problem.

Consequently, the relativistic approach must also be modified, and it is clear that the border region especially for the spirals with their pronounced kinematical effect requires the special attention. But already general considerations (like virial theorem) make one think that since kinetic energy of rotating spirals is comparable to the potential energy of gravitation, the mass and energy equivalence discovered in special relativity causes the effective mass increase. That is, the motion causes gravitation and, therefore, must be also the source of curvature. Curvature is described by metric, thus, the metric must be motion dependent, and this kind of motion has nothing to do with the small corrections that are usually discussed, the division by c^2 relates to another case.

In clusters of galaxies there is no observed ordered motion, but the velocities dispersion is so large that Zwicky [Zwicky (1937)] thought of additional dark masses needed to hold the cluster together. It seems that the additional gravitation potential could be produced by the motion of galaxies itself, the "heating leads to the mass increase".

All this becomes even more intriguing when we think of the linear Hooke's law more precisely. First of all, recollect the simple electrostatics: in the case of the homogeneous ball it is unquestionably equivalent to gravitation (to say nothing of the Maxwell identity approach discussed in the previous Chapter). Suppose, we want to find the electric stress of the simple charged model objects: a zero-dimensional point, a one-dimensional infinite thread, a two-dimensional infinite plane. The stress is given by $E_p \sim 1/r^2; E_{th} \sim 1/r; E_{pl} \sim const$ correspondingly. When we pass from the idealized objects like a point, a thread and a plane to a homogeneous ball of finite radius r_b, infinite homogeneous cylinder of finite radius r_b and infinite homogeneous layer between the parallel planes separated by finite distance r_b, there appear the above mentioned inner and exterior problems. One can immediately discover that the Gauss theorem gives different solutions for the exterior problems, namely, $E_p' \sim 1/r^2; E_{th}' \sim 1/r; E_{pl}' \sim const$, as before, but all the inner solutions have one and the same form

$$E_p' \sim E_{th}' \sim E_{pl}' \sim r \qquad (2.1)$$

where r is the distance $(r < r_b)$ from the point of observation to the center of the ball, to the axis of the cylinder, and, amazingly, to a certain point inside the layer chosen as the origin, at which we postulate $E_{pl}'(\mathbf{r}_0) = E_0 = const,$

for example, $E'_{pl}(\mathbf{r}_0) = 0$. The result for the layer seems paradoxical, but one can check by the direct calculation (integration), that the value of the field in any place inside the layer does not depend on the choice of this point of origin.

This circumstance has a far going consequence: suppose, the density of sources inside the three mentioned bodies is very low and is produced by the homogeneously distributed points. The fixed point of origin is the point where the observer is located, he does not register any stress or any force or any acceleration at his own location. Therefore, he concludes that he moves along the geodesics. But if he is able to measure the stress dependence on the distance from his position, he will find the linear law. It is hardly possible to measure the gravitational stress at a distance, but it could be possible to measure the acceleration of some probe observable bodies. In which way? Use the Doppler effect to find the radial velocities of the luminous probe bodies, measure their velocity dependence on distance, recollect that $dr = cdt$, get the linear law. Hubble did it already, in a sense. Who spoke of a linear law then or before? Hooke did. And not only him. But we must admit that Hooke's achievements are nowadays essentially underestimated. Actually, he also invented the rule for the decrease of gravity being inversely proportional to the square of the distances from the center, and he expected Newton to acknowledge his result in his famous 'Principia'. But Newton claimed that he did the same independently and only wrote [Newton (1687)]:

> I have now explained the two principal cases of attractions: when the centripetal forces decrease as the square of the ratio of the distances, or increase in a simple ratio of the distances, causing the bodies in both cases to revolve in conic sections, and composing spherical bodies whose centripetal forces observe the same law of increase or decrease in the recess from the centre as the forces of the particles themselves do; which is very remarkable.

When Newton in his book came to the description of the existing astronomical observations, he left the linear force term out, probably because at that time there was no evidence for it, neither did he consider the superposition of the two forces. But Laplace already mentioned this possibility and gave the general gravitation force law in the form $F = Ar + B/r^2$, where A and B were constants, and while the trajectory of a planet could be an ellipse or a circle for the first term and any conic section for the second term, the combination of both terms gives a figure of rosette, which became the mystery of Mercury orbit in the end of 19-th century, that is a hundred years after Laplace.

Now if we use this expression for the equation of motion and choose the constants A and B in the form

$$\frac{d^2r}{dt^2} = -\frac{GM}{r^2} + \frac{\Lambda}{3}r \qquad (2.2)$$

we can see that this equation resembles the Friedmann equation eq. (1.103) for the scale factor, $\frac{a''}{a} = -\frac{4\pi G\rho}{3} + \frac{\Lambda}{3}$, written for the case of a homogeneous ball with $\rho = M/V$. In eq. (1.103) Λ is a cosmological term the meaning of which is widely discussed now. Comparing this to eq. (2.1), we can see that the cosmological constant, Λ could be proportional to the total observable mass of the Universe, and this makes one think of the requirement of the ΛCDM-model for the coincidence of the orders of magnitude for the estimated dark energy density and for the mass energy density of the Universe. What could be the origin of this linear and repulsive gravitation force which seems to belong but is still not discovered up to now?

2.1.3 *Newton and Minkowski models for the intuitive space and time*

Investigating the results of laboratory experiments that included the uniform motion, Galileo introduced the notions of rest and straight uniform motion. Galileo's attempt to define forces as the conditions of the straight uniform motion was rejected by Newton, who gave his own definitions, eq. (1.14), relating forces to accelerations, and the acceleration was inversely proportional to the mass defined as the quantity of substance. The question of the existence of the reference frame corresponding to rest or to straight uniform motion was solved by a postulate known as the first Newton law. Applying his definition of forces to the planetary motion description, Newton obtained the gravitation law for the Solar system. But when he later generalized it for all the bodies in the Universe, he didn't pay attention to the inconsistency: from this moment on, no general reference frame could ever correspond to rest or to straight uniform motion. But the approximations to both were extremely good, and that decided the issue. Mach [Mach (1911)] criticized this approach when considering the inertia and trying to answer the question of whether it is inherent to the body itself or is due to the action of "distant stars" that he used instead of Newton's absolute space. He also rejected the idea of the absolute (free) motion and noticed that there always must be something relative to which the motion takes place.

Later, Einstein also missed or neglected this point when, having finished with SRT, he turned to its generalization for the non-inertial motions performed by bodies. At that time the question of physical inertial reference frame existence was discussed in a constructive manner, and the answer obtained by Einstein was negative. (Having rejected ether as a certain substance filling all the Universe and presenting the absolute reference frame, he later made an attempt to introduce a cosmological constant whose properties resembled some of those of ether.) Nevertheless, the bodies moving along geodesics were considered probe ones, i.e. they did not perturb the geometry of the space in which they moved, and the space could remain, for example, flat and, thus, provide the straight uniform motion. It was only when they attempted to describe the physical system consisting of bodies that could not be considered non-disturbing the space geometry, they finally came to the observable contradiction and discovered flat rotation curves in spiral galaxies.

The notion of the absolute empty space endowed with Euclid geometry was introduced by Newton who regarded the motion of bodies in such space, while the interaction between them was distant and instantaneous. Newton did not unite the space and interacting bodies. Therefore, he implicitly assumed that the existence of empty space without bodies was possible. But only God Almighty could be the observer in such space. His properties could not be cognizable and measurable in the same sense as mass and energy. All the other observers including a human equipped with a ruler with the help of which he is intending to perform measurements in order to check if the geometry invented by his brain fits the real world, inevitably change the properties of the system by making the space not empty. As it will soon be clear, this changes the measurable geometry of this space. If we start with "let us imagine", we lose the link with physics and pass to pure mathematics. In physics everything must start with "let us regard" with the meaning of "let us take" or "let us view", and as it was stressed by Niels Bohr, it must be possible to 'take' or to 'view'. All we could legally imagine in physics is the *absence* of some part of this 'taken', and this part must be small enough in order to affect the phenomena in a weak and measurable way.

Thus, Newton did invent a hypothesis. And it was not the interaction at a distance that seemed such a nonsense to him at first, but the existence of the empty space. Later this misleading step revealed itself in the appearance of ether, dark matter, empty space curvature, quintessence etc. The fundamental physical theory created by Newton was the theory of gravitation based on Kepler law obtained from the observations of planetary motion in Solar system. But in Newton gravitation theory, the space did not play any special role remaining just the container.

When the experiments with the fast moving objects (with light) were performed, it became clear that the Galilean structure does not give a satisfactory description of what was going on, and the principle of relativity of the IRFs motions was burdened with the additional sense related to electrodynamics. This measure was forced by the situation, because the observer was not perfect again, possessed only the senses governed by electrodynamics and could not perform the measurements faster than the light moved. This personal grievance was converted into the Lorentz invariance of the laws of Nature, and the inertial reference frames moving in a straight uniform way in the empty space continued their motion, though the velocity addition law changed.

Poincaré was the first to discover that Lorentz invariance in physics corresponded to the invariance of length relative to rotation in the 4D space with the pseudo Euclidean metric, namely, in such a space in which one coordinate is expressed by the imaginary number, this length in physics is now called the interval. From the point of view of algebra, this meant that the feasible coordinate transformations must form a group which is called Lorentz group or rotation group $O(1, 3)$ that preserved the position of the coordinates origin. And then Minkowsky suggested to substitute the habitual combination of space and time which was used to model the physical world by this unified 4D space-time with the metric given by eq. (1.35). The important feature of this substitution was that the physical interpretation of the parametrization of the curve exploited the notion of "time", which meant that a special inherent conversion factor measured in the units of speed was needed. And the speed of light fitted ideally. It was also present in the interpretation of Michelson and Morley experiment results that Lorentz and FitzGerald tried to explain by the contraction of bodies moving relative to ether. New profound and essentially physical idea was the demand to speak not of the intuitive but of the measurable reality.

Demanding that the interval does not change with translations, we get the transformations that form the Poincaré group or non-homogeneous Lorentz group. These transformations is convenient to present in the following form

$$x^{i\prime} = \Lambda^i_k x^k + la^i \qquad (2.3)$$

where Λ^i_k is the matrix of coefficient of the Lorentz group, l is a constant measured in the units of length, and a^i is a set of dimensionless coordinates corresponding to a vector. Therefore, this is a 10-parametric group: three angles of rotation, three boosts corresponding to the IRF velocity, and four translations.

When they speak of the generalization of special relativity with regard of the accelerated motion, that is of the introduction of the non-inertial reference frames (NIRF), they usually mention Einstein's aspiration to formulate the physical laws in the covariant form - as if it was only the aesthetically attractive and logically consistent possibility. But in view of the gravitation existence, it was the inevitability, and the NIRFs should have been considered first of all.

When we perform observations and measurements in our galaxy, we do it from the surface of the rotating planet, rotating around the Sun rotating around the Milky Way center. All these motions can be accounted for at least partly. But when we observe other galaxies, it is impossible to find a reference body whose motion can be described in a justified way. We can interpret the observations only in the indirect way using the effects that we know from laboratory practice (for example, Doppler effect), considering them the same as on the Earth and having no possibility to really account for the effects that are unknown. Therefore, the observable dynamics (physics) at the large (galactic) scales passes on to the region whose geometry should be defined on the base of these very observations, that is of the phenomena of the same scale but not of the smaller scale for which it is sometimes possible to separate "kinematics" from "physics". This was the premise for Lobachevsky's idea, this was what Einstein accounted for when he recognized the common and different in the laboratory scale experiments and Solar system scale experiments.

Clifford's idea of the equivalence between the physical phenomena and geometrical curvature was particularized by Einstein into the equivalence principle dealing with inertial (kinematical) and gravitational masses. This made it possible to generalize SRT onto GRT using the geometrical considerations. In physical terms this corresponded to the pass from IRF to NIRF with the simultaneous account for the ineradicable gravitation. It is essential to notice that in classical GRT none of the two following questions has arisen: what is the character of motion of an arbitrary NIRF? - this is considered unimportant, because the laws are covariant; and does an IRF with straight uniform motion exist? - an IRF is considered to be a trivial case of a NIRF. But taken together, these two issues mean: first, the insufficiency of the Poincaré group for the description of general motion caused by the impossibility to define the character of motion on the galactic scale; second, the identifying of an IRF and a flat space taken together with the passage from the Galileo group to the Poincaré group could lead to the peculiar properties of the degenerate case of the empty space.

The development of science showed that both the possible existence of known or unknown effects and the meaning of the equivalence principle were probably accounted for in an insufficient way, and this led to the contradictions with the observations. The lack of lucidity in the question with IRF caused the introduction into the theory the unobservable components presenting the major part of the Universe.

In order to improve the situation, first of all, we should reject the discussion and use of such notions as empty space and its curvature as the notions that have no physical (testable) meaning in Bohr's sense. In some cases we can speak about space far from bodies which can have this or that (even flat) metric and curvature, but every time these metric and curvature will be not standalone but will be the limiting cases of the metric and curvature of the space with moving bodies. And in the last only the NIRFs can be present.

Thus, the question arises: in which sense or in what limits may the flat space-time with Minkowski metric be used for the modelling of the physical reality? This question is also important because, as it follows from the above said, the introduction of the absolute cosmological constants (particularly those characterizing geometry linked to motions) into the physical theory is inadmissible, and we should limit ourselves to the finite regions. This means that the role of the parameter characterizing these regions and measured in the units of length is still not revealed in full. It has to be introduced in an unequivocal way - similar to the way in which the light speed was introduced. In the last decade, especially because of the SN1a observations, such absolute constant is discussed, and it is related to the empty space and its geometry. For example, DeSitter radius is a constant measured in the units of length. But being correspondent to the geometrical foundation of the GRT, it contradicts its physical meaning. We cannot observe the empty space, but only the space in which the probe bodies move inertially. This also means, that we cannot be sure that this motion observed from the other frame would preserve its inertial character, because of the absence of the absolute reference frame.

It seems that now the situation demands the following. The origination of the SRT corresponded to the expulsion of the preferential reference system. Now we have to map out the motion in the empty space which is understood as the region that is far away from the single central source of gravitation. It's a question, whether there are such regions in the intermediate scale. Even when we speak about the so-called voids in the large scale, they are not the regions far away *from a single body*. The space without bodies is an abstract invention of mind, and, therefore, belongs to mathematics. The empty space with curvature is also an abstraction, and every time we think of it, we should bear in mind that somewhere around there must be bodies that produce this curvature. The attributing of physical properties to a space without bodies (without matter) seems to break down the essence of general relativity in which "the bodies tell the space how to curve", though Einstein himself also made an unsuccessful attempt to introduce a cosmological constant responsible for the uniform gravitation in the Universe, even if the stars were taken away from it.

This means that it is incorrect to start the construction of the theory with the empty space, then fill it in with the massive bodies that will curve it, and then regard a probe. The observable sources of curvature must be present in the theory from the very beginning, or else the theory would be either untestable at all (flat empty space) or provide a variety of predictions broad enough to make the observable phenomena fit various sets of parameters for various theory structures and not vice versa as it should be for a physical theory - that is, the well defined theory must already possess a unique set of parameters and only the choice of their values has to lead to the fit with observations. The theory of theories is closer to metaphysics but further from the physical observations.

From the point of view of the gravitation theory construction, the historically predefined and seemingly natural practice to start with the special relativity and define the fundamental properties of the corresponding empty space is internally inconsistent and could lead to the physically senseless solutions. It worked, when the bearer of the observed property were electromagnetic phenomena, but it cannot work for gravitation, at least, until we believe in the identity of gravitation and curvature. The empty space should not be the starting point but the particular case of the general gravitation theory. The gravitation field equations derived by Einstein must now play the same role as the electromagnetic field equations derived by Maxwell played a century ago.

2.2 Interpretation dilemma, variation principle, equivalence principle

The flat character of the rotation curves of spiral galaxies discussed in the previous Chapter is probably the most simple, not small and statistically verified effect which doesn't find an immediate explanation in GRT. In the same way, it contradicts the simple everyday approximation, that is Newton gravity.

In a sense, the situation with the measurement of flat rotation curves on the galactic scale resembles that of the end of 19th century with Michelson and Morley ether experiment. Michelson and Morley, as well as the physical community of the time, were sure of the simple character of the effect they were trying to register, the accuracy of their instrumentation was far beyond the needed value, and still, the result was not the one they all expected. The repetition of the same measurements with higher accuracy and more precautions changed nothing and brought bewilderment. Simple explanations like ether dragging quickly failed. In order to save the situation, the most profound theory of the time, i.e. electrodynamics, was used to prescribe some inevitable but directly immeasurable effect of the objects contraction. The total scale of the effect was that of a Universe scale and was far beyond the corrections needed for the experiment explanation.

With spiral galaxies, nobody had any doubts that the orbital velocity or orbital velocity squared or the gravitational potential would go to zero with distance from the center of a galaxy. When the first observational results disagreed with the expectations, many other galaxies were tested but with the same result. In order to save the situation, the most profound theory of the time, i.e. the gravitation theory, was used to prescribe the presence of some inevitable but directly immeasurable entity, i.e. massive matter in the quantities far beyond than those that might be considered a small correction like it was with Neptune planet discovery. This invisible but gravitating matter was then supposed to be present everywhere in the Universe and to concentrate in the same places as the visible matter did.

The spiritual effort undertaken by Einstein was to acknowledge the result of Michelson and Morley observation as a given fact and start with this fact. It was an attempt not to underestimate its meaning, not to look for a suitable compromise, not to invent a hypothesis to put this effect to its place. He recognized why this effect was inevitable in the world of high velocities by means of the analysis of what and how was measured.

2.2.1   *Dilemma: new entity or new equations*

In order to realize the problems stemming from the astrophysical observations, let us regard the physical ideas that are now used. The theoretical attempts mentioned briefly in the previous Chapter and dealing with various complications of the GRT formalism do not seem inspiring because of their vague relation to physics. The first and the most popular idea suggests, so to say, to "increase the right-hand side of the known field equations until the obtained solution covers the observed effect". Physically, this corresponds to the introduction of some dark matter whose mass appears to be 3-4 times larger than the mass of the visible luminous baryonic matter. This is how it was done.

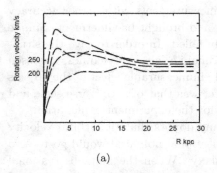
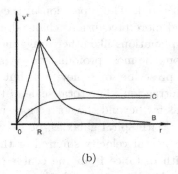

(a)   (b)

Fig. 2.1   (a) Sketch of the experimental data obtained in [Sofue and Rubin (2001)]. (b) Plots for the orbital velocity squared corresponding to the inner and exterior problems for homogeneous ball.

On Fig. 2.1(a) there is a sketch of the rotation curves of spiral galaxies of various types obtained from measurements (drawn after the curves found in [Bekenstein and Milgrom (1984)] where there is a reference to [Sofue and Rubin (2001)]). On Fig. 2.1(b) the plot OAB presents the dependence of the gravitational field stress for the homogeneous ball with radius, R, which is proportional to the satellite orbital velocity squared (the material of the ball is supposed to be rarefied and the satellite is small). The relation between Fig. 2.1(a) and Fig. 2.1(b) is obvious. Since instead of an OAB type curve, the experiment gives OAC and OC type curves, one concludes that the homogeneous ball model is not valid - and for several reasons at once. The non-homogeneity of the matter distribution in a spiral galaxy can be measured and calculated and, thus, the stress dependence on the distance from the center can be accounted for (see the calculations in [Mannheim (2006)]). Nevertheless, it still corresponds to the chosen symmetry of the problem and to the boundary conditions, that is the stress tends to zero with distance from the center. The next specification is to find such parameters of the mass distribution obtained by calculation, that the observational data corresponding to Fig. 2.1(a) would fit the theoretical OAC type curve. It turned out that the fitting calculated values must be the following: the fitting calculated mass must be thrice larger than the total mass estimated by usual means; the fitting calculated radius must be five times larger than the visible galaxy radius. Thus, the larger part of the mass distribution that could provide the observable effect is invisible, i.e. presents a dark matter. Simultaneously, one should suggest that the paradoxical flat curves could correspond not to the exterior problem but rather to the inner one.

The very idea of introduction of the invisible gravitating matter in the spiral galaxies has several logical discrepancies:

- the evaluation of the amount of dark matter in every particular case is performed *a posteriori*, and, therefore, this amount is a fitting parameter which is not stipulated and specified by the theory (GRT);
- there exists an empirically well based relation between the luminous matter and the observed stars' orbital velocity which does not require dark matter, this is Tully-Fisher law for the velocity of stars at the periphery of the spiral galaxy and its luminosity which is related to the luminous and not to the total mass of a spiral galaxy, while no proportionality between these two masses is known;
- the contradiction appears when the observations are compared to the solution of spherically symmetric problem with a point center, that is the difference between the exterior and inner problems mentioned in the previous Section is not taken into consideration.

The existence of the observational data (Fig. 2.1(a)) corresponding to the OC type curve on Fig. 2.1(b) can make one think that the boundary conditions were chosen hastily and did not correspond to the problem in question: before the stress starts to tend to zero, there could appear another effect related to the physical reasons that we don't know yet. In this case the disaccord of the flat rotation curves of the spiral galaxies with the theoretical predictions would not be surprising. These speculations correspond to the attempt of an empirical modification of the dynamics (or of the gravitation) law at large distances and at small accelerations. That is why the second idea suggested by Milgrom [Milgrom (1983)] and now known as MOND was developed particularly to explain the flat rotation curves. It is proposed to modify the dynamics equations or, equivalently, to modify Newton gravity law. Since the intention to fit the experiment was proclaimed as the initial one, many observational data (though not all, see [Aguirre *et al.* (2001)]) can be described in a satisfactory way. But as it was underlined by Fock [Fock (1939)], the dynamics equation (related to the geodesics) are not the independent part of the general relativity theory, they present the condition of solvability of the field equations. That is why they cannot be arbitrarily changed even if to fit the observations. An attempt [Bekenstein (2007)] to make MOND a metric theory led to the introduction of an additional scalar field into the expression for Hilbert-Einstein action. This has the physical meaning of changing the effective gravitational constant from place to place, i.e. suggests to introduce the tunable parameter. At the same time, the present ambient value of the effective gravitation constant must be chosen as a boundary condition. Therefore, the unsatisfactory features of this approach for its phenomenological version are:

- arbitrary choice of the expression for the correction term in the dynamics equations or in the gravitation law;
- introduction of the new empirical world constant (measured in the acceleration units and surprisingly close to cH which was noticed but not explained);

and the drawback of its covariant version is:

- it belongs to the Brans-Dicke type theories, they could be regarded as an alternative for the GRT which is still the best, but they have an extra tunable parameter. Besides, this membership doesn't make the covariant MOND the preferred one.

Thus, we see that the first (dark matter) approach plainly exploits an inconsistent method and is hardly based on a physical idea, the angels or demons would do none the worse; and in the second approach (MOND), the way to a better idea is barely outlined and also seems to lead to a new entity with the property of a tunable parameter. The interpretation dilemma brings to the choice that seems to be misleading in both cases.

Usually they say that the explanation of the large speed of stars close to the galaxy disk edge could be given only by the assumption of the existence of extra mass that reveals itself only through gravitational interaction. The same demand is used for the explanation of the stability of the galactic disk and of the massive galaxies, now the massive halo is usually mentioned and the ratio of its mass to the mass of the visible stars inexplicably varies from case to case. Thus, the dark matter is needed only to produce an additional action, and if we knew an additional type of action, no extra mass would be needed. But currently, no additional type of action is discussed with regard to galaxy dynamics, whence the presence of it in a larger scale is almost generally accepted due to the observations interpreted as the acceleration of the Universe expansion. There they need repulsion, here they need attraction, could it be one and the same force? In electricity there are two types of charges, but in gravitation we haven't yet found a negative mass. But already in magnetism the forces acting in opposite directions appear with the change of current direction. The similarity of gravitation and electrodynamics was drawing interest from the very beginning, and already Maxwell thought of it classically but gave up, and later Lense and Thirring put it in a relativistic way and obtained only a small correction.

The solution of the theory equations which was the base for the observed discrepancy was first obtained approximately by Einstein himself in [Einstein (1916a)] and then exactly by Schwarzschild [Schwarzschild (1916)]. The next step was performed simultaneously and independently by Einstein et al. in [Einstein *et al.* (1938)] and by Fock in [Fock (1939)]. In both papers they constructed the next - second - consecutive approximation of the field equations and stopped on this because of the technical complications. Fock obtained the following expression for the gravitation potential in the point far away from the system of gravitating bodies. In his notation it was

$$U = \frac{G}{r}(M + \frac{E}{c^2}) \qquad (2.4)$$

where M is the total mass of the bodies in a system, and E is a sum of their kinetic and potential energies. Formula eq. (2.4) was obtained from the field equations and is in accord with the idea of mass and energy equivalence stated in the SRT. If E is proportional to the product of the total mass by the average velocity squared, the correction appears to be of the second order in $1/c$.

2.2.2 *Comparison of methods*

Let us pay attention to the comparison of the approach of classical GRT practiced by Einstein and Fock, and to the approach of exact solutions, first demonstrated by Schwarzschild and later by DeSitter, Friedmann and others.

As one can see from Section 1.5, in the first of them, we choose the model of real space in the form of 4D deformed Minkowski space-time and derive the corresponding geodesics equations, put down the field equations that can be obtained from the variation principle, use the simplifying assumptions, the main of them being $g_{ik}(x) = \eta_{ik} + h_{ik}(x); |h_{ik}(x)| << 1$, and at the last but one step the situation is the following

$$\frac{d^2 x^\alpha}{dt^2} = -\frac{1}{2}\frac{\partial h_{00}}{\partial x^\alpha}; \alpha = 1, 2, 3 \qquad (2.5)$$
$$\Delta h_{00} = \kappa\rho$$

$$g_{ik}(x)|_{x\to\infty} \to \eta_{ik} \qquad (2.6)$$

Now we regard the first two equations: the first is the equation of motion (Newton dynamics law), the second is the equation for the field of the force. Notice what happens now. These *two* equations form a *specific boundary problem*: there is a system of differential equations (field equations) that must be supplied with the boundary conditions - that is the given values of the potential on some given border. Since the real heart of the matter is physics and not the abstract mathematics, the border must be such that we are able to perform measurements and compare the result with the mathematical problem solution. So, the border at infinity does hardly suffice. Besides, and this is even more important, we have no instruments to measure the values of the potential at any border at all. On the other hand, we can observe bodies that move according to these potentials. Thus, the equation of motion (geodesics) on this border must be sufficed by the same potentials. It is this transformation of a differential equation with boundary conditions into a system of differential equations that makes the physical problem sensible and solvable. This may be called the Einstein method, and this is the main Einstein's achievement in the geometrization of physics. What he did was not the mere interpretation of the exquisite mathematical solutions. It was the genuine method of posing problems for physical situations.

No general solution of the system of the differential equations for the field is known, thus, Einstein chose the approximation procedure and already in the first order got "what we see", that is Newton gravitation law derived from Kepler observations. Only after that, one can make sure that this solution suffices the third equation in eq. (2.5) which is the usual boundary condition for a mathematical problem. By the way, Einstein didn't even mention this check up - because of its obvious character. Instead, he wrote in the beginning that the flattening of space-time in the infinity means the localization of the source of curvature. Thus, he spoke of the region which was far enough from the source though still possessed some curvature, that is - of an exterior problem, the same that Newton spoke about.

Here it worked. But in the previous case - with SRT - the situation was not that good. And when Einstein decided to stick to "what we see" in exactly the same manner which was just described, he (finally, with Minkowski's help) had to switch to another initial model of real space. This he was obliged to do, if he wanted to discuss physics with its measurement possibilities. Instead of inventing *new* additional "models" he changed the *initial* one. All the others either did not recognize the presence of this initial model or rejected to change it because of its 'obvious' character. They continued with the efforts to preserve the 3D+1 Newtonian model and tried to imagine how the observed results could be possible. However plausible were these efforts based on assumptions and imagination, the most economical and observation based physical approach has rightly won. The new model of real space - flat 4D space-time - appeared to have such profound influence on our understanding of physical reality, because it was totally based on our abilities to get answers to the questions we put. Simultaneously, the new groups of coordinate transformations (Lorentz and Poincaré groups) that already contained everything needed to be in accord with observations took the upper hand, and this gave a new impulse for the mathematical research. (When it later came to the unification of electromagnetism and gravitation, Kaluza [Kaluza (1921)], in a sense, imitated the same approach, though formally, and added an extra dimension to the model of space, that is to space-time. The attempt was successful, but left an unsatisfactory feeling because of the need of the compactification of the extra dimension.)

Now let us have a closer look at what was done, for example, by Schwarzschild, who pursued the traditional mathematical approach and wrote about providing more purity to the result. He also made some simplifying assumptions, and managed to find the ingenious solution of the *usual boundary problem* for the field equations

$$\frac{\partial \Gamma^j_{ik}}{\partial x^j} + \Gamma^j_{im}\Gamma^m_{kj} = 0 \tag{2.7}$$

$$\det g_{ik} = -1; \, g_{ik}(x)|_{x\to\infty} = \eta_{ik}$$

which corresponds to the *last* two equations of the system eq. (2.5). It happened that in the exterior case this solution also corresponded to Newton gravitation law. The model of real space remained the same as before, that is the geodesics equations corresponding to this model also remained the same, but they somehow fell out of the consideration making the problem less physical in the sense of Einstein method. Only after the solution of this abstract mathematical problem was obtained, it was substituted into the geodesics equation which now appeared to be a separate equation, which is absolutely not the case, though the obtained trajectories coincided with observations. It is very natural to trust the exact solutions when they are supported by the observations. Mathematics started to dominate, and the above mentioned Einstein's main achievement was in a sense camouflaged and forgotten.

It was probably this that Fock warned about [Fock (1939)] when speaking of the application of the GRT to cosmology. It was this that Einstein himself noted when he gave the negative reference to the Friedmann's paper when he saw it for the first time and called formal. This is why the only attempt to save the situation when the disagreement with observations appeared was the modification of the field equation or its contents, and, hence, the change of Hilbert-Einstein action. Nothing but this comes to a mathematically oriented mind which is used *to think* that it deals with physics. Whatever solution is obtained on this way, it will again be substituted into the same geodesics which waits its time somewhere aside. This is the road to the adjustments and to the use of imagination which can sometimes lead to the way out but sometimes can be misleading.

At first it led to the notion of black holes that had a physical background since Laplace times, and naturally, as soon as there is a concept, there appear observations that can be interpreted with reference to it. This naturally led to the development of the corresponding model that obtained certain physical features. Fortunately, the black holes present only a small part of the presumable but still not definitely observable physical reality. Then this led to the dark matter, the amount of which was already 3-4 times larger than the amount of the observable matter. Then, when another exactly solved mathematical problem - which has already changed our world view and has made us think that there was a Big Bang and the Universe is expanding - came into the confrontation with observations, there appeared new dark areas in the form of dark energy, and finally, only 4% of the surrounding world was left for physics based on observations. Einstein's method was forgotten, but his name was used.

Unfortunately, when he himself tried to investigate the possibilities of mathematics for cosmology and introduced a mysterious cosmological term, the physical sense of which was inaccessible for measurements, there appeared Hubble's discovery that had a simple interpretation which supported another abstract mathematical solution, the Friedmann's one. The situation was confusing, and probably Einstein had some reasons not to pay enough attention to the interpretation suggested by Hubble himself already in the original paper [Hubble (1929)]: Hubble spoke of the DeSitter Universe [De Sitter (1917)] and the time slowing down there.

At the same time, Friedmann's equations helped Einstein [Einstein (1931)] to realize that the field equation containing the cosmological term had no physical sense, and he did it independently of Hubble's observations. The corresponding solution appeared to be unstable: any fluctuation that brings the system out of the equilibrium would make the system deviate from it further and further with time.

As one could see from Section 1.1, it is not for the first time that on the stage of the leadership of mathematics, we went a little bit too far. Absolute circles transformed into circles with epicycles, regular Platonic polyhedral were inscribed into spheres "corresponding to the planets orbits", bodies in the absolute space were subjected to Lorentz's contraction, etc.

2.2.3   On the variation principle

Let us discuss the connection of what was called Einstein's method with the approach that is considered to be most general in physics, that is with the variation principle method.

The usual variation method suggests to choose a scalar which is usually an integral functional called action, S, "fix the ends", that is demand that the integrand has definite values on these ends, then perform the variation procedure, that is find the conditions that must be fulfilled for the integrand to provide the extremum of the functional. In geometry, the integrand could be the expression for the length of the curve connecting two given points on the surface, and the result of the variation procedure, i.e. the condition to be fulfilled, is the equation of the geodesics. In physics, the integrand could be a Lagrange function and the integral is taken between two moments of time for which the values of the Lagrangian are considered fixed. In general physics this is a natural procedure, since in a sense it models the experimental situation: there are the initial conditions with the known (measured and calculated) values of kinetic and potential energies forming the Lagrangian, then we let the system go and find (measure and calculate) the values of kinetic and potential energies corresponding to the final moment of time. The variation procedure gives the Euler-Lagrange equations that present the differential equations of dynamics for which the measured values at the ends provide the boundary conditions. That is, the "trajectory" is the integral curve for these equations. Due to the definition of kinetic energy, the geometrical geodesics is the same curve as the trajectory of a free particle [Arnold (1974)]. The geodesics on a curved surface corresponds to the trajectory of a particle in a gravitation field. Solving the equations, one could get the expression for the line that is related to the curvature of space in geometry and to the gravitation potential in physics.

But in cosmology this procedure can be inappropriate or even misleading, because there is no observable experimental situation that can be consistently modelled with its help. The reason is that only the last - final point is available for observations, the initial one is not, and, thus, one of the ends is free. Therefore, there could be many different integral curves sufficing the equations and passing through the only given point, this makes the chosen solution arbitrary. That is why for cosmology, the field equations obtained by the variation method cannot have the same significance that they had for planetary system, and one should have a more attentive look at the initial Einstein's intuitive conjecture.

Turning back to the variation principle, one could recollect several modifications that differ from variation principle with the fixed ends. First, the ends might be free, but belong to some surfaces whose equations must be known. Then instead of the vanishing of the integrated term, there appear the transversality conditions, and the solution of these provide the needed boundary conditions for the Euler-Lagrange equation solution. It is the equations of the given surfaces - two in number - that enter the boundary conditions making the problem solvable. The problem is that the equations of such surfaces for cosmological problems are again unknown. Second, there could be some conditional extremum, and in this case, the constraints must be given. They could even be non-holonomic, and in this case the conditions present differential equation. The geodesics equation corresponding to a free motion might be used as such condition. And it is just the case which was "secretly" adopted by Einstein. In this case, it is the condition that must be sufficed first, and only after that one should turn to the boundary conditions and check if the behavior of the function is sensible. Still, it seems that there is no way to avoid arbitrariness completely. The most radical way would be to modify the general form of the least action principle and demand conditional variation. In some cases this would coincide with the results of the use of additional observational back up.

Whatever are the demands to the Lagrange function and whatever form it has, its choice is inevitably arbitrary. There is always a whole class of Lagrange functions that differ from each other by the summand which is a full time derivative of an arbitrary function in classical mechanics or has an equal to zero divergence of an arbitrary function in relativity. Such term vanishes during the integration because of the Gauss theorem, and the equations of motion (or the equation of field) will remain the same. These additive terms would not correspond to any observable value, because it is these equations of motion that can be used for experimental check up, and therefore, this "hidden" function cannot be catched. When the procedure is more complicated, there could be several "hidden" functions, and if we attribute some sense to each of them, there could appear several correlations that could resemble and even "reveal" certain physical properties.

Thus, the variation principle can sometimes lead to the origination of additional physical notions. Performing additional experiments with account for these notions, one could find their traces in the same indirect manner, i.e. the nontrivial results of the experiments can be predicted and observed. For example, this can be frequently seen in various cases in particle physics. But as soon as one tries to perform direct observations, there *must* be a failure, because if there is a success, these notions, being observable, had to be introduced into the initial Lagrangian and change the result of the variation procedure. As long as they are tiny corrections to the general picture, it is quite possible and even reasonable to put up with this situation, but when they start to dominate, this indicates at some crucial defect in the theoretical approach and at the need of the latter's modification.

The demand for the conservation of energy and momentum in an isolated system makes one think of a function whose divergence is equal to zero - this was the guiding line for the energy-momentum tensor introduction. This means that the flow of this function through any closed surface enveloping the system is equal to zero too. This gives us the following options. The function is a constant in the region with the matter inside, which means that the world is stationary. But this somehow contradicts our ideas about the behavior of similar systems like gas, and it was discussed in the previous chapter. We could suggest that this function is present and constant everywhere - inside and outside the surface. This is, actually, the idea of the cosmological constant. But such equilibrium is not stable. We could also suggest that though the function is time dependent and "tries" to produce a flow through the surface, the surface itself moves and expands in the corresponding way, thus, preventing the flow through itself from becoming non-zero. This is an idea of an expanding Universe and time dependent metric.

We can also suggest that this function is a curl of another function. The divergence of a curl is equal to zero, and the curl corresponds to the angular momentum that is also conserved in an isolated system. Such term could be both an alternative and a summand.

The antisymmetric tensor

$$M^{ik} = \sum (x^i p^k - x^k p^i)$$

is the 4-tensor of angular momentum and is conserved for the closed system. Its space components coincide [Landau and Lifshitz (1967)] with the components of the usual three-dimensional vector of angular momentum $\mathbf{M} = \sum [\mathbf{r}, \mathbf{p}]$

$$M^{23} = M_x; \ -M^{13} = M_y; \ M^{12} = M_z$$

while the components M^{01}, M^{02}, M^{03} present the vector $\sum (t\mathbf{p} - \mathcal{E}\mathbf{r}/c^2)$, and the components of tensor M^{ik} can be presented as

$$M^{ik} = \left(\sum (t\mathbf{p} - \mathcal{E}\mathbf{r}/c^2), -\mathbf{M} \right)$$

Since M^{ik} is conserved for a closed system,

$$\sum (t\mathbf{p} - \mathcal{E}\mathbf{r}/c^2) = const$$

But the full energy $\sum \mathcal{E}$ is also conserved, and the last equation can be written as

$$\frac{\sum \mathcal{E}\mathbf{r}}{\sum \mathcal{E}} - t\frac{c^2 \sum \mathbf{p}}{\sum \mathcal{E}} = const$$

One can see that the point with the radius-vector

$$\mathbf{R} = \frac{\sum \mathcal{E} \mathbf{r}}{\sum \mathcal{E}} \qquad (2.8)$$

moves uniformly with the velocity

$$\mathbf{V} = \frac{c^2 \sum \mathbf{P}}{\sum \mathcal{E}}$$

which is the velocity of the system as a whole corresponding to its full momentum and full energy. Equation (2.8) gives the relativistic definition of the center-of-mass coordinates. The important property of this definition is that the components of vector $\mathbf{R}$ do not correspond to the space components of any 4-vector, and, therefore, do not transform as the coordinates of a certain point with the transformation of the reference frame. That is why the center of inertia of one and the same system of bodies occupies different positions relative to different reference frames. It also underlines the role of observations on the cosmological scale.

2.2.4 *On the equivalence principle*

The principle of relativity that is adopted as a guiding principle in physics since Galileo times means the impossibility to distinguish between the phenomena taking place in the inertial reference frames. The equivalence principle formulated by Einstein is the extension of the principle of relativity, and it means that there is no possibility to distinguish between the local phenomena taking place in a non-inertial reference frames and in the gravitation field. The questions arise: how to distinguish between the inertial and non-inertial reference frame when performing measurements on the galactic scale? What does "locality" mean now? What is the criterion of the inertial reference frame? Obviously, it is not the straight uniform motion, because according to the indisputable meaning of the relativity theory, there is no possibility to decide which is it by direct physical measurements and not by mathematical reasoning. The motion - even the motion of light - takes place along geodesics, and the form of the geodesics depends on the presence and intensity of the gravitation sources that curve the space, and the observable gravitational lenses are good illustrations for that. Therefore, the equivalence principle dominates, it is the main one, because the principle of relativity is its particular case corresponding to the absence of sources or, strictly speaking, to the region far from sources. Then how far?

Where is the border of the region beyond which the inertial motion differs little enough from the straight uniform one? It is beyond this border that the usual Lorentz invariance must be demanded from a theory. But what is inside this region? Whatever there is, the radius of the region, l, must enter the speculations on the same grounds as the extremal value of the available velocity, c. The discussion of this question will be continued in the subsequent section.

Together with the inertial forces related to the rectilinear acceleration of the frame (change of the velocity magnitude), there are also the inertial forces depending on the velocities of the body and of the frame (change of the velocity direction), for example, Coriolis force. Then the following proposition must be true

Proposition 2.1. *The literal meaning of the equivalence principle suggests that the gravitation force and, hence, the gravitational potential depend on the relative velocity of a body.*

Apart from the formal reason, this suggestion has an additional one related to the ideas mentioned above. As long as the observer is inside the system of moving bodies and takes part in their motion, there is no experimental way to distinguish the observable effect which is due to inertia from the effect due to gravitation. The expression for the inertial force has to include the velocity dependent term, therefore, the expressions for the gravitation force and, hence, for the gravitational potential also have to include such dependence. This means that the metric tensor which *is* the gravitational potential starts to depend on velocities, i.e. from the geometric point of view, the space becomes anisotropic. What about the meaning of locality and the possibility to reveal this meaning experimentally? When it comes to the experimental check up, the passage to the infinity where the Coriolis type forces become infinite requires special consideration, and it will be given in the next section. This makes the situation in a galaxy different from the situation in a planetary system in which the phenomena that take place in the vicinity of a star, in the vicinity of a planet and in the laboratory can be easily separated according to their scale, and the reference frame with the definite properties of motion can be chosen with regard to these phenomena.

Besides, it is well known that any theory that consistently unifies the inverse square law for the forces and Lorentz invariance has to contain the field originating from currents (see the discussion of Maxwell identities in the previous chapter and in [Siparov (2008a)]). In electrodynamics where the interaction constant q is the electric charge, this field is magnetic vector field, and it produces the Lorentz force which depends on the velocities of charges and enters the right-hand side of the geodesics equation. The expression for the Lorentz force is well known

$$\mathbf{F}_L = q\mathbf{E} + q[\mathbf{v}, \mathbf{B}] \qquad (2.9)$$

In the approach known as gravitoelectromagnetism (GEM) [Lense and Thirring (1918); Mashhoon (2001); Ruggiero and Tartaglia (2002)], the gravitation is described with the help of scalar and vector potentials analogously to electrodynamics. This approach was used to calculate the corrections to the classical solutions obtained with regard to the rotation of the massive gravitation source. The calculations did not contradict the measurements, but this analogy is too straightforward and cannot be applied to gravitation directly. GEM theory copies electrodynamics formally, but in the latter the velocity dependent Lorentz force is produced by the additional (external) field and by the corresponding sources (electric charges) [Landau and Lifshitz (1967)]:

$$mc\left(\frac{du^i}{ds} + \Gamma^i_{kl} u^k u^l\right) = \frac{e}{c} F^{ik} u_k \qquad (2.10)$$

where u^i is the 4-velocity and F^{ik} is electromagnetic tensor.

In general relativity, the ineradicable gravitation force cannot be attributed to an external (additional) field; similarly, no part of it can be arbitrarily regarded as an additional field as in GEM. Being the cause of the geometrical curvature, the gravitation field is identified with the metric itself, this means that the GEM theory is only an ingenious trick which must be substituted by the consideration of the anisotropic space. The same refers to an attempt to apply to gravitation the theory of retarded (or Lienard-Wiechert) potentials. And of course, the account for the metric tensor velocity dependence in the next orders of asymptotic expansion of the usual GRT solutions is also not the case, these corrections are very small and do not infer a principally new approach able to repair the situation.

The conclusion is that the metric corresponding to the potentials of the gravitation forces must include the vector fields related to the velocities of the sources and of the probe particle. The velocity dependent metric characterizes the anisotropic space whose geometry is no longer Riemannian one. We see that a hundred years old situation is repeated. Then it was the result of Michelson and Morley experiment, now it is the rotation curves of spiral galaxies. Then the physical theory and formalism that everybody wanted to preserve was the electrodynamics and Maxwell equations, now the physical theory and formalism that everybody wants to preserve is the gravitation theory and Einstein equations. The proposed solution is the same - the recognized necessity to change the geometry of the model used to describe the physical reality. Then it was the passage from the Euclidean 3D space plus 1D time to the 4D Minkowski space-time. What now? At first glance, the anisotropic space seems to be a serious complication.

In order to realize what does it really mean and what should be considered, notice the following:

- every point of this space-time (main manifold) must be endowed with the vector obtained as a derivative along the curve (trajectory) with respect to the natural parameter, thus, forming the tangent bundle;

- the structure resembling the direct sum of the 4D main manifold and 4D fiber that constitutes an 8D analogue to the usual 6D phase space of general physics is sufficiently simple;

- Einstein equations for the gravitation field formulated for the 4D space-time are supposed to preserve their form in the modified theory.

As it was mentioned, the geodesics equation is not independent, and the equation of motion in the Riemannian space takes Newtonian form only in the lowest approximation of the metric expansion in the vicinity of the Minkowski metric. Even in the next approximation for the solution in the Riemannian space, the obtained corrections must be considered in the equation for the geodesics [Fock (1939)]. This is also true for the space which we are going to consider and in which the geodesics equations are supposed to take another form because of the velocity dependence of the metric. Thus, the physical considerations has brought us close to the mathematical formalism which can be used for the construction of the theory.

Finally, the set of the principal postulates forming the base for the construction of the gravitation theory applicable for the intermediate (galactic) scale is the following:

(1) the equivalence principle is true in its full meaning, that is the observable gravitation interaction (and, therefore, the gravitation potential expressed by the metric tensor) depends not only on the coordinates of the bodies but on their velocities as well;

(2) the appropriate model for the description of the real physical world is the anisotropic space equivalent to the 8D phase space-time whose geometry corresponds to the tangent bundle that includes main manifold (space-time) and fiber;

(3) the only origin of the geometrical curvature of the model space is the distribution of the moving masses;

(4) the principle of relativity is the corollary of the equivalence principle and involves the inertial motion along geodesics; in the situations where the influence of the gravitation source is negligible, it coincides with Galileo-Einstein principle of relativity;

(5) the boundary conditions for a physical problem must account for a border separating the inner problem from the exterior one;

(6) the invariant group of coordinate transformations must account for the existence of the mentioned border.

The solution of the field equation must be performed with regard to the modified expression for the geodesics, and the result must be comparable with the observations.

2.3 Construction of the formalism

2.3.1 *Space and metric*

Let $M = R^4$ be a differentiable 4-dimensional manifold of class C^∞, TM be its tangent bundle, with coordinates $(x, y) = (x^i, y^i)$, $i = 0–3$. If c is a curve on M, described by $c : [a, b] \to M$, $t \mapsto (x^i(t))$, then its natural extension to TM is $\tilde{c} : [a, b] \to TM$, $t \mapsto (x^i(t), y^i(t))$ where $y^i = \dfrac{\partial x^i}{\partial t}$. The arclength, s, which is usually taken as a natural parameter, therefore, is equal to $s = \int\limits_0^t \sqrt{g_{ij} y^i y^j} \, d\tau$, where g_{ij} is the metric tensor, and $\sqrt{g_{ij} y^i y^j} \equiv F$.

Let the metric depend on y introduced above, i.e. $g_{ij} = g_{ij}(x, y)$. In general case such metric is a generalized Lagrange one, $g_{ij} = g_{ij}(x, y)$; $i, j = 0 - 3$, which is a 2-covariant, symmetric tensor on TM with the only restrictions: a) $\det(g_{ij}) \neq 0$ for all (x, y) on TM and b) with respect to the coordinate changes on TM induced by the coordinate changes on M, its components transform by the same rule as (0,2)-type tensor components on the base manifold M. This means that TM turns to be an eight-dimensional Riemannian manifold (the obvious analogy with a well-known six-dimensional phase space in physics has a deep meaning and will be discussed in the subsequent sections). The 4D base manifold M can be regarded as embedded into the 8D tangent bundle, TM, and the tangent vectors for all possible curves on M compose a 4D vector space, that is a tangent space or a fiber. Tangent bundle can be regarded as isomorphic to the direct sum of the base manifold and the tangent space. Strictly speaking, the geometry of this 8D "phase space-time" is a bit more complicated and requires some specific ingredients, such as Ehresmann (nonlinear) connections. But if we limit ourselves to linear coordinate transformations $x^i = \Lambda^i{}_j x^j$ (where $\Lambda^i{}_j$ are constants) and weakly curved space with metric like $g_{ij}(x,y) = \eta_{ij} + \varepsilon_{ij}(x,y)$, then this geometry simplifies a lot. The definition of y^i makes it possible, when needed, to use the simplest case - the one of the Sasaki lift [Bao *et al.* (2000)] - to raise and lower indices corresponding to both "horizontal" (x-) and "vertical" (y-) quantities on the tangent bundle with the help of the same metric tensor $g_{ij}(x,y)$. In order to insure the curve length invariance with regard to the change of the parameter on the curve, tensor g_{ij} should be 0-homogeneous in y, that is $g_{ij}(x, \lambda y) = g_{ij}(x,y)$, $\lambda > 0$, i.e. the metric depends only on the direction of y, but not on its magnitude. The last is equivalent to the relation

$$\frac{\partial g_{ij}}{\partial y^k} y^k = 0; i,j,k = 0 - 3 \qquad (2.11)$$

If it also holds $\frac{\partial g_{ij}}{\partial y^k} y^j = 0$, the introduced metric would become a Finsler one as it is defined in [Rund (1981)], but now we don't demand this. Various Finsler extensions of the relativity theory were investigated by Pimenov in [Pimenov (1987)].

Let us once again emphasize the way in which the notion of space-time established in physics. The absolute Newtonian space and time possess no intrinsic parameters. The same is true for an abstract four-dimensional geometric space (x^i), $i = 0, 1, 2, 3$ with a specified metric, for example, Minkowski one $\eta_{ij} = diag\{1, -1, -1, -1\}$. But when we use such space for physical interpretation, and appoint the first coordinate x^0 to be 'time' (i.e. seconds) while the rest three are appointed to be 'space' (i.e. meters), there appears the need for a dimensional conversion factor which makes all the four coordinates equivalent. Obviously, it has the meaning of some fundamental 'speed' (meters per second). Fortunately, in physics there exists a very convenient speed - the speed of light. Light plays the basic role for most observations, since the majority of them is performed with the help of electromagnetic signals. The convenience is due to the fact that its value is very high and it is rarely actually measured or needed in the everyday scientific practice. Therefore, it is postulated to be a constant of our world and characterizes the four-dimensional space-time which is used to model the reality.

For the same reasons, when the introduced eight-dimensional space is used for physical interpretations, there must appear an extra conversion factor with the dimensionality of length, *fundamental length*. Really, if x^i are positional variables, their measurement units correspond to length. The units for the directional variable y^i depend on the used definition of y^i. If $y^i = \frac{dx^i}{ds}$ where s is an arclength, y^i are dimensionless. Since x and y are the coordinates in the eight-dimensional space and must be treated equivalently, there must appear a dimensional conversion factor - now in the definition of y^i - chosen such that the measurement units of y^i are the same as those of x^i :

$$y^i = \frac{dx^i}{ds} \Rightarrow y^i = l\frac{dx^i}{ds}; \ [l] = length \tag{2.12}$$

If t is a parameter with the meaning of time and is used instead of an arclength, $ds = cdt$, where c is a constant with the dimensionality of speed, then

$$y^i = l\frac{dx^i}{cdt} = \frac{1}{H}\frac{dx^i}{dt} = \frac{1}{H}v^i; [H] = (time)^{-1}; v^0 = c \tag{2.13}$$

Thus, when passing from the mathematical coordinates to the physical ones, the following correspondence should be used

$$(x^0, x^1, x^2, x^3, y^0, y^1, y^2, y^3) \longleftrightarrow (ct, x, y, z, \frac{c}{H}, \frac{v_x}{H}, \frac{v_y}{H}, \frac{v_z}{H}) \tag{2.14}$$

One can immediately notice that $y^0 = \frac{c}{H} = const$. It means that all the points in the newly defined space that present the states of a physical system belong to a 7D surface imbedded into the 8D phase space-time. In mathematical transformations, it is sometimes convenient to use the system of units where $c = H = 1$. But when we come back to physics, both constants must be accounted for and chosen in a most convenient or natural way.

Let us regard a weakly curved space and choose the metric in the simplest form: $g_{ij}(x, y) = \eta_{ij} + \varepsilon_{ij}(x, y), \forall (x, y) \in TM$, where $\eta_{ij} = diag\{1, -1, -1, -1\}$ is Minkowski metric tensor on M, and $\varepsilon_{ij}(x, y) = \sigma\epsilon_{ij}(x, y), (\sigma << 1)$ is a small (linearly approximable) anisotropic deformation of η such that only the terms proportional to $\varepsilon_{ij}, \frac{\partial \varepsilon_{ij}}{\partial x^k}, \frac{\partial \varepsilon_{ij}}{\partial y^k}$ and $\frac{\partial^2 \varepsilon_{ij}}{\partial x^l \partial y^k}$ should be retained in calculations.

2.3.2 *Generalized geodesics*

Geodesics are defined as extremal curves for the arclength $s = \int\limits_{0}^{t} F d\tau$, and now $F = \sqrt{(\eta_{ij} + \varepsilon_{ij}(x, y))y^i y^j}$. The variation procedure applied to this expression similarly to [Siparov and Brinzei (2008a)] results in the Euler-Lagrange equations of the form

$$\frac{\partial F}{\partial x^k} - \frac{d}{ds}\left(\frac{\partial F}{\partial y^k}\right) = 0.$$

Multiply the above equalities by $2F$:

$$2F\frac{\partial F}{\partial x^k} - 2F\frac{d}{ds}\left(\frac{\partial F}{\partial y^k}\right) = 0.$$

and take into account that $\frac{dF}{ds} = 0$ along the curve. Then we get

$$2F\frac{\partial F}{\partial x^k} = \frac{\partial F^2}{\partial x^k}$$

$$2F\frac{d}{ds}\left(\frac{\partial F}{\partial y^k}\right) = \frac{d}{ds}\left(2F\frac{\partial F}{\partial y^k}\right) = \frac{d}{ds}\frac{\partial F^2}{\partial y^k}.$$

and the Euler-Lagrange equations become

$$\frac{\partial F^2}{\partial x^k} - \frac{d}{ds}\left(\frac{\partial F^2}{\partial y^k}\right) = 0, \tag{2.15}$$

where F^2 can be related to the kinetic energy as usual. Let $,k$ be the notation for $\frac{\partial}{\partial x^k}$, and $\cdot k$ be the notation for $\frac{\partial}{\partial y^k}$. Then $\frac{\partial F^2}{\partial x^k} = g_{ij,k}y^i y^j$ and $\frac{\partial F^2}{\partial y^k} = g_{ij\cdot k}y^i y^j + 2g_{kj}y^j$. Therefore,

$$\frac{d}{ds}\left(\frac{\partial F^2}{\partial y^k}\right) = \frac{\partial^2 F^2}{\partial x^l \partial y^k}y^l + \frac{\partial^2 F^2}{\partial y^l \partial y^k}\frac{dy^l}{ds}$$

$$= \frac{\partial}{\partial x^l}(g_{ij\cdot k}y^i y^j + 2g_{kj}y^j)y^l + \frac{\partial}{\partial y^l}(g_{ij\cdot k}y^i y^j + 2g_{kj}y^j)\frac{dy^l}{ds}$$

$$= g_{ij\cdot k,l}y^i y^j y^l + 2g_{kj,l}y^j y^l + (g_{ij\cdot k\cdot l}y^i y^j + 2g_{jl\cdot k}y^j + 2g_{kj\cdot l}y^j)\frac{dy^l}{ds}.$$

By grouping the terms and denoting

$$\Gamma_{klj} = \frac{1}{2}(g_{kj,l} + g_{kl,j} - g_{lj,k}),$$

we have

$$(g_{ij\cdot k\cdot l}y^iy^j + 2g_{jl\cdot k}y^j + 2g_{kj\cdot l}y^j + 2g_{kl})\frac{dy^l}{ds} + (2\Gamma_{klj} + g_{hj\cdot k,l}y^h)y^ly^j = 0.$$

Notice, that Γ_{klj} depends also on y. Now introduce $2g^*_{kl} \equiv g_{ij\cdot k\cdot l}y^iy^j + 2g_{jl\cdot k}y^j + 2g_{kj\cdot l}y^j + 2g_{kl}$, where the above defined g^*_{kl} is a (0,2)-type symmetric tensor. Thus, the geodesic equations are

$$g^*_{kl}\frac{dy^l}{ds} + (2\Gamma_{klj} + g_{hj\cdot k,l}y^h)y^ly^j = 0,$$

and if g^* is invertible and $(g^{*ik}) = (g^*_{ik})^{-1}$, we get

$$\frac{dy^i}{ds} + g^{*ik}(2\Gamma_{klj} + g_{hj\cdot k,l}y^h)y^ly^j = 0.$$

Taking into account the linear approximation, we get the expression for the generalized geodesics in the anisotropic space

$$\frac{dy^i}{ds} + (\Gamma^i_{jk} + \frac{1}{2}\eta^{ih}\frac{\partial\varepsilon_{jk}}{\partial x^l\partial y^h}y^l)y^jy^k = 0 \qquad (2.16)$$

Contrary to the usual geodesics equation, $\frac{dy^i}{ds} + \Gamma^i_{jk}y^jy^k = 0$, valid in the Riemannian space, eq. (3.95) shows that in an anisotropic space the use of coordinate transformation such that all Γ^i_{jk} become equal to zero does not make the frame locally inertial. This feature represents the situation when the observation of the motion of bodies belonging to a system is performed from one of such bodies.

Locality in the usual sense would here mean that an observer is able to have a look at this system from aside and register that all the bodies move inertially, for example, planets are orbiting around a star (a point static source of the gravitational potential) along their geodesics. Global character in the usual sense would mean that there is a global reference frame whose state of motion is somehow known and accounted for during the observations, for example, the center of the rotating frame is at the star and its axis points at a planet whose center oscillates near the mean position suffering the action of gravitational and inertial forces. Now both these locality and global character are broken. We see what we see. Neither additional invisible motion (epicycles), nor additional invisible entity (dark matter) should be invented as long as we stay with the equivalence principle and acknowledge the indistinguishability of geometry (observations) and gravitation ('force at a distance' as Newton unwillingly put it). But it doesn't seem that remaining consistent, we have a choice. In terms of formal logics, this is called a tautology.

2.3.3 *Anisotropic potential*

Let us transform the obtained expression for the geodesics eq. (3.95) and use two standard simplifying assumptions formulated already by Einstein in [Einstein (1916a)], and one additional assumption characteristic to the space we regard now:

(1) The velocities are considered much less than the fundamental velocity, c. This means that the components $y^1 = \frac{dx^1}{ds}, y^2 = \frac{dx^2}{ds}$ and $y^3 = \frac{dx^3}{ds}$ can be neglected in comparison with $y^0 = \frac{dx^0}{ds}$ which is equal to unity within the accuracy of the second order;

(2) Since the velocities are small, the time derivative of metric $\frac{\partial \varepsilon_{jk}}{\partial x^0}$ can be neglected in comparison with the space derivatives $\frac{\partial \varepsilon_{jk}}{\partial x^\alpha}; \alpha = 1, 2, 3;$

(3) The same is taken true for the corresponding accelerations, i.e. the y^0-derivative can be neglected in comparison to the y^1-, y^2-, and y^3-derivatives. It is obviously justified, when we pass to $x^0 = ct$, and $y^0 = c = const$ which is not a variable parameter.

With these assumptions, the only terms that remain in eq. (3.95) in linear approximation are those with $j = k = 0$, that is the only ε_{jk} which is present there is ε_{00}, while the corresponding $y^j = y^k = 1$. Let us denote

$$\frac{1}{2} \frac{\partial \varepsilon_{00}}{\partial y^i} \equiv w_i \qquad (2.17)$$

If the coordinate transformations are limited only to those that preserve $\varepsilon_{00} = \tilde{\varepsilon}_{00}$, (for example, only spatial rotations and translations are regarded), eq. (2.17) defines a 1-form (covector) with the components w_i which can be called the 'proper 4-potential of the anisotropic space'. With this notation the linear approximation for the generalized geodesics given by eq. (3.95) yields

$$\frac{dy^i}{ds} + \Gamma^i_{00} + \eta^{ik} \frac{\partial w_k}{\partial x^j} y^j = 0 \qquad (2.18)$$

The third term in eq. (2.18) does not vanish, since, though $\frac{\partial w^i}{\partial x^0} << \frac{\partial w^i}{\partial x^\alpha}$, $y^0 >> y^\alpha$, $\alpha = 1, 2, 3$. Let us now add and subtract the same value $\eta^{ik} \frac{\partial w_j}{\partial x^k} y^j$ and obtain

$$\frac{dy^i}{ds} + \Gamma^i_{00} + \eta^{ik} [\left(\frac{\partial w_k}{\partial x^j} - \frac{\partial w_j}{\partial x^k} \right) + \frac{\partial w_j}{\partial x^k}] y^j = 0 \qquad (2.19)$$

where the expression

$$\left(\frac{\partial w_k}{\partial x^j} - \frac{\partial w_j}{\partial x^k}\right) \equiv D_{jk} \tag{2.20}$$

presents the components of a 'proper antisymmetric tensor of the anisotropic space'. The expressions $\left(\frac{\partial w_\alpha}{\partial x^\beta} - \frac{\partial w_\beta}{\partial x^\alpha}\right)$, $\alpha, \beta = 1, 2, 3$ are the components of the $curl \mathbf{w}$.

The goal of these manipulations is to construct a geometrical object related to the anisotropic metric that suffices Maxwell identity eq. (1.39). This will eventually lead to the mass current description and to the corresponding forces in a geometrical way which is consistent with the spirit of relativity and not in an artificial way used in GEM theory.

In isotropic spaces there can be terms that involve the curl-like factors such as $\partial_j h_{0k}(x) - \partial_k h_{0j}(x)$, where h corresponds to the small isotropic perturbation of the Minkowski metric, $g_{ij}(x) = \eta_{ij} + h_{ij}(x)$. They describe the vorticity of fields (in this case the field that would have vorticity is $h_{\alpha\beta}$) and correspond to the choice of the reference frame. In isotropic spaces such terms differ from zero only if the frame is not a pure free-fall one, for example, it could be a rotating frame. Now in the anisotropic space the $curl$-term is provided by the $\varepsilon_{00}(x, y)$ part of the metric, while the frame even being a free-fall one cannot escape containing fields with vorticity.

Substituting the proper tensor into eq. (2.19), we get

$$\frac{dy^i}{ds} = -\Gamma^i_{00} + \eta^{ik} D_{kj} y^j - \eta^{ik} \frac{\partial w_j}{\partial x^k} y^j. \tag{2.21}$$

The third term on the right-hand side of eq. (2.21) can be transformed with regard to the second assumption

$$-\eta^{\alpha k} \frac{\partial w_j}{\partial x^k} y^j = -(-\delta^{\alpha k})[\frac{\partial}{\partial x^k}(w_j y^j) - w_j \frac{\partial y^j}{\partial x^k}] = \left(\nabla(w_j y^j)\right)^\alpha \tag{2.22}$$

where $w_j \frac{\partial y^j}{\partial x^k}$ vanishes because x and y are independent variables. Recollecting that initially the right-hand side of eq. (2.21) was multiplied by $y^0 y^0$, and $y^0 = 1$ *unit of length*, (according to the chosen system of units), we introduce all the dimensional factors into eq. (2.21) explicitly, consider $y^\alpha = \frac{1}{H} v^\alpha$ for the 3D-space and $y^0 = c/H$, and get the equation of motion in the 3D-space in the form

$$\frac{d\mathbf{v}}{dt} = \frac{c^2}{2}\left\{-\nabla\varepsilon_{00} + 2[\mathbf{v}, curl\mathbf{w}] + 2\nabla(\mathbf{v}, \mathbf{w})\right\} = -\nabla\Phi_{(a)}, \qquad (2.23)$$

where

$$\Phi_{(a)} = \frac{c^2}{2}\left\{\varepsilon_{00} - 2(\mathbf{v}, \mathbf{w}) - 2\varphi_{(a)}\right\} \qquad (2.24)$$

$$\nabla\varphi_{(a)} = [\mathbf{v}, curl\mathbf{w}] \qquad (2.25)$$

is the anisotropic potential. If the metric is y-independent, the equation of motion coincides with the known one [Einstein (1916a)]. The gravitation force acting on a particle with mass m is, therefore,

$$\mathbf{F}^{(g)} = -m\nabla\Phi_{(a)} \qquad (2.26)$$

and now it also depends on the particle velocity, $\mathbf{v}$, and on the velocities of the sources of the gravitation field.

2.3.4 *Field equations*

It is essential to understand what kind of modification we are trying to perform here. The main feature of it is the aspiration not to change but to preserve the mathematical structure that appeared to have such a deep physical meaning, and this is the Einstein field equations. We would like once again to point at the parallelism of the approach we use here and the approach used by Einstein when he tried to preserve Maxwell field equations as the mathematical structure with the deepest physical meaning at that time. What he did then did not disturb their form. Is it possible to see that the use of the anisotropic space or of the phase space-time as a model of the physical reality not only preserves the form of gravitational field equations but provides some new perspectives?

The gravitation field equations in the isotropic space have the form

$$G_{ik} = 8\pi\frac{G}{c^4}T_{ik}, \qquad (2.27)$$

where $G_{ik} = \frac{\partial\Gamma_{ik}^j}{\partial x^j} + \Gamma_{im}^j\Gamma_{kj}^m$ is Einstein tensor, T_{ik} is the energy-momentum tensor and G is the gravitation constant. The field equations can be obtained by the variation of the expression for the corresponding Hilbert-Einstein action. Because of the anisotropy, the connection coefficients, the simplest scalar (Ricci one), and the variation *procedure* itself in general case would demand modification. But as it was mentioned in the beginning of this section, for the linearly approximable case the geometrical problems simplify a lot. In [Voicu (2010)] it is shown that for the weak (linearly approximable) anisotropic perturbation of the metric, the appearing additional terms are negligibly small, and the field equations formally remain the same (no new derivatives are present), though the terms they include are now y-dependent.

Sticking to this approximation, we have only $i = k = 0$ as before on the left-hand side of eq. (2.27), and the x^0-derivatives can be neglected because of the linear character of the approximation. Then the left-hand side becomes

$$\frac{\partial \Gamma_{ik}^j}{\partial x^j} + \Gamma_{im}^j \Gamma_{kj}^m = -\frac{1}{2}\Delta \varepsilon_{00}, \qquad (2.28)$$

where Δ is the Laplace operator.

On the right-hand side of eq. (2.27), the energy-momentum tensor now cannot be transformed into the diagonal form, because the space is anisotropic and there is no suitable reference frame - one and the same for all parts of the system. But in view of our assumptions, we should again regard only the T_{00} component of the energy-momentum tensor. Notice, that now we also have the specific geometrical structures originated from the anisotropy, i.e. the proper 4-potential, w_i, and the proper tensor, $D_{ik} = \frac{\partial w_k}{\partial x^i} - \frac{\partial w_i}{\partial x^k}$. They may be seen as related to an additional 'background' field produced by the motion of the distribution of the moving sources (in terms of classical GRT it corresponds to the additional mass of the system originating from its internal or interaction energy related to the defect of mass). It is this background field that makes the straight uniform motion impossible as it was mentioned in the end of the previous section. Let us obtain the characteristics of this background field directly in the following way.

Let us use [Einstein (1916a); Siparov (2008a)] the geometrical identity

$$\frac{\partial D_{ik}}{\partial x^l} + \frac{\partial D_{kl}}{\partial x^i} + \frac{\partial D_{li}}{\partial x^k} = 0 \qquad (2.29)$$

which is true for any antisymmetric tensor of the form $D_{ik} = \frac{\partial w_k}{\partial x^i} - \frac{\partial w_i}{\partial x^k}$, to describe the sources of curvature and anisotropy of the anisotropic space. Introduce new designations for the components of the proper tensor

$$D_{ik} = \begin{pmatrix} 0 & E_{(a)x} & E_{(a)y} & E_{(a)z} \\ -E_{(a)x} & 0 & -B_{(a)z} & B_{(a)y} \\ -E_{(a)y} & B_{(a)z} & 0 & -B_{(a)x} \\ -E_{(a)z} & -B_{(a)y} & B_{(a)x} & 0 \end{pmatrix} ; D^{ik} = \begin{pmatrix} 0 & -E_{(a)x} & -E_{(a)y} & -E_{(a)z} \\ E_{(a)x} & 0 & -B_{(a)z} & B_{(a)y} \\ E_{(a)y} & B_{(a)z} & 0 & -B_{(a)x} \\ E_{(a)z} & -B_{(a)y} & B_{(a)x} & 0 \end{pmatrix}$$

$$(2.30)$$

Then eqs. (2.29-2.30) immediately lead to the first pair of "Maxwell equations":

$$\frac{\partial \mathbf{B}_{(a)}}{\partial t} + rot\mathbf{E}_{(a)} = 0 \qquad (2.31)$$

$$div\mathbf{B}_{(a)} = 0$$

If we introduce

$$J^i \equiv \frac{\partial D^{ik}}{\partial x^k}; D^{ik} = g^{il}g^{km}D_{lm} \qquad (2.32)$$

then the second pair of "Maxwell equations" appears

$$curl\mathbf{B}_{(a)} - \frac{\partial \mathbf{E}_{(a)}}{\partial t} = \mathbf{j}_{(a)} \qquad (2.33)$$

$$div\mathbf{E}_{(a)} = \rho_{(a)}$$

where the new designations

$$J^0 \equiv \rho_{(a)}; J^1 \equiv j_{(a)x}; J^2 \equiv j_{(a)y}; J^3 \equiv j_{(a)z}; \mathbf{j}_{(a)} = \rho_{(a)}\mathbf{u} \qquad (2.34)$$

were used with $\mathbf{u}$ being the effective velocity of the mass density current. Thus, the curvature and the anisotropy described by the metric $g_{ik}(x, y)$ originate from the background mass density distribution, $\rho_{(a)}$, and from the background mass current density, $\mathbf{j}_{(a)}$, that characterize the moving distribution of masses present somewhere in this space.

Therefore, the T_{00} component of the energy-momentum tensor could contain the density, $\rho_{(a)} = div\mathbf{E}_{(a)}$ and the energy density, $W_{(a)} = \frac{E_{(a)}^2 + B_{(a)}^2}{8\pi}$ of the proper potential of the anisotropic space where

$$\mathbf{E}_{(a)} = -\nabla \int \frac{\rho_{(a)}(r)}{|r - r_0|} dV; \mathbf{B}_{(a)} = curl \int \frac{\mathbf{j}_{(a)}(r)}{|r - r_0|} dV. \qquad (2.35)$$

But since the field is considered weak (linearly approximable), the terms $\rho_{(a)}$ and $W_{(a)}$ should be omitted in T_{00}, because as it can be seen from their definitions they have the next asymptotic order.

Thus, in case of the linearly approximable anisotropic space there are no new terms that could be present in the energy-momentum tensor, and the T_{00} component of it still contains only the terms related to the matter *imbedded* into the anisotropic space. This means that the relativistic increase of mass that is due to the internal motion in the system does not enter the stress-energy tensor in the linear approximation for the gravitation potential though it can affect the observable motion.

In our approximation, the field equation (2.27) gives the Poisson equation with regard to x, as usual. But the y-dependence of metric imparts some specific features to its solution:

- the integration constant may now be a function of velocity and contain a certain $f(v/c)$. Therefore, if the perturbation of the flat metric is considered to be localized and to be fading to zero with distance from the place of its localization, the regions of localization for x and y may not coincide and may not be equal in size. Thus, the fading could be not monotonous;
- the existence of the intrinsic characteristic length, $l = c/H$, in the anisotropic space (compare with the existence of the intrinsic characteristic speed, c, in Minkowski and Riemann spaces) suggests that alongside with the exterior problem (with the solution of Poisson equation proportional to $\frac{1}{r}$), the inner problem (with the solution of Poisson equation proportional to $\frac{r^2}{l^2}$) should be regarded. Both solutions coincide at $r = l$, thus, giving the corresponding boundary condition for both problems in accord with the modified variation principle. Hence, $f(v/c)$ function must be the same for both of them.

As it was underlined by Fock [Fock (1939)] and Infeld [Einstein *et al.* (1938)] (without mentioning the variation principle problem), the solutions of the geodesics equation (2.23) and of the field equation (2.27) are not independent. For the above said, the immediate consequence of this is that the solution of the Poisson equation in both cases includes such function $f(v/c)$ that provides the correct form of eq. (2.23). But the meaning of this remark is deeper and we will come back to it in the next subsection.

The motion of a probe particle in the anisotropic space in the non-relativistic limit is now described by the following Lagrangian and action

$$L = -mc^2 + \frac{mv^2}{2} - m\Phi_{(a)} \qquad (2.36)$$

$$S = \int L dt = -mc \int (c - \frac{v^2}{2c} + \frac{\Phi_{(a)}}{c}) dt = -mc \int ds$$

The line invariant or the interval takes the form

$$ds^2 = (c^2 + 2\Phi_{(a)}) dt^2 - dr^2 - \frac{1}{H^2} dv^2 \qquad (2.37)$$

or

$$ds^2 = (c^2 + 2\Phi_{(a)}) dt^2 - dr^2 - l^2 d\beta^2 \qquad (2.38)$$

Here $\beta = v/c$, v is the velocity of a probe body (e.g., a star) relative to a frame whose origin could be taken on the one of the gravitation sources that form the regarded system of the moving bodies (e.g., on another star of the same galaxy). Therefore, $v \sim V$, where V could be the relative velocity of the sources, so that V is related to a kind of temperature characterizing the system. Then $l = c/H$ corresponding to $v^0 = c$ is the fundamental length or characteristic scale which is a new parameter of the theory. The appearance of this parameter is required by the geometrical considerations, that is by the manner of construction of the space which is supposed to become the model of the physical world. The choice of this parameter could influence the results of the possible calculations during the interpretation on this or that scale. One can also notice that the expression $(dr^2 + \frac{1}{H^2} dv^2)$ represents the sum of the radial and tangential 3D parts of the line invariant.

The problems that motivated this research belong to cosmology. Therefore, the choice of Hubble constant, $H = 72(\text{km/s})/\text{Mpc} \sim 10^{-18}\text{s}^{-1}$ as a value of the new geometric parameter seems reasonable. On the other hand, we saw recently that for the vast and rarefied distribution of matter - like that of the Universe - the corresponding Schwarzschild radius $l = r_S = \frac{2MG}{c^2}$ could also be acceptable, especially, with regard to the fact that it includes such important characteristics as the total mass of the system and the gravitational constant introduced earlier. The use of $l = r_S$ transforms the theory into the mass dependent one which seems attractive. But then we see that this step is not consistent, because, instead of the new world constant as it should be for a new model of space-time, we come back to an old scheme with the fitting parameter in the same manner as it was done with the introduction of the dark matter. Maybe now this scheme would be better, because there is hope that all the mass present in the theory is supposed to be observable and can only influence the parameter of the model describing physical space-time. On this way, the new approach would be good only at the Universe scale with the total mass of the Universe as the parameter, thus, making the theory rather metaphysical than physical. But the problems already began on the galaxy scale. In MOND there was introduced a new world constant which was related to acceleration and chosen to fit the observations, that is it was an empirical constant sufficing the problem under discussion. But the accompanying expression was of little sense, while the covariant MOND contained an unknown field. Now the situation is different, and the need for the constant comes from the basics, like it was with the speed parameter when Minkowski introduced the space-time. The choice of it must account for observations but not stem from them.

Since β in these problems remains small, one can find the distance at which the last term in eq. (2.38) starts to play. It also seems clear how to account for the scale and velocity separately and track the influence of the order of the limit transitions $c \to \infty$ and $l \to \infty$. The simultaneous growth of c and l does not tell us anything about the relation between cdt and dr. For example, in an abstract empty space, if we have $\beta cdt/dr << 1$ but $\beta dr/cdt \sim 1$, that is if $\beta << dr/cdt$ but $cdt/dr \sim \beta << 1$, we get into the case which is called Carrol kinematics where the roles of time and space are in a sense interchanged. It corresponds to the situation when space intervals are much larger than time intervals and, thus, no true motion and no causal relations are possible, all the events are independent and isolated. Abstract as it is, this scheme seems to be in accord with what we observe in astronomy. The presence of the last term in eq. (2.38) widens the number of such possibilities.

For a planetary system scale the third term is so large in comparison with the first two that the only correction that must be accounted for is the one in the expression for the anisotropic potential, $\Phi_{(a)}$, and it would be expectedly small (though measurable) and negligible for most applications. The g_{00} component of the metric tensor in the limit case has the form

$$g_{00} = 1 + \frac{2}{c^2}\Phi_{(a)} \qquad (2.39)$$

which is different from the usual Newtonian one: it contains the velocity of the probe particle and accounts for the motion of the sources in accord with eq. (2.24). When there is only one source which is static relative to an observer and is spherically symmetric, we come back to the Newton gravitation law, and Schwarzschild exterior solution is valid.

For a galaxy scale, the third term remains too large, but there are no reasons to neglect the velocity dependence in the anisotropic potential, $\Phi_{(a)}$, that now defines the geometry suitable for modeling the observable world in a more pronounced way. Notice the difference between this formalism and the so-called post-Newtonian approximation in the isotropic space when the velocity dependence of the metric has the next asymptotic order and is very small - this is definitely not the case of the flat rotation curves.

Finally, for the Universe scale the 8D domain of the phase space-time separates into two: the first resembles the usual 4D space-time with the only account for anisotropy caused by the motion of sources - now they present the clusters and super-clusters of galaxies; the second 4D region is characterized by a constant value of one of the coordinates, and this refers to DeSitter and Lobachevsky models that contain the corresponding constant values. Interestingly, we can also distribute the eight coordinates into the following two sets: (x_1, x_2, x_3) and $(x_0, y_0, y_1, y_2, y_3)$ or (x, y, z) and $(ct, l, l\beta_x, l\beta_y, l\beta_z)$ that is Newtonian absolute 3D space and Lobachevsky "space" and time in the velocities domain, the last is a familiar place for the relativistic particles theory. These combinations provide the model of physical space with the experimental arena, - that is with the observations of "motions" of gravitating bodies in astronomy.

But this could not be the end. The model we discuss does not seem foreign to the micro world. If we go for this, the choice for l could broaden and start to include such alternatives as the radius of atom, the radius of atomic nucleus, the radius of a proton, the Planck's length, $L_P = \sqrt{\frac{G\hbar}{c^3}} \sim 10^{-35}$m. Among the time meaning parameters like H, there could appear Planck's time, $T_P = \frac{L_P}{c} = \sqrt{\frac{G\hbar}{c^5}} \sim 10^{-43}$s. All these values would be meaningful and able to provide not only qualitative but quantitative results too.

2.3.5   *Back to Einstein method*

Let us underline the relation of the approach used here to what was done before. The remark made by Fock hints that the way to obtain the exact solution briefly discussed in section 1-6 is not good, because the equation (1.92) did not take part in the solution of the field equation (1.93) supplemented by eq. (1.94) and by the set of 4 conditions. It was only after the usual "boundary problem" for the field equation was solved, that the solution was used in eq. (1.92) to obtain a trajectory of motion. That is, a body was prescribed how to move by the space-time curvature, but it was deprived of the possibility to tell the space-time how to curve. And it was deprived of it not because the body was small and probe one, but because the method of solution contradicted the spirit of relativity theory established by Einstein already in the SRT: it is "what we see" that tells us which model of the reality to choose, and not the chosen model that tells us what we should see and have. The last opportunity could be and sometimes is a successful guiding method - actually, it is the main method of theoretical physics - but none of the models can be the everlasting dogma. One can see what can happen otherwise, if one recollects the "paradoxical" flat rotation curves of spiral galaxies that resulted later from the beauty of the exact solution of a mathematical problem. Similarly, De Sitter and Friedmann did not bother about bodies curving the space, but regarded the homogeneous (mass free) gravitation field equations and expanding space-time - the *correct mathematical problems* that became a kind of mines under the building of physics waiting their time to explode.

Comparing the AGD approach to those discussed before, we can say that the situation similar to the SRT appearance repeated. It is important to realize that from the moment the idea of covariance of physical laws has established, it is not the physical law which has already proved its fundamental character and applicability that must be changed when the disagreement with the observations appears. It is the model which is unavoidably present in any theory and presents the object liable to changes caused by this disagreement. And in case when the physical law expressed by the gravitation equations proved its consistency and applicability, the model we speak about, i.e. unavoidably present in the theory, is the model of real space and its geometry.

Since the flat rotation curves is "what we see", the AGD suggests to modify the model of real space and time. As a result, the problem in the linear approximation is now described by the system

$$\frac{d\mathbf{v}}{dt} = \frac{c^2}{2}\left\{-\nabla\varepsilon_{00} + [\mathbf{v}, curl\frac{\partial\varepsilon_{00}}{\partial\mathbf{v}}] + \nabla(\mathbf{v}, \frac{\partial\varepsilon_{00}}{\partial\mathbf{v}})\right\} \tag{2.40}$$

$$\Delta\varepsilon_{00} = \kappa\rho$$

Lacing condition for the inner and exterior problems

What is going to be done now following the receipt of Einstein's paper

[Einstein (1916a)], is the solution of the equations in the system eq. (2.40) with regard to "what we see". Only then we should turn to the boundary condition. Thus, the style of the AGD approach is the unification of the "SRT style" (model correction) and of the "GRT style" (the solution of the field equation is first compared with the what-we-see geodesics) used by Einstein, and not the style of the solution of the traditional boundary problem which is probably more appealing for mathematicians but is not always suitable for physics.

2.4   Gravitation force in anisotropic geometrodynamics

Let us discuss the physical meaning of the components of the gravitation force present in eq. (2.26). Rewriting it in more detail, we obtain

$$\mathbf{F}^{(g)} = \frac{mc^2}{2}\left\{-\nabla\varepsilon_{00} + [\mathbf{v}, curl\frac{\partial\varepsilon_{00}}{\partial\mathbf{v}}] + \nabla(\frac{\partial\varepsilon_{00}}{\partial\mathbf{v}}, \mathbf{v})\right\} \qquad (2.41)$$

The first term can be naturally related to the usual expression of the Newtonian force, $\mathbf{F}_N^{(g)}$ acting on a particle with mass m. For a single stationary spherically symmetric gravitation source with mass M, the solution of Poisson equation will lead to the familiar expressions, and we can introduce

$$\varepsilon_{00}(x) = \frac{r_S}{r}$$

where

$$r_S = \frac{2GM}{c^2}$$

is the Schwarzschild radius. Then the result will be the usual Newton gravitation law

$$\mathbf{F}^{(g)} = G\frac{Mm}{r^2}\frac{\mathbf{r}}{r} \qquad (2.42)$$

If we introduce

$$\mathbf{\Omega}(x) = \frac{c^2}{4}curl\frac{\partial\varepsilon_{00}}{\partial\mathbf{v}} \qquad (2.43)$$

then the second term in eq. (2.41) can be recognized as an analogue of "Coriolis force"

$$\mathbf{F}_C^{(g)} = 2m[\mathbf{v}, \mathbf{\Omega}(x)] \qquad (2.44)$$

which in the usual case appears in the non-inertial (rotating) reference frame. It is proportional to the velocity, $\mathbf{v}$, of the probe particle and also depends on the frame motion. Now when we observe the "motion of atoms sitting on one of them", we can never be sure of the character of motion of any observed object including our own. Therefore, the clear distinction between non-inertial and inertial frames cannot be used, if we intend to test the theory by observations.

On the other hand, we can introduce

$$\mathbf{B}^{(g)}(x) = \frac{c^2}{2} curl \frac{\partial \varepsilon_{00}}{\partial \mathbf{v}} = 2 curl \, \mathbf{u} \qquad (2.45)$$

$$u^\alpha \equiv \delta^{\alpha\beta} \frac{c^2}{4} \frac{\partial \varepsilon_{00}}{\partial v^\beta}$$

where $\mathbf{u}$ is the effective velocity of the mass current in the given point. Then we can speak of the gravitational analogue of Lorentz force,

$$\mathbf{F}_L^{(g)} = m[\mathbf{v}, \mathbf{B}^{(g)}(x)] \qquad (2.46)$$

This could be convenient from the point of view of calculations and will be used later.

Notice that since the action produced by the gravitation force component $\mathbf{F}_C^{(g)}$ (or $\mathbf{F}_L^{(g)}$) is related to the motion of the body and to the motion of the gravitation sources, it could be attraction or repulsion or tangent action depending on the angle between $\mathbf{v}$ and $\mathbf{\Omega}(x)$. If in a certain region vector $\mathbf{\Omega}(x)$ preserves constant direction, then the component of the body's velocity, $\mathbf{v}$, which is parallel to $\mathbf{\Omega}$ doesn't suffer the action of the second term in eq. (2.41).

This explains the observational paradox known alongside with the flat rotation curves [Aguirre *et al.* (2001)]. Contrary to the orbital motion of stars and gas on the periphery of a spiral galaxy, the motion of the globular star clusters in the plane orthogonal to the galaxy plane is in better agreement with the calculations based on Kepler law [Dauphole and Colin (1995); Bridges (1997)]. The idea of the additional component of the gravitational force $\mathbf{F}_C^{(g)}$ acting on the stars moving in the galaxy plane but not acting on the clusters moving orthogonally to it explains this paradox.

The third term in eq. (2.41) will present an additional force of attraction or repulsion acting on a moving particle when the system of gravitation sources extends radially (explosion) or contracts radially (collapse). This could be accounted for in the description of the following astrophysical situations.

1.   The velocity of the mass current, $\mathbf{u}$, has no radial component. This means that we deal with the stationary spiral galaxy. (The corresponding applications of the AGD theory are given in the next section.)

a)   The radial component of the probe velocity, $\mathbf{v}$, does not affect the situation.

b)   If the tangent component of the probe velocity, $\mathbf{v}$, gives positive scalar product, an additional repulsive force appears and pushes the probe outwards.

c)   If the tangent component of the probe velocity, $\mathbf{v}$, gives negative scalar product, an additional attraction force appears and pulls the probe inwards.

2.   The velocity, $\mathbf{u}$, of the mass current has radial component directed outwards with regard to the distribution center (expansion, explosion). This means, that together with the consequences b) and c) mentioned in item 1, the following will also take place:

•   If the radial component of the probe velocity, $\mathbf{v}$, is directed outwards, the additional repulsive force appears.

•   If the radial component of the probe velocity, $\mathbf{v}$, is directed inwards, the additional attraction force appears.

3.   The velocity, $\mathbf{u}$, of the mass current has radial component directed inwards with regard to the distribution center (contraction, collapse). This means that besides the consequences b) and c) mentioned in item 1, the following will also take place:

•   If the radial component of the probe velocity, $\mathbf{v}$, is directed outwards, the additional attraction force appears.

•   If the radial component of the probe velocity, $\mathbf{v}$, is directed inwards, the additional repulsive force appears. Both periods correspond to the point No. 120 of the total number of points for which the calculation of the Fourier coefficient was performed. This means that the period is 83 minutes. Table 3.2 contains the characteristics of the spectra details of the radio sources The corresponding calculations can be performed in a straightforward way for any concrete case.

Let us transform the expression for the gravitation force eq. (2.41) obtained for the space with anisotropic metric depending on the velocities of the distributed sources and, first, write it in the following form

$$\mathbf{F}^{(g)} = \frac{mc^2}{2}\left\{-\nabla\varepsilon_{00} + \frac{2}{c^2}2[\mathbf{v}, curl\mathbf{u}] + \frac{2}{c^2}2\nabla(\mathbf{u}, \mathbf{v})\right\} \qquad (2.47)$$

Notice that despite the $1/c^2$-factor in the second and third terms, they have the same order as the first one, because the definition of $\mathbf{u}$, eq. (2.45), includes the c^2-factor. Now unite the first and the third components in eq. (2.47) under the common gradient symbol

$$\mathbf{F}^{(g)} = \frac{mc^2}{2}\left\{\nabla\left(-\varepsilon_{00} + \frac{4}{c^2}(\mathbf{u}, \mathbf{v})\right) + \frac{4}{c^2}[\mathbf{v}, curl\mathbf{u}]\right\} \qquad (2.48)$$

In order to get the expression for the force in the form more convenient for the experimental testing in the spiral galaxy case, let us regard a system of moving sources which has a symmetry of a disk, take a reference body in the center and calculate the needed parameters with account to the rotation of all the other bodies moving around the central one. As it was underlined above and in [Siparov (2008a)], both field theories (electromagnetism and gravitation) have the same geometric origin which means that the calculation of $\mathbf{u}(x)$ and other notions of anisotropic geometrodynamics can be performed similarly to electrodynamics by switching the notation, it seems clear that $curl\mathbf{u} \sim \mathbf{B} \sim \boldsymbol{\Omega}$ mentioned above have the same meaning. The corresponding mathematical formulas can be used for both cases. For example, the magnetic induction formula for the gravitation case gives

$$\boldsymbol{\Omega} = curl \int \frac{\mathbf{j}(\mathbf{r})}{|\mathbf{r} - \mathbf{r}_0|}dV \qquad (2.49)$$

where $\mathbf{j}(\mathbf{r})$ is the mass current density, and $\mathbf{r}_0$ is the observation point. It is essential to notice that $\mathbf{j}(\mathbf{r})$ characterizes the given distribution of mass currents with reference to the origin (observer) and has no properties related to the distance from $\mathbf{r}_0$, i.e. it neither decreases nor increases with distance but is a given function. Therefore, all the dependence of $\boldsymbol{\Omega}$ on the distance is described by the denominator.

In this case $curl\mathbf{u} = const$ and $\mathbf{u} = [\boldsymbol{\Omega}, \mathbf{r}]$, where $\mathbf{r}$ is a radius vector, and the expression eq. (2.48) can be further simplified. Notice two useful relations

$$[\mathbf{a}, curl\mathbf{b}] = \nabla([\mathbf{a}, curl\mathbf{b}], \mathbf{r}) \qquad (2.50)$$

$$(\mathbf{a}, \mathbf{b}) = (\mathbf{a}, [curl\mathbf{b}, \mathbf{r}]) = ([\mathbf{a}, curl\mathbf{b}], \mathbf{r}) \qquad (2.51)$$

and introduce the scalar potential $([\mathbf{v}, curl\,\mathbf{u}], \mathbf{r})$ for which the second term in eq. (2.47) will also be a gradient and finally get the expression

$$\mathbf{F}^{(g)} = -\frac{mc^2}{2} \nabla \left\{ \varepsilon_{00} - \frac{8}{c^2}(\mathbf{u}, \mathbf{v}) \right\} \tag{2.52}$$

The gravitational potential, eq. (2.24), can be written now as

$$\mathbf{\Phi}_{(a)} = \frac{c^2}{2} \left\{ \varepsilon_{00} - 2(\frac{\partial \varepsilon_{00}}{\partial \mathbf{v}}, \mathbf{v}) \right\} \tag{2.53}$$

This is the gravitation law obtained from the dynamics equation (geodesics equation) which is the condition of solvability of the field equations in the general case. When calculating, one should consider all the bodies taking part in motion. Alongside with the usual term related to Newton potential, in the new equation there is a *convection term* corresponding to the anisotropic scalar potential whose role cannot be neglected in certain conditions. Notice, that similarly to [Einstein *et al.* (1938)], no suggestions about the concrete form of the metric were made, and the only assumption is the usual weak field approximation. Notice also that all this could look as if there is a preferred point in space - the chosen reference body - for which the gravitation law eq. (2.52) works. But this is not so. If the frame origin is moved on another reference body, all the motions and accelerations of sources and probe body will be described in a different way, but all the observable features like finite motions or collisions will remain the same. In a sense, this is similar to those seeming paradoxes of the SRT. In a case without any specific symmetry, one just has to use eqs. (2.48, 2.49) in a straightforward way.

If there is only one body producing gravitation, the natural place for the frame origin location is this body, there is no velocity dependence, and the character of the gravitation law must become the Newtonian one. Therefore, this limit case suggests that the expression for the force must become

$$\mathbf{F}^{(g)} = -\frac{mc^2}{2} \nabla \left\{ \sum_n \left(\frac{r_{S,n}}{r_n} - \frac{8}{c^2}(\mathbf{u}_n, \mathbf{v}) \right) \right\} \tag{2.54}$$

Despite the formal similarity, the expression (2.54) can't be regarded as a post-Newtonian approximation for the two obvious reasons. First, when we speak about post-Newtonian approximation, we usually mean a correction to the gravitation around a center, while there could be no central symmetry in the convection problem in question. Second, in a post-Newtonian approximation, the correction to Newton law starts to play when the point approaches the center, while in our case the second term becomes important at a distance specified by the scale of the distribution of matter and could be important when the Newtonian term has already faded away.

The new circumstance is the possibility of the negative correction to Newton potential – up to the predominance of (observable) repulsion over attraction forces. But this is of no surprise, because it is related to the equivalence principle and to the identifying of the gravitational and inertial forces. The latter include, for example, the centrifugal forces. The additional attraction forces (with regard to Newton ones) that are due to the motion can also be found. But this is also in agreement with the GRT ideas that can be seen in eq. (2.4).

The alternating-sign character of the second term in eq. (2.54) means that in case of the approximate spherical symmetry of the distribution of randomly moving sources of gravitation, for example, stars in elliptical galaxies, the mean force would become Newtonian as soon as the scale of averaging permits. But when there is any kind of order in the motion, for example, a finite total angular momentum like that of a spiral galaxy, the terms won't cancel out and the second term in eq. (2.54) cannot be neglected. In some cases, it can obtain the expected form, for example, in spiral galaxies, it will correspond to the logarithmic potential as will be shown in the next section.

Let us compare the obtained expressions for the gravitation force eqs. (2.48, 2.52) with known expressions of the similar character. First of all, there is the expression for the force in the non-inertial system. Taking for simplicity the uniformly rotating frame, we see that the force is equal to

$$\mathbf{F}_{nI} = m\nabla(\Phi + \frac{V^2}{2}) + m[\mathbf{v}, curl\mathbf{V}] \tag{2.55}$$

where Φ is the force potential, $\mathbf{v}$ is the probe velocity, $\mathbf{V} = [\omega, \mathbf{r}]$ is the linear velocity at distance r from the rotation axis measured in the motionless frame. The second term in eq. (2.55) is proportional to Coriolis force; the second term in brackets under gradient is called the centrifugal potential energy. As r approaches infinity, this energy approaches infinity too, this makes one speak of the fictitious character of this energy and of the formal character of this formula use. But this ambiguous attitude is neither neglected when it comes to practical applications, nor it can survive, when we come to the impossibility to tell which frame is the accelerated one. The expression eq. (2.55) looks much alike that of eq. (2.48), but in the last one there could be no talk of infinite values however far the regarded point could be. The new part of the gravitation force caused by motion will decrease and finally vanish with distance because of the integration over angles in eq. (2.49). This makes it different from the centripetal force in the rotating reference frame and makes one think of its relation to the "Lorentz force" in which $\mathbf{B}^{(g)}(x)$ decreases with distance.

The "inertial" convection term may correspond to various observable types of behavior. Thinking of spiral galaxies in these terms, one can see the possible origin of the additional forces. If inertia is really indistinguishable from gravitation with regard to the motion, as we assume, the whole picture starts to simplify.

Second, let us compare the potential in formula (2.54) and Fock's result eq. (2.4). The difference is due to the fact that eq. (2.4) describes an external problem and deals with the potential in the point which is far away from the bound system of bodies. In our case the situation is different. The probe body could be one of the bodies of the bound system, and the potential is defined in the point which the probe occupies moving inside or outside the distribution of moving masses.

Third, we can have a look at the expression obtained for the gravitation force in the isotropic space of the classical GRT [Landau and Lifshitz (1967)] for the probe body moving in constant gravitation field

$$\mathbf{F} = \frac{mc^2}{\sqrt{1 - \frac{v^2}{c^2}}} \left(-\nabla \ln \sqrt{g_{00}} + \sqrt{g_{00}} [\frac{\mathbf{v}}{c}, curl\mathbf{g}] \right) ; g_\alpha = -\frac{g_{0\alpha}}{g_{00}} ; \alpha = 1, 2, 3$$

The second term here is also analogous to the Coriolis force when the velocity of the body is not large. This Coriolis type force is proportional to $mc\sqrt{g_{00}}[\mathbf{v}, curl\mathbf{g}]$, and corresponds to the one that appears in the gravitation free situation in a frame rotating with angular velocity $\Omega^* = \frac{c}{2}\sqrt{g_{00}}curl\mathbf{g}$. Thus, in the isotropic space of the classical GRT, we can speak of the potential only for the motionless body; the *curl*-type term depends on the 0α-components of the metric and can be eliminated by the proper choice of the frame; the angular velocity of the corresponding rotating frame is an order in c less than the term we have in eq. (2.43).

In [Landau and Lifshitz (1967)], there is an essential remark:

"in order to completely describe the distribution and motion of matter in case of the gravitation field, Einstein equations must be supplemented by the equation of state (which doesn't enter the field equations), i.e. by the equation relating pressure and density. This equation must be given alongside with the field equations. [Actually, the equation of state relates not two but three thermodynamical parameters, for example, pressure, density and temperature of the substance. In application to the gravitation theory, this circumstance is usually not essential, because the equations of state used here do not contain temperature (for example, $p = 0$ for rarefied substance, limiting relativistic equation $p = \varepsilon/3$ for strongly compressed substance, etc.)]".

But one can argue, that there exists a scale for which the definition of temperature which is the mean kinetic energy might include the rotation motion of "molecules", i.e. the type of motion that doesn't take part in pressure. And if we regard, for example, spiral galaxies as giant molecules as it was mentioned above, then their rotation might correspond to an additional degree of freedom according to the Boltzmann energy distribution theorem, which might not be negligible. In this case, the equation of state that includes temperature must be accounted for alongside with the gravitation field equations. The use of the anisotropic metric seems to be quite an applicable method to account for that.

Fourth, the most evident analogue which comes to mind when speaking about the motion dependent gravitation is the GEM theory. Here are the corresponding formulas originating from the GEM approach

$$\mathbf{F}^{(G)} = m\mathbf{E}_G + m[\frac{\mathbf{v}}{c}, \mathbf{B}_G]$$

$$\mathbf{E}_G = -\nabla\Phi_N - \frac{1}{2c}\frac{\partial\mathbf{A}}{\partial t}; \mathbf{B}_G = curl\mathbf{A}; \Phi_N = -\frac{GM}{r};$$

$$A^i = \frac{G}{c}\frac{S^n x^k}{r^3}\epsilon^i_{nk}; M = \int \rho d^3 x; S^i = 2\int \epsilon^i_{jk} x'^j \frac{T^{k0}}{c} d^3 x'$$

$$\mathbf{B}_G = -4\frac{G}{c}\frac{3\mathbf{r}(\mathbf{r},\mathbf{S}) - \mathbf{S}r^2}{2r^5}; j^i = \frac{T^{i0}}{c}$$

where ρ is mass density, j^i is mass-current density, S^i is the total angular momentum of the system, and ϵ_{ijk} is the three-dimensional completely antisymmetric Levi-Civita tensor. The metric tensor used in GEM theory can be read from the corresponding space-time invariant [Mashhoon (2001)]

$$ds^2 = (1 + 2\frac{\Phi}{c^2})c^2 dt^2 + 4dt(\mathbf{dr}, \frac{\mathbf{A}}{c}) - (1 - 2\frac{\Phi}{c^2})\delta_{\alpha\beta}dx^\alpha dx^\beta \qquad (2.56)$$

As it was already discussed, the straightforward introduction and usage of the gravitational scalar and vector potentials similar to those in electrodynamics is inconsistent, because the gravitation tensor becomes divided into two parts: one is responsible for the curvature and has the geometrical meaning, the other is responsible for the Lorentz type force related to the velocities as in eq. (2.10) which is the most appealing gravitational analogue.

Finally, let us compare the expressions for the gravitation force presented by eq. (2.48) and eq. (2.44) and the phenomenological additional term that was introduced in MOND in order to satisfy the observations of the flat rotation curves in spiral galaxies. This term contained an empirical constant measured in the acceleration units whose value was chosen to fit the observations and appeared to be close to the value of cH where c is the speed of light and H is the Hubble parameter. Let us introduce dimensionless $\beta = \mathbf{v}/c$ and $\mathbf{\Theta} = \mathbf{\Omega}(x)/H$, where c and H are the fundamental constants of the geometrical theory (AGD) based on the notion of phase space-time. Then the acceleration which is present in eq. (2.44) is given by

$$\mathbf{a}_C = 2cH[\beta, \mathbf{\Theta}] \qquad (2.57)$$

The corresponding *a priori* choice of the fundamental constants and the corresponding physical conditions described by the vector product $[\beta, \mathbf{\Theta}]$ points at the obvious relation of this general expression to the empirical MOND value. For example [Siparov (2010a)], the period of galactic rotation is about 10^{15}s, the orbital velocity of stars is about 10^5m/s, thus, for $H \sim 10^{-18}$s^{-1} which corresponds to Hubble parameter, $|[\beta, \mathbf{\Theta}]| \sim 10^0$, and $|\mathbf{a}_C| \sim 10^{-10}$m/s^2 which has the same order as in MOND but has no fitting character. Notice that when a probe body moves in the system of gravitational sources taking part in rotation around the common center, the additional radial acceleration - attraction to the center or repulsion from it - depends on the tangential velocity of the probe body.

2.5   Model of the gravitation source and its applications

2.5.1   *Center plus current model*

As it was already mentioned, the point source of gravitation which in a sense was the basic model of the gravitation source in the classical GRT, now is not enough. Generally, this is due to the lack of the usual notions of locality and globality in the anisotropic space. From the formal point of view, this is due to the fact that the gravitational potential now depends not only on coordinates but also on the vector field that could have a solenoidal part. This means that in general case, the singularities of the solutions may possess more complicated properties than points. Therefore, if such singularities are interpreted as the sources of the gravitation field, the simplest basic model of the source will not be the point with its spherical symmetry as in the GRT, but a system like a "center plus current" (CPC), i.e. the central body surrounded by the effective mass current $J_{(m)}$. The spiral galaxies are obviously the objects suitable to be described by the CPC model. In order to describe more complicated systems of moving masses, the model has to be more complicated too.

(a)   (b)

Fig. 2.2 Spiral galaxies: (a) M-104. (b) NGC-7742. (Figures made after the images obtained by Hubble telescope, NASA/ESA.)

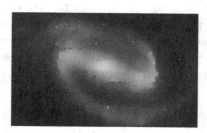

Fig. 2.3 NGC-1300. Figure made after the images obtained by Hubble telescope, NASA/ESA.

For such galaxies as M-104 (Sombrero) or NGC-7742 Fig. 2.2(a),(b) that have rings, the CPC model can be applied directly. For other spiral galaxies – with more pronounced arm structure – the effective values can be introduced. They are the contour radius, R_{eff}, constant angular velocity, Ω_{eff}, and linear velocity of the mass current density along the contour, $V_{eff} = \Omega_{eff} R_{eff}$. This can be done, for example, in the following way

$$I_{eff} = \sum I_n \equiv M R_{eff}^2 \Rightarrow R_{eff} = \sqrt{\frac{I_{eff}}{M}} \qquad (2.58)$$

where I_{eff} is the moment of inertia of the galaxy with the total mass M, and I_n is an estimated moment of inertia for a selected part of it (star or a group of stars) with regard to the galaxy center. The effective angular velocity, Ω_{eff}, can be obtained from the condition

$$I_{eff} \Omega_{eff} \equiv L_{eff} = \sum L_n \qquad (2.59)$$

where L_n is the angular momentum of the component of the system. Thus, we get

$$\Omega_{eff} = \frac{L_{eff}}{I_{eff}} \qquad (2.60)$$

These parameters can be obtained for a concrete galaxy from the astronomical observations. Notice the relation between the phenomenological value Ω_{eff} and the general expression $\Omega(x)$ introduced earlier.

Let us find the conditions for which the regular GRT model of the gravitation source (point) is not enough to describe the physical system, and one has to use the CPC model of the AGD. These conditions take place when both terms in eq. (2.54) start to play comparable roles. Let us suppose that the probe body, i.e. one of the stars on the periphery of the spiral galaxy, moves in the same way as the bodies that we consider belonging to the circular current, so that $u \sim v$. Then the formula (2.54) gives the required condition accounting for the distance from the star to the center of galaxy and for the star velocity

$$rv^2 \sim \frac{1}{4} GM \qquad (2.61)$$

For a galaxy with mass M, with the radius of its visible disk r, and for the velocity v of the stars moving at its periphery, this condition is in accord with the values measured in astrophysics on galactic scale. Therefore, from the point of view of this theory, the discrepancy between Newton-Einstein-Schwarzschild predictions and the observations is not surprising.

It is very convenient to use the consequence of the common geometric origin ("Maxwell identities") of electromagnetism and gravitation discussed in Chapter 1 and in [Siparov (2008a, 2009)]: the mathematical results obtained in electrodynamics can be used in calculations of the motion of the gravitational systems and vice versa. The gravitational CPC model is analogous to an electromagnetic model that consists of a circular electric current and a charge in the center of it.

2.5.2   *Flat rotation curves of spiral galaxies*

Let us, first, speak for simplicity about the electric version of the CPC model and regard a positive charge surrounded by a circular current J, and an electron circling around in the plane of the contour and close to the wire, (from the point of view of mathematical formalism, this will be just the same as speaking of the corresponding gravitation problem). Strictly speaking, an electron in such a system cannot be in a finite motion because of the presence of the magnetic field and has either to fly away or to fall on the center. But the number of rotations performed before that could be large enough. The value of $B_z(r)$ component of the magnetic induction produced by the contour with radius, R_{eff} , can be found with the help of Biot-Savart law, and according to [Landau and Lifshits (1967)] with $c = 1$ as before, it is equal to

$$B_z(r) = J\frac{2}{\sqrt{(R_{eff}+r)^2+z^2}}\left[K + \frac{R_{eff}^2 - r^2 - z^2}{(R_{eff}-r)^2+z^2}E\right] \quad (2.62)$$

$$K = \int_0^{\pi/2}\frac{d\theta}{\sqrt{1-k^2\sin^2\theta}}; E = \int_0^{\pi/2}\sqrt{1-k^2\sin^2\theta}\,d\theta$$

$$k^2 = \frac{4R_{eff}r}{(R_{eff}+r)^2+z^2}$$

where K and E are elliptic integrals. Taking $z = 0$ and introducing the notation, $b = r/R_{eff}$, one gets

$$B_z(r) = \frac{2J}{R_{eff}}\left(\frac{K}{1+b}+\frac{E}{1-b}\right) \quad (2.63)$$

The inner region close to the charge corresponds to $b << 1$ and to the constant value $B_z(r) \to J/2R_{eff}$; the remote region corresponds to $b >> 1$ and to $B_z(r) \to 0$; the intermediate region to which the contour belongs and in which the electron moves corresponds to $b = O(1)$ and

$$B_z(r) \sim J/r \qquad (2.64)$$

The centrifugal force $m\frac{v_{orb}^2}{r}$, acting on an electron orbiting around with velocity v_{orb} is equal to the sum of the Coulomb attraction, $F_{Cl} = qC_1/r^2$ produced by the central charge and of the Lorentz force

$$m\frac{v_{orb}^2}{r} = \frac{qC_1}{r^2} \pm qv_{orb}B_z(r) \qquad (2.65)$$

Consider that in the intermediate region (not far from the wire in the electromagnetic case and at the periphery of a galaxy in the gravitation case) $B_z^{(g)}(r) \sim J/r$. Let us now pass to the gravitational version of the CPC model, change the electric charge to gravitational one, i.e. set $q = m_g$, and use the equivalence principle, i.e. $m_g = m$. For a spiral galaxy its effective radius does not differ much from the radius of the visible disk where the measurements of the stars orbital velocities are possible. Therefore, we can neglect the change of the character of $B_z(r)$ decrease with r. Then we designate $J \equiv C_2$ and instead of eq. (2.65) get the dynamics equation in the form

$$v_{orb}^2 = \frac{C_1}{r} \pm v_{orb}C_2 \qquad (2.66)$$

where C_1 and C_2 are constants characterizing the system and units, and the sign corresponds to the direction of the current and to the location of the probe inside or outside the contour. The smaller root of the quadratic equation (2.66) gives the expression

$$v_{orb} = \frac{C_2}{2}\left(1 - \sqrt{1 \pm \frac{4C_1}{rC_2^2}}\right)$$

corresponding to Newton law, that is to the decrease of the velocity with distance (the sign corresponds to the direction of the probe motion). In the larger root of the quadratic equation (2.66), i.e. in the expression

$$v_{orb} = \frac{C_2}{2}\left(1 + \sqrt{1 \pm \frac{4C_1}{rC_2^2}}\right)$$

the small term under the square root can be neglected, and we see that the velocity remains constant with distance, i.e.

$$v_{orb} \sim C_2 \qquad (2.67)$$

This result corresponds to the observable flat rotation curve at the periphery of a spiral galaxy.

2.5.3 *Tully-Fisher and Faber-Jackson relations*

In any course of observational astronomy (for convenience, see e.g. [Myers (1999)]) there is a discussion of the measurable physical parameters of a galaxy with mass M, visual radius R and luminosity L_{lum}. These are given by the relations

$$\theta = \frac{R}{d}$$

$$f = \frac{L_{lum}}{4\pi d^2}$$

$$v^2 = \frac{GM}{R}$$

where θ is a small angle at which the luminous spot of a galaxy is observed, d is the distance to it, f and L_{lum} are flux and luminosity, and v is a characteristic velocity of the objects belonging to a galaxy. The distance, d, can be eliminated by the use of the surface brightness, I, defined as

$$I = \frac{f}{\theta^2} = \frac{L_{lum}v^4}{4\pi G^2 M^2}$$

This can be rearranged as

$$L_{lum} = \frac{v^4}{I} \frac{1}{4\pi G^2 (M/L_{lum})^2}$$

where M/L_{lum} is the mass-to-light ratio which is an important characteristic parameter in astronomical reasoning. The basis of the most useful distance indicators in cosmology is the assumption that for a given class of galaxies the central surface brightness, I, and the mass-to-light ratio, M/L_{lum}, are constant. Then the following relation is true

$$L_{lum} \sim v^4$$

For example, for spiral galaxies, the appropriate velocity would be the characteristic constant velocity on the rotation curve which is determined with the help of spectral analysis and Doppler effect. Then, we get

$$L_{lum} \sim v_{orb}^4 \tag{2.68}$$

which is known as the Tully-Fisher relation. For elliptical galaxies, the appropriate velocity is the central velocity dispersion

$$L_{lum} \sim \sigma_v^4 \qquad (2.69)$$

which is known as the Faber-Jackson relation. The velocity dispersion σ, is the range of velocities close to the mean velocity for a group of objects, such as a cluster of stars in a galaxy. For example, in a galaxy, a typical value for the velocity dispersion of the objects orbiting the galactic center is about 10^5m/s. For the Faber-Jackson relation the index in eq. (2.69) can differ from 4 and depend on the range of galaxy luminosities that is fitted. The fact that massive galaxies originate from homologous merging, and the fainter ones from dissipation must be taken into account, and the assumption of constant surface brightness can fail. Both Tully-Fisher and Faber-Jackson relations were first discovered empirically and only later explained in the manner given above.

Let us now estimate the value of the constant C_2 present in eq. (2.67) which corresponded to the CPC model. This constant was introduced as $C_2(R_{eff}) = J_{(m)}(R_{eff})$. In the simplest case, the mass current can be evaluated as $J_{(m)}(R_{eff}) \sim \frac{M}{T}$, the spiral galaxy mass, M is proportional to its area, i.e. to R_{eff}^2, and the period can be evaluated according to Kepler law as $T \sim R_{eff}^{3/2}$. This leads to

$$J_{(m)}(R_{eff}) \sim \sqrt{R_{eff}}$$

Since the integral luminosity, L_{lum} is also proportional to the area of the galactic disk, we get

$$R_{eff} \sim \sqrt{L_{lum}}$$

Therefore

$$J_{(m)}(R_{eff}) \sim \sqrt{R_{eff}} \sim L_{lum}^{1/4}$$

and finally

$$v_{orb} \sim L_{lum}^{1/4} \qquad (2.70)$$

which is in accord with Tully-Fisher relation eq. (2.68). Notice the assumption of the constant mass-to-light ratio that was used, the same assumption is commonly used in astronomy. In order to describe the motion of stars in elliptical galaxy the simple model including only one current is not enough.

Reasonably improved for any concrete case, the CPC model would give the result that could be compared not only with the speculative Faber-Jackson relation given above but with the concrete measurements of the real galaxies.

Similarly to SRT which doesn't tell us if there is any ether or if there isn't - it simply does without it and exploits the notion of space-time for predictions, the AGD doesn't tell us if there is any dark matter or if there isn't - it also simply does without it and exploits the notion of phase space-time for predictions.

2.5.4 *Logarithmic potential in spiral galaxies*

Regarding a spiral galaxy, one can see that the logarithmic gravitational potential is an appealing idea for many reasons. First of all, the matter seems to belong mostly to the galaxy plane and, therefore, the Poisson equation in 2D *has* the solution proportional to $\ln r$. The problem is that comparing it with the geodesics, we find no usual Newton law which works so well. Second, it has an observational support represented by the OC-type curves on Fig. 2.1. Now the problem is that there are also OAC type curves that do not correspond to the logarithm. Third, it seems to provide the possibility to do without dark matter. For example, logarithmic potential was suggested in [Kinney and Brisudova (2000)] from heuristic considerations just in order to describe the gravitation in the spiral galaxy (actually, the flat rotation curves) without the use of dark matter. This worked well, but the authors eventually confronted problems with the interpretation of the gravitational lenses observations. Something was missing, and they expressed the hope that such potential could be somehow obtained from the relativistic considerations. The same potential was also discussed in [Dvali *et al.* (2001)] where it was introduced to describe a certain (2+1)-dimensional scalar field in the cosmological domain wall whose plane contains the planes of spiral galaxies, this potential also produced the flat rotation curves.

It is interesting to notice that the logarithmic potential can be neither guessed as in [Kinney and Brisudova (2000)] nor be the result of the additional unknown field action, but can be obtained in frames of general physics, that is directly from observations. If the Earth were the only celestial body rotating around the Sun, Kepler would have been unable to obtain his third law, $\frac{R_{orb,i}^3}{T_i^2} = const$ for planets' orbits. But it was just this law that led Newton to the expression for the gravitation potential in the

Solar system. For two planets' accelerations, we get

$$\frac{a_1}{a_2} = \frac{v_1^2}{R_1}\frac{R_2}{v_2^2} = \frac{4\pi^2 R_1^2}{T_1^2 R_1}\frac{T_2^2 R_2}{4\pi^2 R_2^2} = \left(\frac{R_{orb,i}^3}{T_i^2} = const\right) = \frac{1/R_1^2}{1/R_2^2}$$

and with regard to the definition of force, i.e. to the dynamics law $F = ma$, and to the demand $F = \nabla\Phi$, we obtain $\Phi \sim \frac{1}{R}$. But on the solitary Earth, an observer who would be able to discover that the stars' motion in spiral galaxies give $v_{orb} = const$, that is $\frac{R_{orb,i}}{T_i} = const$, would by the same simple calculations come to

$$\frac{a_1}{a_2} = \frac{v_1^2}{R_1}\frac{R_2}{v_2^2} = \frac{4\pi^2 R_1^2}{T_1^2 R_1}\frac{T_2^2 R_2}{4\pi^2 R_2^2} = \left(\frac{R_{orb,i}}{T_i} = const\right) = \frac{1/R_1}{1/R_2}$$

and the phenomenological gravitation potential for the far away galaxies appears to be $\Phi \sim \ln R$. Thus, the flat parts of the rotation curves are evidently provided by the logarithmic gravitation potential.

In the geometrical approach of AGD developed here, this feature appears as follows. Now the metric depends on velocity, and we account for it when solving the field equation with regard to the generalized geodesics, and obtain the gravitation force in the form of eq. (2.54). Taking the CPT model for a galaxy and calculating the gradient with regard to eq. (2.49), we see that the second term of the force is proportional to $1/R$. The corresponding potential gives the logarithm, and the first term gives the usual Newton gravitation that tends to vanish.

Thus, in the AGD the logarithm for a spiral galaxy naturally appears from the anisotropic metric. There is no need for observations nor for the introduction of any additional field. When the moving masses are distributed in a more complicated way, the potentials will obtain a related form in accord with eq. (2.49).

2.5.5 *Classical tests on the galaxy scale*

Let us use the CPC model in order to calculate the results of the classical GRT tests on the galactic scale. They include orbit precession, gravitational redshift and light beam bending. The visual results can be obtained with the help of numerical calculations which is due mostly to the presence of the elliptic integrals in the functions.

2.5.5.1 *Orbit precession*

In order to describe the motion of a star in a spiral galaxy around the galaxy nucleus, let us use the CPC model and regard a particle orbiting in the contour plane. Choosing various initial conditions, one can obtain scattering, swift fall down on the center and long enough orbiting (see examples on Fig. 2.4). If the initial conditions lead to the large enough number of particle rotations around the CPC center, then Fig. 2.8a presents what could be called quasi-precession. This type of a star motion in a galaxy is the anisotropic space analogue of the GRT planet orbit precession in the Solar system in the isotropic space. The part of a circle line on the plot represents the position of the effective current.

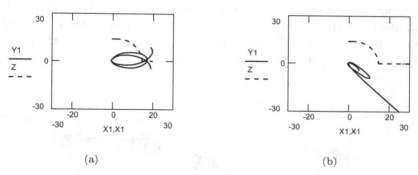

Fig. 2.4   Trajectories obtained with the help of the CPC model of the source: (a) Quasi-precession. (b) Motion of globular clusters that disagree with Kepler law. (Part-of-the-circle line corresponds to the part of a circular current).

According to Kepler, the planets' orbits were elliptical and this was the basis of Newton gravitation theory. Then the major part of their lifetime the orbiting bodies spend at the periphery, far from the gravitation center vicinity. The paradoxical feature of the globular clusters motion is that there are too many of them found in the vicinity of the galaxy center and not in the periphery. And the explanation of this feature was also mentioned in [Aguirre *et al.* (2001)] as one of the demands to be sufficed by any modification of gravitation theory. The trajectory given on Fig. 2.4b suggests an example of such explanation. In AGD instead of an attracting point center, there is a CPC, the orbits are not elliptical, and if the plane of a cluster orbit is close to the galaxy plane, the influence of the mass current cannot be neglected and it can hold the cluster close to the center. That is why the globular star clusters can be frequently found in the vicinity of the galaxy center and not on the periphery.

In [Lin and Shu (1964)], there was developed the theory of the galactic arms origination based on the idea of density waves. They are produced by the precession of the elliptical orbits of the stars. But this theory does not give any suggestion about the origination of galaxies' bars, while 2/3 of the total number of spiral galaxies possess them. The problem of bars origination is not solved up to now. A bar usually looks like a dense elongated feature consisting of stars and interstellar gas. It is considered to be the main importer of the interstellar gas to the center of a galaxy. But as a matter of fact, the 3D computer simulation [Fux (1999)] shows that the gas doesn't reach this center.

The famous Sandage list of 23 astronomical problems for the 21st century contained two problems concerning this issue: No.1 What causes the appearance of Hubble sequence (Fig. 2.5) in the galaxy formation: initial conditions or the evolution of galaxies?; and No.4 What is the role of rotation in the spiral structure of galaxies?

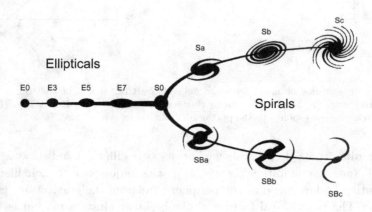

Fig. 2.5   Hubble classification scheme.

The numerical modeling of the CPC system based on the phase space-time approach contributes to all this in the following way. Let us suppose that the nucleus of the rotating galaxy that belongs to E or S Hubble classes explodes and throws two approximately equal giant pieces of matter in the opposite directions with approximately equal velocities in the plane of galaxy rotation. This can be modeled as a corresponding burst of the nucleus of the CPC model. Figure 2.6 represents the comparison of the

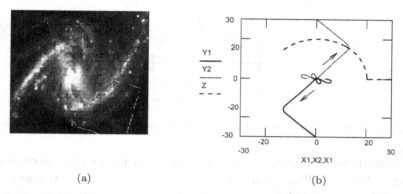

(a)   (b)

Fig. 2.6   (a) NGC-1365, Hubble image of a spiral galaxy (NASA/ESA) and (b) the CPC calculation of the explosion.

image of spiral galaxy NGC-1365 and the result of the corresponding numerical solution. The forms of the small loops that can be seen on the trajectories close to the center of the plot may vary with the step of calculation but the loops remain and demonstrate how a fragment first moves away, then it is caught and pulled back by the attraction of the center but then the field of the current interferes and drives it away.

On Fig. 2.7, there is the reverse situation corresponding to the symmetrical inward motion of equal masses towards the center of the CPC system.

Notice, that what we see in the real image are the positions of different particles in roughly the "same" moment of time. It is the "same" in comparison with the time it takes the light to reach the Earth, but one should not forget the De Sitter's objections to the Ritz hypothesis mentioned in Section 1.2 and consider that the scale of the binary star system is much smaller than the scale of a galaxy and, thus, the view of one and the same galaxy could be completely different from the different distances to it. On the contrary, the plots of the calculated trajectories given on Figs. 2.6, 2.7 present the consequent positions of one and the same body. Therefore, the calculated plots and the real image given on Figs. 2.6, 2.7 present different situations.

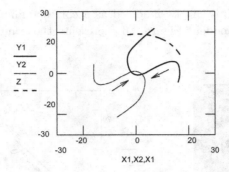

Fig. 2.7   CPC model of the motion of the two symmetrical pieces moving inwards.

Nevertheless, the pattern coincidence is remarkable and in a sense could be regarded as an illustration to the ergodic theorem. Varying the parameters of the chosen CPC model and initial conditions, one can obtain a lot of patterns resembling the observed ones for various observable galaxies. The same can be performed when modeling the interacting galaxies observed in space with the help of a couple of CPCs located close to each other.

2.5.5.2   *Gravitational redshift*

In the observations on the galactic scale, the kinematical Doppler type effect is indistinguishable from the gravitational redshift effect, i.e. if the gravitational potential is related to the dependence of metric on (here: tangential) velocity, then it is not obligatory the Universe expansion that causes the frequency shift.

When we pass to the consideration of the gravitational potential far from the observer in frames of AGD, there appears a principal feature which characterizes the anisotropic potential introduced above and underlines the difference between the point-like source characteristic to the GRT and the CPC model of the source inherent to AGD. The point-like source or the isle model suggests that there is a finite region occupied by matter where the curvature differs from zero, and this region is surrounded by the empty space stretching up to infinity. The system under consideration is close to the border of this region where the curvature is small. Therefore, when we calculate the gravitational potential at distances approaching infinity, the potential tends to zero. But this situation is only imaginary. It does not contradict the pure logic but it is not observable and, thus, testable already on the galaxy scale. It does not mean that the idea is inconsistent, but it means that there are scales at which the isle-source model is not applicable. Really, what we measure in observations (rotation curves) contradicts its prediction. And this is not because something unknown is missing and should be added, but because the model is not appropriate. For the CPC model, one can choose as origin any point (star) in a spiral galaxy, calculate (or try to measure) the radial dependence of the gravitational potential on the distance from the origin, and discover that it grows both inwards and outwards. The same can be done for the Universe scale, only the origin would coincide rather with a galaxy and not with a star. Is there any "inwards" or "outwards" there? Presumably, no, but it doesn't matter because the problem remains the inner one and not exterior, thus, the linear growth of the potential with distance from the origin pertains, and it will lead to the observation of the gravitational redshift which will grow with distance from the selected source to the observer. It sends us back to the Mach's ideas and to the notion of the infinite world mass critically discussed by De Sitter in his paper [De Sitter (1917)]. Now we see that no "infinite world mass" is needed as De Sitter himself was sure of. The effect stems from the observation, and the character of it is characteristic to the AGD in the phase space-time. Actually, the observation of Hubble redshift can be interpreted as a manifestation of this very effect inherent to the AGD. As it was already mentioned, Hubble realized [Hubble (1929)] that the Doppler effect could not be the only reason of the redshift he observed. He thought of attributing it to De Sitter's solution which presumed metrical effects.

This means that the redshift can be due not only to the radial expansion of the Universe, but also to the tangential motion of its parts. Notice that in accord with the expression for the gravitation potential in eq. (2.54), the first term tends to zero when the observer moves away from the center, but the second term grows as $(\mathbf{u}, \mathbf{v}) \rightarrow v^2 \sim r^2$. Therefore, the time interval will be inversely proportional to the first power of the distance from the origin to the point close to the (tangentially) moving source, it will give the Hooke's law, and this corresponds to the linear Hubble law. Data scattering and the observed anisotropy of the Hubble law could be due to the various values of

Such interpretation has an indirect observational support presented by the measurements of the tangential velocities of the far away quasars given on Fig. 2.8 on which the velocities appear to be surprisingly high [MacMillan (2003)].

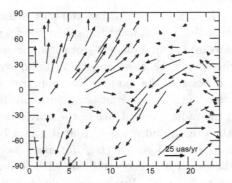

Fig. 2.8   Observed tangential motion of quasars (sketch of the experimental data obtained in [MacMillan (2003)]).

2.5.5.3   *Light bending*

Figure 2.9 presents the comparison of the scattering of the probe body on a center with Newtonian gravitation and on a CPC as suggested by the AGD. Since the expression for the metric which has to be used for the calculation of the light bending angle now depends the proper velocities of the sources, these trajectories can be related to the light beams trajectories.

The comparison of Fig. 2.9 (a) and Fig. 2.9 (b) explains why the observed values of the light bending by the convex gravitational lenses can

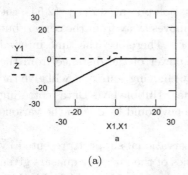

(a)

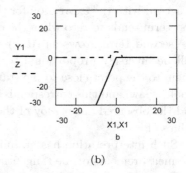

(b)

Fig. 2.9   Trajectories of a probe body scattering: (a) on the Newtonian center; (b) on the CPC.

be essentially larger than those predicted by the GRT, see the subsection below. These results also don't have any need of the dark matter. Moreover, Fig. 2.10 illustrates a probe body scattering on a CPC along the trajectory which bends twice in the opposite directions. This result was obtained

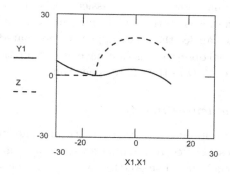

Fig. 2.10   Trajectory of a probe body scattering on the CPC: double bending with opposite signs.

when the direction of the mass current present on Fig. 2.9 (b) was changed to the opposite one, and the starting point of the probe was put close to the contour. If instead of a probe body there is a light beam starting from a source located in the same place at the same direction, we can say that from the point of view of light refraction, it means that the AGD predicts the existence of the concave gravitational lenses that diminish the angular sizes of the objects located behind them. It means that if such a lens is located between the object and the observer on the Earth and is oriented in a due manner, then the estimated value of the distance to the object could be higher than it really is. If this object is an astronomical standard candle, this situation could be the reason of the misinterpretations of the last decade observations of the type 1a supernovae presumably showing the break down of Hubble linear law [Reiss (1998)]. These observations and interpretations led to the idea of the acceleration of the Universe expansion and to the introduction of the notion of dark energy of repulsion providing the acceleration. It caused the revival of the cosmological constant in Einstein equation.

Now we see that the interpretation of the SN1a observations with regard to the possible existence of concave gravitational lenses provides the possibility not to demand dark energy existence from the Universe. In frames of the AGD, the gravitational potential in eq. (2.54) contains an additional term which can have different signs depending on the velocity of the object and on the initial conditions. This makes the introduction of an absolute constant less natural than the use of the local corrections due to the relative motion of the parts of the Universe.

Thus, the phenomena known as the classical tests of the GRT do not only correspond to the ideas of the AGD and can be calculated but are directly observed. One can see that they appear to be related just to those paradoxical astrophysical observations that have no explanation in the GRT but are natural for the AGD in the phase space-time.

2.5.6 *Gravitational lenses in AGD*

Let us now apply the obtained results to the theory and observations of the gravitational lenses. The corresponding formulas were given in Section 1.5 by eqs. (1.88, 1.89, 1.90). The AGD gravitation potential in classical approximation is given by

$$\Phi_{(a)} = \Phi_N + \frac{8}{c^2}(\mathbf{u}, \mathbf{v}) \tag{2.71}$$

where $u = V_{eff}$ is the effective value of the mass density current and v is the velocity of a probe body. In order to simplify the evaluations that could be compared with the observation, let us make some approximations. First, we will use $V_{eff} = \Omega_{eff} R_{eff}$ similarly to above, and second, we will take the light speed as a probe body velocity. The possible errors could affect quantitative values, but hardly more than the errors in the evaluation of the astronomical parameters that are usually present in such calculations. The qualitative aspect would not change. It is clear that when a light beam belongs to the plane of the lens which is a spiral galaxy, the second term in eq. (2.71) would enter with different signs depending on which side from the center does the beam pass. Then the light bending angle will be evaluated not as $\theta = \frac{2r_S}{\xi} = 2\Phi_N$ as usual but as

$$\theta = 2\Phi_{(a)} = 2\Phi_N + \frac{16}{c^2}(\mathbf{V}_{eff}, \mathbf{c}) \tag{2.72}$$

On Fig. 2.11 there are two orientations of the spiral galaxy that could be a gravitational lens for the light source.

To the left is the en face orientation, and the lens will be described by the usual formulas, to the left is the en profile orientation, and the AGD

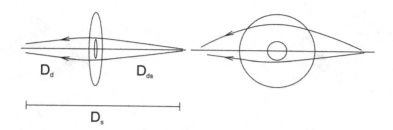

Fig. 2.11 Orientations of a gravitational lens-spiral galaxy.

theory might be used to evaluate the corrections. In the last case, the AGD theory gives the following formulas. The gravitational lens equation will become

$$\frac{1}{\xi}[\eta + D_{ds}(\frac{2r_S}{\xi} + 16\frac{V_{eff}}{c})] = \frac{D_s}{D_d} \tag{2.73}$$

and the role of Chwolson-Einstein radii when the light passes by different paths will be played by two unequal segments the lengths of which are given by

$$\xi_{1,2} = 4\xi_0 \frac{\xi_0}{r_S} \frac{V_{eff}}{c} \left[\pm 1 + \sqrt{1 + \frac{c^2}{V_{eff}^2} \frac{r_S D_s}{32 D_{ds} D_d}} \right] \tag{2.74}$$

where $\xi_0 = \sqrt{2r_S \frac{D_d D_{ds}}{D_s}}$. The ratio ξ_1/ξ_2 that could be compared with the observations is

$$\frac{\xi_1}{\xi_2} = \frac{1 + \sqrt{1 + \frac{c^2}{V_{eff}^2} \frac{r_S D_s}{32 D_{ds} D_d}}}{-1 + \sqrt{1 + \frac{c^2}{V_{eff}^2} \frac{r_S D_s}{32 D_{ds} D_d}}} \tag{2.75}$$

In order to make the estimations, the effective velocity V_{eff} could be taken of the order of 10^5 m/s, which gives $\frac{c^2}{V_{eff}^2} \sim 10^6$ while D_s, D_{ds} and D_d are galactic distances and r_S for a galaxy is about a light year. Thus, the estimation suggested by eq. (2.75) seems reasonable.

For example, let us take the famous Einstein Cross, Figs. 2.12, and 2.13. The quasar 2237+030 is at the distance of $8 \cdot 10^9$ light years [NASA/ESA (1990)], the lensing galaxy is the spiral ZW 2237+030 at the distance of $4 \cdot 10^8$ light years [NOAO (1999)], and as can be seen from its elliptical form

(a)   (b)

Fig. 2.12   Sketches of (a) quadra-image of the quasar in Einstein cross (b) lens - spiral galaxy.

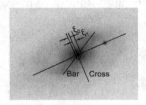

Fig. 2.13   Einstein cross, lens and segments.

on the image with the bar, its orientation at least partly provides the needed en profile tilt. The arrows point at the segments whose ratio we are going to find. Substituting numbers into eq. (2.75), we get the order of 10^0, and this value is in accord with what we see. Of course, higher accuracy is hardly achievable, because of the astronomical parameters evaluations accuracy, but qualitatively we can say that the AGD result does not contradict the observations and can be used for interpretations.

How many of the total number of galactic gravitational lenses could reveal such property? Crude estimate could be made as follows. Spirals constitute about one half of the total number of all the galaxies, therefore, of the lenses too. Two thirds of them will be within the AGD assumption, because of the orientation. Thus, one third of all the gravitation lenses might give the anomalously large effect, because of the anisotropic properties of the gravitation potential.If a lens presents another type of galaxy or a cluster of galaxies, then the inner motion of its components should be taken into consideration.

2.5.7 *Pioneer anomaly*

Another obvious attempt to apply the AGD approach is the well-known Pioneer anomaly [Anderson *et al.* (1998)]. It constitutes the additional constant sunward accelerations measured for two probes Pioneer-10 and Pioneer-11 that appeared equal to $(8.74 \pm 1.33) \cdot 10^{-10}$ m/s^2 for both spacecrafts. Notice that the value is again close to the cH product, the reason of that being still unknown. The probes were launched with the mission to leave the Solar system, and the anomaly revealed itself when they were already at about 20 a.u. from the Sun. For other missions like Voyager, Galileo, Ulysses etc. there is no solid evidence of the same effect: some of these probes were not spin stabilized and use thrusters too often, some burned in the atmospheres of planets before the effect was noticed. Pioneer-10 and Pioneer-11 are flying away in the opposite directions almost radially with the speeds of about $1.4 \cdot 10^4$m/s obtained from Jupiter and Saturn as a result of gravitational manoeuvre, Fig. 2.14.

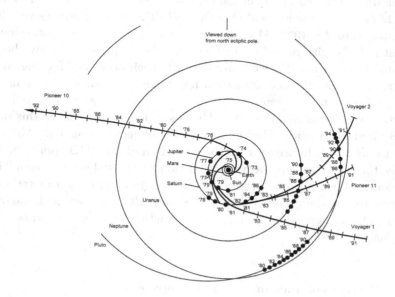

Fig. 2.14   Pioneer missions. (NASA image)

The data concerning the flight details at distances closer to the Sun is now taken out from the storage devices in order to be processed and analyzed to see if the effect shows up in the inner region of the Solar system.

Explanations for the discrepancy that have been considered include observational errors with account for measurements and computations; statistical errors; gravitational forces from unidentified sources such as the Kuiper belt; drag from the interplanetary medium; gas leaks; radiation pressure of sunlight, the spacecraft's radio transmissions, or thermal radiation pressure from the radioisotope thermoelectric generators. All these appeared to be either irrelevant (acceleration does not show up in the orbits of the outer planets) or insufficient to provide the needed value. Some new physics was also suggested as a possible explanation of the Pioneer anomaly. It included: clock acceleration between coordinate or Ephemeris time and International Atomic Time; MOND induced explanations based on the deviation of the gravitation force from the traditional Newtonian value to a different force at low accelerations or on a modification of inertia; down-scaling of photon frequency; Hubble law extension and others.

From the point of view of AGD, the additional acceleration is defined by the product of the constants of the theory, i.e. cH, and the value of $[\beta, \Theta]$, eq. (2.57), where $\beta = v/c$ and $\Theta = \Omega(x)/H$. The value of probe's velocity corresponds to $\beta \sim 10^{-4}$. Moving along a hyperbola arm, the probe moves almost radially with regard to the Sun, but when it passes a massive body like Jupiter or Saturn it moves tangentially to their orbits and has the same velocity. That is why the estimation for Ω can be taken as a ratio of its measured sunward acceleration to its measured velocity. According to [Iorio (2008)], this ratio is equal to $7.3 \cdot 10^{14}$rad· s^{-1}, and we see that in this case $|[\beta, \Theta]|$ is close to unity. Therefore, the value of acceleration, eq. (2.57), is close to cH and is in accord with the prediction of the AGD approach.

By the way, the qualitative explanation of the flyby effect could be the same. When a probe passes a massive object during a gravitational manoeuvre, the change of the expected spacecraft trajectory could be due to the velocity dependent gravitation interaction between the probe and the planet.

2.6   Electrodynamics in anisotropic space

The results of the previous section show that the perspectives in application of the new theoretical approach to the interpretation of the observable astrophysical situations governed by the gravitational interaction are promising. At the same time, in case we absolutely choose to go this way, we will have to reform a great amount of our knowledge concerning the world structure and performance. This situation again repeats the situation that was a century ago, when Poincaré rejected the idea of reforming science of that time on these very grounds - the range of the forthcoming reforms seemed to him too huge. But Einstein and Minkowski took the challenge, and the results appeared to be worth the efforts.

As it was already underlined, we try to follow Einstein method, and when the result of the observations - flat rotation curves - starts to contradict the theory, there appears the AGD which is based on the 8D phase space-time model instead of involving an unobservable entity - dark matter. When Einstein did the same with the result of Michelson and Morley experiment, and there appeared the SRT which was based on the 4D space-time model instead of involving an unobservable entity - ether, he had to reform mechanics in order to preserve electrodynamic observations and Maxwell equations. Mechanics was based then on Euclidean geometry. It seems that now we have to reform electrodynamics in order to preserve gravitational observations and Einstein equations. Thus, new geometry demands the changes in electrodynamics which is based on Riemannian (Minkowskian) geometry. And in the same way as then there appeared the "relativistic mass", now the "anisotropic space currents" are going to appear.

Studying anisotropic spaces has an obvious meaning with regard to the physical interpretations [Voicu and Siparov (2008, 2010)]. The direction dependence of the metric could cause the appearance of the motion dependent forces [Siparov (2010a)] associated with inertial forces in the accelerated frames. In case there is a physical vector field - electromagnetic one - in the anisotropic space, this may lead to the appearance of the extra Lorentz type forces or extra currents that could reveal themselves in the special laboratory environment or even in Nature. From the mathematical point of view, it is possible to treat the problem in the purely Finslerian setting when $g_{ij} = \dfrac{1}{2}\dfrac{\partial^2 \mathcal{F}^2}{\partial y^i \partial y^j}$ for some 2-homogeneous in y function $\mathcal{F} = \mathcal{F}(x, y)$ or introduce a more general type of an anisotropic metric that could explicitly give extra terms in the equations for geodesics.

In isotropic (pseudo-Riemannian) spaces with $R_{ij} = 0$, the components of the free electromagnetic potential 4-vector $A^i = A^i(x)$ obey de Rham equations:

$$A^{i;j}_{\;\;\;;j} = 0$$

or

$$g^{ij}(x,y)A^k_{\;\;;ij} = 0.$$

When passing to anisotropic spaces with metric $g_{ij} = g_{ij}(x,y)$, the corresponding solution of such an equation would generally depend on directional variables. So, it is meaningful to take into consideration the case when the potential 4-vector depends on directional variables $y = (y^i)$,

$$A^i = A^i(x,y),$$

and

$$A_i = A_i(x,y).$$

Let us investigate the following simple case of an anisotropic space. Let its metric be a small (linearly approximable) deformation of metric tensor whose components do not depend on the positional variables (locally Minkowski metrics), and let the coordinate changes preserve the positional independence of the metric.

2.6.1   *Weak deformation of locally Minkowski metrics*

Let us consider the space $\mathbb{R}^4$ endowed with linear coordinate changes. Let $(x,y) = (x^i, y^a)_{i,a=\overline{0,3}}$, $y^i = \dfrac{\partial x^i}{\partial t}$ (t is a parameter), $i = 0 - 3$ be the coordinates in a local frame of $T\mathbb{R}^4 \equiv \mathbb{R}^8$.

Let g be a small (linearly approximable) deformation of a locally Minkowski metric:

$$g_{ij}(x,y) = \gamma_{ij}(y) + \varepsilon_{ij}(x,y). \tag{2.76}$$

This type of metrics includes the small deformations of Minkowski metric regarded earlier as particular case.

We suppose that this metric tensor is equal to $g_{ij} = \dfrac{1}{2}\dfrac{\partial^2 \mathcal{F}^2}{\partial y^i \partial y^j}$ for some 2-homogeneous in y function $\mathcal{F} = \mathcal{F}(x,y)$ (according to [Rashevsky (2003)], if such metric tensor is non-degenerate, it is Finslerian). We denote by , $_k$ (with commas) partial derivation with regard to x^k and with dots $._a$, partial derivation by the directional variable y^a. Whenever convenient - and just in order to point out the difference, we will denote the indices corresponding to y with $a, b, c, \ldots$ and those corresponding to x with $i, j, k, \ldots$, (though, they run over the same set $\{0,1,2,3\}$). Let $\Gamma^i{}_{jk} = \dfrac{1}{2}g^{ih}(g_{hj,k} + g_{hk,j} - g_{jk,h})$ denote the usual Christoffel symbols (with respect to x) of metric g. In our case, $\Gamma^i{}_{jk}$ depend both on x and y. Let $_{|k}$ denote covariant derivation with respect to x^k :

$$X^i{}_{|k} = X^i{}_{,k} + \Gamma^i{}_{jk}X^j \tag{2.77}$$

In the following, we shall also need the Cartan tensor C_{ijk} defined as

$$C_{ijk} = \frac{1}{2}(g_{ij \cdot k} + g_{ik \cdot j} - g_{jk \cdot i}) = \frac{1}{2}g_{ij \cdot k}. \qquad (2.78)$$

Due to the 0-homogeneity of g_{ij}, the Cartan tensor satisfies:

$$C_{ijk}y^k = C_{ijk}y^j = C_{ijk}y^i = 0. \qquad (2.79)$$

Let $|_a$ denote covariant derivation with respect to y^a :

$$X^i|_a = X^i{}_{\cdot a} + C^i{}_{ja}X^j. \qquad (2.80)$$

If we take into account the y-dependence of the fundamental metric tensor, it follows that in the anisotropic spaces, the components of an electromagnetic-type tensor F_{ij}, $F^i{}_j$, F^{ij} in general case depend on the directional variables. So, it is natural to suggest that either electromagnetic potential A_i or its alternative A^i could also depend on direction.

2.6.2 *Lorentz force*

2.6.2.1 *Variational principle*

The equations of electrodynamics can be obtained from the variational procedure applied to a Lagrangian. In isotropic spaces, the Lagrangian is

$$L(x, y) = \frac{1}{2}g_{ij}(x)y^i y^j + \frac{q}{c}A_i(x)y^i$$

where q is the electric charge, and $A_i(x)$ are the covariant components of the 4-vector potential. In order to obtain Lorentz force in our case let us consider the Lagrangian

$$L(x, y) = L_0 + \frac{q}{c}L_1 \qquad (2.81)$$

where g_{ij} is a 0-homogeneous anisotropic metric tensor (2.76), $L_0 = \frac{1}{2}g_{ij}(x, y)y^i y^j$ and $L_1(x, y)$ is a scalar function which is 1-homogeneous in the directional variable. Let

$$A_i = \frac{\partial L_1}{\partial y^i}$$

then

$$L_1 = A_j(x, y)y^j$$

and

$$L(x, y) = \frac{1}{2}g_{ij}(x)y^i y^j + \frac{q}{c}A_i(x, y)y^i$$

where $A_i(x, y)$ is the direction dependent potential. The components of the covector field $A_i = A_i(x, y)$ are 0-homogeneous functions in y, and possess the property

$$A^i_{i \cdot k}y = 0 \qquad (2.82)$$

Thus,

$$\frac{d}{dt}\left(\frac{\partial L}{\partial y^k}\right) = g_{kj,h}y^j y^h + g_{kh}\frac{dy^h}{dt} + \frac{q}{c}\left(A_{k,h}y^h + A_{k \cdot h}\frac{dy^h}{dt}\right)$$

and the Euler-Lagrange equation

$$\frac{\partial L}{\partial x^k} - \frac{d}{dt}\left(\frac{\partial L}{\partial y^k}\right) = 0$$

becomes

$$g_{kh}\left(\frac{dy^h}{dt} + \Gamma^h_{jk}y^j y^k\right) + \frac{q}{c}(A_{k,h} - A_{h,k})y^h + \frac{q}{c}A_{k \cdot h}\frac{dy^h}{dt} = 0. \qquad (2.83)$$

Let us designate

$$F_{kh} = A_{h,k} - A_{k,h} \qquad (2.84)$$

then the extremal curves $t \mapsto (x^i(t)) : [0, 1] \to \mathbb{R}^4$ of the Lagrangian (2.81), (2.82) are given by

$$\frac{dy^i}{dt} + \Gamma^i_{jk}y^j y^k = \frac{q}{c}g^{ik}\left(F_{kh}y^h - A_{k \cdot h}\frac{dy^h}{dt}\right), \qquad (2.85)$$

The usual interpretation of the extremal curves is the equation of motion. Therefore, the expression on the right-hand side of eq.(2.85) presents the Lorentz force in the anisotropic space. We see that its first term which is common with the isotropic case is proportional to velocity, while the second term is proportional to acceleration which brings to mind the idea of an "inertial force" in the accelerated reference frame.

Let us designate

$$\tilde{F}_{ia} := -A_{i \cdot a}, \quad \tilde{F}_{ai} = A_{i \cdot a}. \qquad (2.86)$$

Then $\tilde{F}_{ia}$ is (-1)-homogeneous in the directional variables, that is

$$\tilde{F}_{ia}(x, \lambda y) = \frac{1}{\lambda} \tilde{F}_{ia}(x, y)$$

Then the relation between $\tilde{F}_{ia}$ and the new term in eq. (2.85) is

$$\tilde{F}^i = \frac{q}{c} \tilde{F}^i{}_a \frac{dy^a}{dt},$$

and, thus, we have obtained an antisymmetric 2-form on $T\mathbb{R}^4$:

$$F = F_{ij} dx^i \wedge dx^j + \tilde{F}_{ia} dx^i \wedge dy^a. \tag{2.87}$$

Mathematically speaking, the above is the exterior derivative of the 1-form $A = A_i(x, y) dx^i + 0 \cdot dy^a$ on $T\mathbb{R}^4$:

$$F = dA. \tag{2.88}$$

We see that the direction dependent electromagnetic potentials cause the natural correction of the expression of the electromagnetic tensor.

2.6.2.2 *Example f-1*

In the particular case when the covariant components of A do not depend on direction, $A_i = A_i(x)$, we get $\tilde{F}_{ia} = 0$, which is the regular expression of Lorentz force.

2.6.2.3 *Example f-2*

If the observations show that the interpretation should be performed with the help of contravariant components of 4-vector potential, and it turns out that these components do not depend on direction, $A^i = A^i(x)$, we should take into account that the perturbed metric tensor does depend on y, and the covariant components will be direction dependent

$$A_i = g_{ij}(x, y) A^j(x) \Rightarrow A_i = A_i(x, y)$$

Then the new term appears in Lorentz force

$$\tilde{F}^i = \frac{q}{c} \tilde{F}^i{}_a \frac{dy^a}{dt} = -\frac{q}{c} g^{ih} A_{h \cdot a} \frac{dy^a}{dt}$$

which leads to

$$\tilde{F}^i = -2 \frac{q}{c} C^i_{ja} A^j \frac{dy^a}{dt}$$

and

$$\tilde{F}_{ia} = -2 C_{ija} A^j$$

2.6.2.4 *Example f-3*

If the weak deformation has happened to the flat Minkowski metric, $\eta_{ij} = diag\{1, -1, -1, -1\}$, and $g_{ij} = \eta_{ij} + \varepsilon_{ij}(x, y)$, then the above is

$$\tilde{F}_{ia} = -\varepsilon_{ij \cdot a} A^j \tag{2.89}$$

and the new term is small. But this is not necessarily so for other locally Minkowski metrics.

2.6.2.5 *Example f-4*

Let $\gamma_{ij}(y) = \dfrac{1}{2} \dfrac{\partial^2 \mathcal{F}^2}{\partial y^i \partial y^j}$ be the Berwald-Moor Finslerian metric with $\mathcal{F}^2 = \sqrt{y^0 y^1 y^2 y^3}$, and we consider $\varepsilon_{ij} = 0$, and $A^i = A^i(x)$. Then $g_{ij} = \dfrac{1}{2} \dfrac{\partial^2 \mathcal{F}^2}{\partial y^i \partial y^j}$, and the correction $\tilde{F}_{ia}$ is

$$\tilde{F}_{ia} = -A_{i \cdot a} = -\frac{\partial}{\partial y^a}(g_{il} A^l) = -2C_{ila} A^l$$

where the Cartan tensor $C_{ila} = \frac{1}{4} \dfrac{\partial^3 \mathcal{F}^2}{\partial y^i \partial y^l \partial y^a}$ has the explicit values

$$C_{ila} = \alpha \frac{\partial^3 \mathcal{F}^2}{\partial y^i \partial y^l \partial y^a}; \alpha = -\begin{array}{ll} \frac{3}{32} & \text{if } i = j = a \\ \frac{1}{32} & \text{if } i = j \neq a \\ \frac{1}{32} & \text{if } i \neq j \neq k \neq i \end{array}$$

Here we see that C_{ila} are not necessarily small.

2.6.3 *New term - "electromagnetic" vs. "metric"*

Let us now have a look at eq. (2.83):

$$g_{kh}\frac{dy^h}{dt} + \Gamma_{khl}y^h y^l = \frac{q}{c}F_{kh}y^h - \frac{q}{c}A_{k \cdot h}\frac{dy^h}{dt}$$

From the mathematical point of view, we can formally interpret the last term in two ways

- Since it appears multiplied by the acceleration $\dfrac{dy^h}{dt}$ and moreover, since $A_{k \cdot h} = A_{h \cdot k}$, we can "stick" it to the metric:

$$(g_{kh} + \frac{q}{c}A_{k \cdot h})\frac{dy^h}{dt} + \Gamma_{khl}y^h y^l = \frac{q}{c}F_{kh}y^h.$$

and get a new metric tensor

$$\tilde{g}_{kh} = g_{kh} + \frac{q}{c} A_{k \cdot h} \tag{2.90}$$

(if the matrix $(\tilde{g}_{kh})$ is invertible) with the property

$$\tilde{g}_{kh} y^k y^h = \mathcal{F}^2$$

With this, we can write the equation of motion as

$$\frac{dy^i}{dt} + \Gamma_{khl} y^h y^l = \tilde{g}^{ik} \frac{q}{c} F_{kh} y^h \tag{2.91}$$

and the obtained expression for Lorentz force $\tilde{g}^{ik} \frac{q}{c} F_{kh} \dot{x}^h$ differs from the case of isotropic perturbation $\frac{q}{c} g^{ik} F_{kh} \dot{x}^h$ due to the new metric (2.90).

- Also, we might leave the metric as it is, preserve the third term in the right-hand side and interpret it as a new term added to Lorentz force:

$$g_{kh} \left(\frac{dy^h}{dt} + \Gamma_{khl} y^h y^l \right) = \frac{q}{c} F_{kh} y^h - \frac{q}{c} A_{k \cdot h} \frac{dy^h}{dt}.$$

This would yield

$$\frac{dy^i}{dt} + \Gamma^i{}_{jk} y^j y^k = \frac{q}{c} \left(F^i{}_h y^h + \tilde{F}^i{}_a \frac{dy^a}{dt} \right). \tag{2.92}$$

with the influence of the anisotropy given by the second term on the right-hand side. Notice that $\frac{q}{c} \left(F^i{}_h y^h + \tilde{F}^i{}_a \frac{dy^a}{dt} \right)$ is equal to $\frac{q}{c} g^{ik} \left(F_{kh} y^h - A_{k \cdot h} \frac{dy^h}{dt} \right)$ given by eq. (2.85) since $F^i{}_h = g^{ik} F_{kh}$, $\tilde{F}^i{}_a = g^{ik} A_{k \cdot h}$. In eq. (2.91) the term $\frac{q}{c} g^{ik} A_{k \cdot h} \frac{dy^h}{dt}$ was brought to the left-hand side of the equation of motion and absorbed by the metric - the new "metric" was denoted by $\tilde{g}_{ik}$.

From the physical point of view, the second variant seems preferable, since we agree that it is only mass - inertial or gravitational - that makes a space curved. Therefore, the correction to the force (to Lorentz force) which is due to the anisotropy of space caused by the motion of masses seems natural. On the other hand, we notice that the AGD approach brings the descriptions of electromagnetism and gravitation more close to each other, because they stem from the same source which is geometrical in origin, that is from Maxwell identities. And the passage to the anisotropic space that we discuss here does not bring away from it. Really, it can be checked that there holds

$$F_{ij,k} + F_{ki,j} + F_{jk,i} = 0 \qquad (2.93)$$

or, in terms of covariant derivatives defined by eq. (2.80),

$$F_{ij|k} + F_{ki|j} + F_{jk|i} = 0.$$

Besides,

$$F_{ia,k} + F_{ki\cdot a} + F_{ak,i} = 0$$
$$F_{ia\cdot b} + F_{bi\cdot a} = 0$$

where $F_{ia\cdot b} = \dfrac{\partial F_{ia}}{\partial y^b}$. The last two relations, together with (2.93) mean actually that the exterior derivative of F is equal to zero

$$dF = 0. \qquad (2.94)$$

This means that we could act exactly as in Chapter 1 when we obtained Maxwell identities and start with some $F = F(x,y)$ such that $F_{ij,k} + F_{ki,j} + F_{jk,i} = 0$, then prove that this implies the existence of some A with $F_{ij} = A_{j,i} - A_{i,j}$ and then define $\tilde{F}_{ia}$ such that $dF = 0$ (or $F = dA$) remains true.

Thus, this discussion of electromagnetism in the anisotropic spaces is not the construction of the unified theory of fields yet, but we move in this direction with no special effort. That is why the first variant of the new term treatment should not be thrown away forever but should just wait for its time to come.

2.6.4 *Currents in anisotropic spaces*

In the classical Riemannian case, the non-homogeneous Maxwell equations can be obtained [Raigorodski *et al.* (1999)] by means of the variational principle applied to the action in the form

$$S = \int (\alpha F_{ij} F^{ij} - \beta j^k A_k) \sqrt{-g} d\Omega$$

where α and β are constants, $g = \det(g_{ij})$ and $\Omega = dx^0 dx^1 dx^2 dx^3$. Taking into account that in our case, at least one of the quantities F_{ij}, F^{ij} depends on y, the whole integrand depends on y, and is actually defined on some domain in $\mathbb{R}^8$.

$$S = \int (\alpha F_{\lambda\mu} F^{\lambda\mu} - \beta j^k A_k) \sqrt{G} d\Omega, \qquad (2.95)$$

where $\lambda, \mu \in \{i, j, a, b\}$, $G = \det(G_{\alpha\beta})$ is the Sasaki lift of g to $T\mathbb{R}^4 \equiv \mathbb{R}^8$, [Bao *et al.* (2000)]:

$$G_{\alpha\beta}(x, u) = g_{ij}(x, y) dx^i \otimes dx^j + g_{ij}(x, y) dy^i \otimes dy^j$$

and $\Omega = \prod_{i,a} dx^i dy^a$ gives the volume form on $\mathbb{R}^8$. The product $F_{\lambda\mu} F^{\lambda\mu}$ is in our case

$$F_{\lambda\mu} F^{\lambda\mu} = F_{ij} F^{ij} + \tilde{F}_{ia} \tilde{F}^{ia}.$$

Performing variation with respect to A_k in eq. (2.95), we get

$$\frac{1}{\sqrt{G}} (F^{ki} \sqrt{G})_{,i} + \zeta^k = J^k, \qquad (2.96)$$

where

$$J^k = \frac{-\beta}{4\alpha} j^k$$

and

$$\zeta^k = \frac{1}{\sqrt{G}} (\tilde{F}^{ka} \sqrt{G})_{\cdot a}. \qquad (2.97)$$

In comparison to the case of isotropic spaces, there appears a new term in the expression for the current, namely

$$\zeta^k = \tilde{F}^{ka}|_a \qquad (2.98)$$

This means that in an anisotropic space the measured fields would correspond to an effective current consisting of two terms: one is a current provided by the experimental environment, the other is a current corresponding to the anisotropy of space. This also seems consistent, since the initial AGD agreement was that the anisotropy is produced by the motion of the distributed masses.

2.6.4.1 *Example c-1*

If $A_i = A_i(x)$, then $\tilde{F}_{ia} = 0$, and $\zeta^k = 0$

2.6.4.2 *Example c-2*

If $A^i = A^i(x)$, then we saw that $\tilde{F}_{ia} = -2C_{ija}A^j$ which means that $\tilde{F}^{ia} = -2C_j^{ia}A^j$ and

$$\zeta^k = -2(C_j^{ka}A^j)|_a = -2C_j^{ka}|_a A^j - 2C_j^{ka}C_{ha}^j A^h$$

The presence of the last current in the experimental situation can be noticed if $|\tilde{F}^{ka}|_a| \sim |F_{|i}^{ki}|$.

2.6.4.3 *Example c-3*

For the weak deformation of the flat Minkowski metric, we get

$$\zeta^k = -2\varepsilon_{\cdot ja}^{ka}A^j \tag{2.99}$$

Consequently, the current originating from the anisotropy of space can be noticed when the "regular" current is small and $|2\varepsilon_{\cdot ja}^{ka}A^j| \sim |F_{|i}^{ki}|$.

2.6.4.4 *Example c-4*

In case of the deformed Berwald-Moor metric, the Cartan tensor components are no longer small, and hence the corrections ζ^k are also no longer small, so that they could be noticed even if the "regular" current is not small. Therefore, if there exists a physical system whose geometrical properties can be described by the Berwald-Moor metric, its electromagnetic properties would be specific to this case.

We conclude that an anisotropic space with electromagnetic field possesses inherent currents that could produce observable effects. Thus, the formal scheme given in Section 2.3 is developed for the case when there is an additional physical field in the anisotropic space. We see that the notions of Lorentz force and electric current were redefined with regard to the mentioned circumstances and took the forms given in eqs. (2.92) and (2.98). It should be underlined that it is important in which way - covariant or contravariant - do the measurable physical variables transform with the coordinate transformation. The calculations given in Examples f-3 and c-3 for the weakly deformed Minkowski metric lead to the concrete expressions eq. (2.89) and eq. (2.99) that can be compared with observations. The similar estimations were given [Voicu and Siparov (2008, 2010)] for another locally Minkowskian metric - Berwald-Moor one. In the last case it could be easier to notice the effect of anisotropy, but first, there is a need for an appropriate environment in which it could be possible to perform the experimental check up.

2.7 Approaching phase space-time

2.7.1 *Coordinate-free dynamics*

The introduction of the 8D phase space-time as a model of physical world demands not only the new geometry but also several definitions and approaches that are not widely used in physics now. Let us outline the directions needed for the construction of the formalism that seems relevant to these approaches.

The idea of a phase space is simple and the corresponding formalism is well developed [Godbillon (1969); Arnold (1974); Sardanashvili (1996); Vershik and Faddeev (1975); Besse (1978)]. In spite of the impression that in macro world we deal with the usual - configuration - space in which the initial and current positions of particles are realized, the classical mechanics actually performs in another place, namely, in a space whose points are states of particles represented by pairs (position, velocity). This fact being simple in itself follows from the Newton-Lagrange determinacy principle [Mornev (2001)] and points at the not-so-simple geometrical aspect of reality - the performance of dynamics takes place in a space S which is the direct product of the positional space Q and velocity space ς, that is $S = Q \times \varsigma$, and, besides, locally S is a tangent bundle TQ of the configuration space Q. Thus, if we speak of a Law of Nature and attribute an onthological status to this Law, the same status must be attributed to the space of states. If we agree that a "law" is just a suitable mathematical construction, a consistent definition able to provide a consistent picture and give verifiable predictions, we have even more reasons to accept the space of states as its natural surroundings [Siparov (2010b)], its base and its base manifold [Mornev (2001)]. It is natural to consider S to be a smooth manifold and regard a TS which is its tangent bundle. Every fiber of this bundle is a linear vector space tangent to S which is a storage for all the vectors tangent to S at a given point. Passing from point to point on S and (smoothly) selecting a certain vector in every such storage on TS, we obtain a section of tangent bundle, that is a certain vector field of selected vectors. From this point of view, a dynamics law prescribes how to choose the tangent vector in every fiber. Therefore, the coordinate-free geometrodynamic approach seems to be a natural generalization of Newtonian mechanics, and the main equation of dynamics will include the Lie derivative operator along the vector field on TQ. It can be shown [Mornev (2001)] that there appear the *fields of currents* that interact with the *field of momenta* and cause the appearance of the *generalized Lorentz forces*. It can also be noticed that in all the formulations the notion of metric is actually not used. The metric appears to be *dependent on momenta* and plays a secondary role in the presented scheme. Thus, we see that the approach presented in this book can be further developed with the help of the formalism that is independently coming into being in abstract mathematics.

On the other hand, speaking of geometry in the sense of Klein's Erlangen program, one has to pay attention to the group aspect of the AGD theory. There are two principal features that have to be discussed now: the first is the generalization of the main notion of the relativity, that is the Lorentz group of transformations with regard to the presence of a characteristic length; the second is the possibility to regard the whole 8D phase space-time as the unification of subspaces that can possess different geometries allowing various simplifications. These geometries correspond to groups and groups allow contractions. These last areas in the recent years have attracted attention of the researches, and in view of the approach suggested here, new perspectives can appear.

2.7.2 *Generalized Lorentz transformations*

Earlier, we have already discussed the meaning of the inertial motion and underlined the inconsistency in regarding it as a straight uniform one. The idea is well known, but the attention that was paid to it and its consequences is hardly enough (exceptions, for example, are [Fock (1964); Kerner (1976); Manida (1999)]). In view of the modification of the gravitation theory performed above and of the introduction of the space with a characteristic length scale, the fundamental concept of an IRF understood as a frame in which a free body preserves its state requires some modification.

The transformations that provide the possibility to construct a set of IRFs form groups, and for the linear transformations these groups are Galilean or Lorentz ones. As we know, they work well in their fields of applicability, and besides, for the limit $c \to \infty$, Lorentz group becomes the Galilean one. Therefore, if some considerations require the introduction of a fundamental length, l, in such a way that the group properties of the transformations are preserved, the limit $l \to \infty$ should bring us back to Lorentz group. The question of the order of such limit transitions and the meaning of the obtained results immediately arises. It was mentioned already in Section 2.3, and we will turn to it in the next subsection. The problem of introduction of a fundamental length into the mathematical apparatus of special relativity is usually treated within the quantum field theory approach (one could mention the so called doubly special relativity, scale relativity etc.) and has different grounds, but here the reasons for such introduction are different.

Let us regard the simplest case of transformations and use eq. (1.1), that is

$$x = x_0 + vt \tag{2.100}$$

Usually, the transformations $x \to x'(x,t); t \to t'(x,t)$ are given by the linear functions, so that the linear character of $x = x(t)$ dependence given by eq.(2.100) is preserved for $x' = x'(t')$ dependence. The same goal can be reached [Fock (1964)] by the use of the fractional-linear transformations of the following form $x \to f_1(x',t')/f(x',t'); t \to f_2(x',t')/f(x',t')$ where f, f_1, f_2 are also linear functions. The dependence $x' = x'(t')$ will be also linear because the function f is the same for both coordinates. It can be checked that the velocity $v' = \frac{dx'}{dt'}$ now depends not only on v but on x_0 too.

At first glance it seems strange, but we have already discussed this seeming preference attributed to point x_0 in several places in this book when we spoke of observational possibilities. In spite of the arbitrary choice of the origin, all the physical processes won't change, but will be only observed from different places and states none of which is preferable. The observation of the straight uniform motion presumes the existence of an observer, but his own reference frame can move in a way which he cannot perceive. So, we have to speak of the straight uniform motion from the point of view of the observer and investigate the transformations that preserve this type of *observation*, while the *absolute character of motion* should not be discussed. It also means that v does no longer mean the relative velocity of the frames but is a parameter with the dimensionality of speed which could have some analogues in hydrodynamics of the non-uniform flows.

Then, instead of the usual Poincaré or Lorentz formulas

$$x'^\mu = L^\mu_\nu x^\nu + a^\mu \tag{2.101}$$

$$x'^\mu = L^\mu_\nu x^\nu \tag{2.102}$$

the new group transformation rule can be expressed by

$$x'^{\mu} = \frac{L_{\nu}^{\mu} x^{\nu} + a^{\mu}}{1 + l^{-1} b_{\mu} x^{\mu}} \tag{2.103}$$

where $L \in SO(1,3)$, and $a^{\mu} = lb^{\mu}$ are the components of the translation vector, l is a length scale, b^{μ} are dimensionless parameters. The homogeneous transformations

$$x'^{\mu} = \frac{L_{\nu}^{\mu} x^{\nu}}{1 + l^{-1} b_{\mu} x^{\mu}} \tag{2.104}$$

also form a group. The physical meaning of this result is that for both frames the "free" particle is seen as moving uniformly along the straight line. At the same time, when the particle moves not freely, and we calculate the corresponding accelerations $d^2 r/dt$ and $d^2 r'/dt'$ in both frames, they will be proportional to each other. Since v has lost its simple character, there are reasons to ask about c which was a constant factor of the theory and obtained the physical meaning of the light speed through interpretation. It must remain a constant, but in order to verify its relation with light, the electrodynamics itself has to be revised with regard to the changes caused by the length scale existence. Hopefully, it will be possible to preserve it undamaged with the help of the reasonable length scale parameter choice. It can also be noticed [Kerner (1976)], that for the particles passing through the common origin of eq. (2.102) and eq. (2.104) transformations, the motion $r = vt$ implies $r' = v't'$ with v' obtained by the ordinary Lorentz velocity addition rule, no dependence on the origin position will be present in this case. In particular, if $v = c$ then $v' = c$, and light from the common origin would be registered invariably in both descriptions. Thus, to this extent the special relativity holds. If we check the influence of the order of limit transitions, we see that for $l \to \infty$ first and $c \to \infty$ second, the Galilean limit is reached in two steps, but if $c \to \infty$ goes first, we get to the Galilean transformation directly. In [Kerner (1976)] the generalization of this approach for the non-homogeneous case was also performed and the transformations appeared to be a particular form of fractional-linear (or projective) transformations. They contain 10 parameters corresponding to space-rotations, boosts (or frame changes to relatively moving systems), and space and time translations. All the parameters of the group fall together in a simple and unified way in a type of overall single rotation, in contrast to the usual situation when the space-rotations and boosts come together but the translations stand rather apart.

2.7.3 *Geometry, groups and their contractions*

From the point of view of classical geometry of the pseudo-Euclidean plane, the circles that have an arbitrary non-zero (real or imaginary) radius are the hyperbolas. Analogously, in the 3D pseudo-Euclidean space with signature (2,1), the spheres that have a non-zero real radius are one-sheet hyperboloids and the spheres that have a non-zero imaginary radius are parted hyperboloids. The analogy works in the spaces with higher dimensions, for example, in a 4D space with signature (3,1). The geometrical properties of each half of a hypersphere with imaginary radius in the $(n + 1)$D pseudo-Euclidean space with signature $(n,1)$ are the same with those of nD Lobachevsky space. In Lobachevsky space, its subspaces of dimension k (k varies from 0 to $n - 1$) correspond to the $(k + 1)$D subspaces of the initial pseudo-Euclidean space passing through the origin of coordinates, and its motions correspond to Lorentz transformations.

The 8D phase space-time (x, y) of the AGD which is an analogue of the usual phase space is a combination of a 4D space whose coordinates describe position and a 4D space whose coordinates describe direction. The set of eight coordinates of the phase space-time can be split into two sets: (x, y, z) and $(ct, l, l\beta_x, l\beta_y, l\beta_z)$ where x, y, z can be related to the space coordinates of a probe body, $l\beta_x, l\beta_y, l\beta_z$ can be related to the normalized components of velocity of the moving probe body, and $l = c/H = const$ is the characteristic scale. Then the second set corresponds to the anti-De Sitter space. We can imagine the picture of a body moving in a slightly curved space that is approximated by the deformed Minkowski space, while the second set of coordinates has a relation to the one-sheet hyperboloid mentioned above. In Fig. 2.15, there is a visual aid to this picture drawn for the smaller number of dimensions.

In Riemannian space, the group of coordinate transformation preserving the inertial motion is the 10-parametric Poincaré group. But on the hyperboloid with its characteristic parameter with the dimension of length, it will be DeSitter group which relies on the fraction-linear transformations. If we use the Beltrami coordinates for the anti-DeSitter space and perform a limit transition $c \to \infty$, the result will be the flat Newton-Hooke space. But this means that in the non-relativistic case the bodies in DeSitter space have an extra "repulsion" from the origin and tend to move away from the center with an outward pointing *fictitious* force proportional to their distance from the origin. It resembles the *fictitious* centrifugal force in the rotating frame, but it does not coincide with it and has a different origin

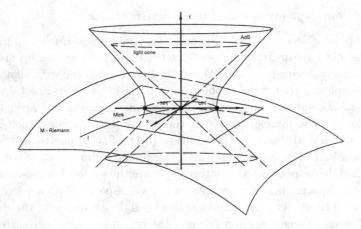

Fig. 2.15   Illustration of a phase space-time in case of less number of dimensions.

of the geometrical nature. It looks like as if it picks out a preferred point in space - the center of repulsion, but actually it is isotropic for the reasons discussed in Section 2.1 that concerned the difference between an inner and an exterior problems. If we move to another point, we should transform to the uniformly accelerated frame of reference of an observer at this point, and this changes all accelerations in such a way that the repulsion center moves to the new origin of coordinates. This means that in a space-time with non-vanishing curvature, gravitation is different from Newtonian gravity, and this is just the case of AGD. At distances comparable to the fundamental length parameter of this space, the bodies feel an additional linear repulsion from the center of coordinates. Besides, the Newton-Hooke space relates to the space-like part of the Minkowski space corresponding to the given point. As it was mentioned before, this is in agreement with the general observational situation in astronomy.

The genuine group theoretical analysis was performed in the fundamental paper by Bacry and Levy-Leblond [Bacry and Levy-Leblond (1968)], (subsequently, it was developed in more formal way in [Gromov (1990)]). It was shown that there exists only one way to generalize classical SRT, and this way is to endow it with length scale factor or endow the space-time with a kind of constant curvature. This is in accord with what was done by DeSitter half a century before, though not on the group theory grounds but by solving mass-free Einstein equation.

The symmetry group of the SRT is the ten-parametric Poincaré group. Ten basic elements of its Lie algebra are the following: the generator H of time translations; three generators P_i of space translations along the ith axis; three generators J_i of spatial rotations; and three generators K_i of boosts (inertial transformations) along ith axis. The commutation relations involving spatial rotations have the form

$$[J_i, H] = 0; [J_i, P_j] = \epsilon_{ijk}P_k; [J_i, J_j] = \epsilon_{ijk}J_k; [J_i, K_j] = \epsilon_{ijk}K_k \quad (2.105)$$

which demonstrates that H is a scalar and P_j, J_j, K_j are vectors. These generators (H, P_i, K_i, J_i) of the group described by eq. (2.101) are given by

$$H = \partial_t; P_i = \partial_i; K_i = (t\partial_i + \frac{1}{c^2}x^i\partial_i); J_i = \epsilon_{ij}^k x^j \partial_k \quad (2.106)$$

In order to find a similar structure such that the transformation properties remain the same, and it can have reasonable physical interpretation, one can turn to the results obtained in [Huang (2009); Guo *et al.* (2009, 2008); Huang *et al.* (2007)].

Preserving the same notation for coordinates as in eq. (2.106), another set of generators can be written as

$$H' = -c^2 l^{-2}tx^k\partial_k(= cP_0'); P_i' = l^{-2}x^i x^k\partial_k; K_i = (t\partial_i + \frac{1}{c^2}x^i\partial_i); J_i = \epsilon_{ij}^k x^j \partial_k$$
$$(2.107)$$

It corresponds to the group described by eq. (2.104) and also has 10-parameters. This set does not generate the isometry of the Minkowski space-time, but preserves the light cone at its own origin which is essential for applications. The details of this geometry with this type of symmetry and a lot of meaningful discussions can be found in the references mentioned above. Here we mention only the coordinate transformation

$$x^0 = l^2\eta^{-1}\cosh(r/l) \quad (2.108)$$
$$x^1 = l^2\eta^{-1}\sinh(r/l)\sin\theta\cos\phi$$
$$x^0 = l^2\eta^{-1}\sinh(r/l)\sin\theta\sin\phi$$
$$x^0 = l^2\eta^{-1}\sinh(r/l)\cos\theta$$

where

$$\eta = \det \eta_{ij}; \eta_{ij} = diag\{1, -1, -1, -1\}$$

with which we come to the expressions

$$g = g_{\mu\nu}dx^{\mu} \otimes dx^{\nu} = -(dr^2 + l^2 \sinh^2(r/l)d\Omega_2^2)$$
$$h = h^{\mu\nu}\partial_{\mu} \otimes \partial_{\nu} = (\partial_{\mu})^2$$

that define the non-degenerate 3D metric g and 1D contravariant metric h. The expressions eq. (2.108) can be recognized as the introduction of the Beltrami coordinates while contracting the De Sitter group into Newton-Hooke one related to the possible applications in frames of AGD.

In the beginning of Chapter 1, there were already mentioned the possible applications of the Euclidean geometry for the description of the regions of the physical world where no cause and effect can be distinguished. The geometry based on the Newton-Hooke group can also be applied to derive a certain form of quantum mechanical equations. For example, in [Liu and Tian (2008); Tian *et al.* (2005)], Schrödinger equations were obtained and discussed from the algebraic viewpoint for Newton-Hooke space-time. From the point of view of 8D phase space-time of the AGD, it means that various subspaces of it can deliver possibilities for various branches of fundamental science to be developed.

Sometimes they also argue, that the special relativity is no longer correct in the presence of gravity and should be replaced by general relativity. The usual counter argument is that the interpretation of gravity as a pseudo-Riemannian metric of curved space-time is not the only interpretation [Silagadze (2008)], nor is it always the best one, because this interpretation sets gravity apart from other interactions. But the approach developed in this book is an attempt to show that if special relativity is really incorrect in the presence of gravity, it should be replaced not by the usual general relativity which has serious drawbacks on the large scales, and, therefore, cannot be the base of the generalized approach, but by anisotropic geometrodynamics. The last is free from these drawbacks and, therefore can pretend to become the starting point for the modified gravitation theory. The new model of physical reality suggested here gives new perspectives to the description of phenomena observable on the scale of the Universe, besides, the same could also be true for the micro and macro worlds.

2.8 Cosmological picture

Not long ago, actually, less than hundred years, the cosmological picture that everyone was aware of presented the stationary homogeneous and isotropic distribution of stars surrounded by the infinite empty space. The stars were so far from each other that their inverse square attraction forces were absolutely negligible and they could move freely. The space scale of this distribution was that of our galaxy, its time scale was infinite in the past, and the existence of such a system was hard to explain in terms of physics of that time. In order not to let the stars fly away from each other during the infinite time of their existence, and, thus, enable people to observe them, Einstein tried to introduce an omnipresent force described by the cosmological constant, and let the Newton potential affect only the vicinities of stars. But later he realized that such constant could not save the situation because it led not to stationary but to an unstable equilibrium. It was the time when the astronomers finally agreed that the tiny nebulae that they observed with the telescopes consisted of billions of stars, presented the distributions similar to the Milky Way, had the similar dimensions and were located at the distances from each other comparable to the characteristic dimension of Milky Way, i.e. the furthest of the newly recognized galaxies were essentially further from us than astronomers thought before. The scale of the Universe increased by several orders of magnitude.

In the 30th the Hubble discovery of the redshift of radiation coming from the far galaxies and the linear character of this shift dependence on distance from the Earth supported the model of expanding Universe suggested earlier by Friedmann. Hubble himself mentioned such possibility, but in the original paper he was rather inclined to interpret his result as an evidence in favor of the De Sitter model which presumed that the empty Universe had a finite though large characteristic radius. This model suggested that the time at the periphery was slowing down, and this explained the redshift. But the majority of the scientific community preferred the expansion interpretation. Not everyone agreed to that, and the attempts to find new proofs of the expansions were undertaken even in the end of 20th and in the very beginning of the 21st centuries.

Usually they mention [Reshetnikov (2003)] three observational tests suggested by Sandage and aimed at the measurements that could prove the expansion. The first of them is Tolman test which is based on the measurement of the surface brightness of galaxies as a function of their redshift related to parameter z of the expansion theory. It is considered to provide the possibility to find out whether the Universe is expanding or static. The static character of the Universe suggests that the surface brightness is independent of the distance, because the intensity of light received from a distant object drops inversely as the square of this distance, while the apparent area of the object under observation also drops inversely with the square of the distance. But if the Universe expands, the measured radiation power is reduced at least for two reasons: the rate of the received photons is lower, because each photon has to travel a longer distance than its predecessor; there appears the redshift due to Doppler effect. On the other hand, distant objects might seem larger than they really are, because there come additional photons that were emitted when the object was closer. The calculation showed that the surface brightness in a flat and uniformly expanding Universe should decrease as $(1 + z)^4$. Sandage and Lubin [Sandage and Lubin (2001)] published a series of papers in which the results of the investigation of the relationship between surface brightness and redshift were presented. They found that the exponent is not exactly 4 as it was expected in the simplest model, but is 2.6 or 3.4, depending on the frequency band. Nevertheless, the authors concluded that the Tolman surface brightness test is consistent with the reality of the expansion, though the model of it could be more complicated.

The second test is based on the measurement of the time slowing down with distance - if the Universe is expanding, the further is the object the faster it runs away. Then according to SRT, the duration of the similar processes must be different at different distances, and for the redshift parameter equal to z, the processes are slower by $(1 + z)$. In [Goldhaber (2001)] they measured the light curves of the type 1a supernovae. The process of type 1a supernovae explosion is well understood and calculated and is broadly used in astronomy. The measurements seemed to support the uniform expansion, though the measurement of the date of explosion was model dependent.

The third test suggests to measure the relic microwave background temperature for various z, i.e. for various epochs of the Universe evolution. According to the expansion model, the temperature should grow linearly with z as $(1 + z)$. The results obtained in [Molaro (2002)] show this growth though its character seems to be more complicated than linear.

Thus, all the three Sandage tests were performed and could serve as a wishful indication of the real character of cosmological expansion though the place for doubts still remains. The proof of expansion makes meaningful the great variety of ingenious theories regarding the cosmological evolution that started with the singularity state and Big Bang followed by inflation which is now considered the standard cosmological model. Even stronger proof of the Big Bang theory comes from the prediction and subsequent discovery and measurement of the cosmic microwave background radiation in 1964, the event that Hawking called "the final nail in the coffin of the steady-state theory". The last is a model developed in the 50s F. Hoyle, T. Gold, H. Bondi and others as an alternative to the Big Bang theory. In steady-state theory the universe is also expanding, but the theory presumes that the new matter is continuously created. The calculations showed that a static universe was impossible under general relativity. Still, even in 1993 F. Hoyle, G. Burbidge, and J. Narlikar proposed a quasi-steady state cosmology in which some additional features like minibangs or mini-creation events appeared. The observations that were interpreted as the acceleration of the Universe expansion led to further modifications of the steady-state model: the main feature of them is that though the Universe is expanding, it never changes its appearance with time and has neither beginning nor end.

Together with the principal idea of the Universe as it is, there are several secondary cosmological issues dealing with the contents of it like matter and its forms. Among them there are the problem of unification of the interactions, the problem of dark matter, the problem of dark energy, the problem of galaxies origination, structure and distribution and many others that are more concrete.

What kind of impact could be produced on the cosmological picture by the introduction of anisotropic geometrodynamics, that is by the introduction of the phase space-time as a model of physical world and by the velocity dependent gravitation? First of all, there is a hope to get rid of the dark notions that seem foreign to the theory that works well, that is to GRT. These notions draw attention and activity that could be used in a more fruitful way. The current principal ideas of the Universe evolution can do without them and probably would be retained. But some new possibilities also come to mind. Instead of dark masses and energies, there is now tangential motion that plays an important role in the functioning of the Universe machine. Considering a galaxy as an element of the Universe construction, we can regard it as a turbulent whirl that can have a form of a disk or of a ball or of their remnants. It can be imagined as a turbulent whirl in hydrodynamics or as a ball lightning in magnetoplasmadynamics The evolution of such a whirl is a result of a hydrodynamic process involving the concentration of the moving rarefied matter - gas of atoms suffering the action of velocity dependent gravitation forces, the origination of a whirl, the period of stability that includes the motion of it through the rarefied medium, the interaction with the other whirls if they are present around, and the subsequent dissipation. The rotation velocity is defined by Newton gravitation, by the radial gradient of the pressure of interstellar gas and by the velocity dependent gravitational forces discussed above. This process is simultaneously going on at a smaller scale for the stars evolution and on a larger scale for galaxy clusters evolution.

If we look at a galaxy as gas of stars, we can see that in such a medium there are no collisions, and, therefore, the system is far from equilibrium and the velocity distribution of stars must be chaotic and not Maxwellian. The characteristic time of the relaxation of this state to the equilibrium must be much longer than a star free path time, and for a Solar type star, the free path time is three orders of magnitude larger than the Universe age. But the measurements show [Zasov and Postnov (2006)] that the majority of stars with exception of the most young ones presents the result of a partial relaxation: the velocity distribution is maxwellian but has different dispersions for different axes. Moreover, one and the same space region demonstrates a systematical growth of the random velocities of old stars. This can be interpreted as the heating of the star disk with time, the growth of kinetic energy corresponds to the growth of temperature. This process is the same with what we can see on a certain stage of a turbulent whirl behavior.

The interrelation between physics and mathematics in AGD approach seems to be more balanced, and the last uses geometry as a tool that can fit this or that situation and give this or that prediction. The only demand is that in order to make the observed cosmological picture less paradoxical, we should speak of observable situations and possibilities in the first place. The existence of giant voids and the presence of galaxies mainly in the walls, edges and vortices of these voids makes one think of a foam which is less abstract than foam walls discussed in brane theories that grew up from the fictitious but fruitful extra dimensions introduced by Kaluza. As it was shown in [Trujillo *et al.* (2006)], not only galaxies are arranged in an intricate "cosmic web" of filaments and walls surrounding bubble-like voids, but even the axes of spiral galaxies are oriented in a coherent manner.

In our galaxy, we also live in such a wall and observe the surroundings and the Universe as a whole from this position. The geometries used to describe these observations could be different and depend on scale. First it was Euclidean geometry of 3D space, the observations on a planetary system scale made it Minkowski geometry of 4D space-time, the observations on a galaxy scale led to AGD with the 8D geometry of phase space-time. It could be interesting to think of geometry suitable to describe the Universe observed from the center of a void. The Universe would be seen as a kind of anisotropic crystal seen from inside with visible consolidations of matter at the edges formed by filaments and vortices. The space-time in the center of the void will be flat, but closer to the borders, the curvature would appear, enter the metric and correspond to a type of Finsler geometry with Berwald-Moor metric that was introduced and investigated by Bogoslovsky [Bogoslovsky (1973); Bogoslovsky and Goenner (1998)] and Pavlov and Garas'ko [Pavlov and Garas'ko (2007)]. This geometry has attractive mathematical features and is based on algebra H4 of the hyper complex numbers. The last authors discussed the simple geometrical figure - rhombic dodecahedron - as characteristic for the symmetry of the real world. It can be noticed that rhombic dodecahedron is an example of the regular 3D polyhedra that can be used for filling the space by translations, the so called 3D parquet, and can be used for modeling the observable geometry of a void in the foam. The results of the observations and calculations presented in [Luminet (2003)] suggest that there are observable geometrical patterns like regular figures in the distribution of luminous objects, and this is in accord with the foam concept. But it seems that the voids in a real foam hardly have a unique and fixed form. At the same time they can have a characteristic scale depending on the properties of the "liquid" they are made from. It means that the values of the world constants in our Universe are such that the bubbles-voids can exist and are observed on the cosmological scale.

When averaging on scales much larger than voids, the matter distribution is homogeneous and isotropic. This is not changed with the introduction of the phase space-time. All those anisotropic interactions are local and the average distribution of mass currents should also be homogeneous and isotropic.

As it was shown, the gravitational or, better say, gravito-inertial redshift could be an equivalent interpretation of the Hubble law which means that the whirl model is applicable to the Universe as a whole. It also naturally refers to the dark energy problem, now there is no need for any specific unknown interaction or essence, while the new constant of the theory has a natural meaning that does not demand any physical properties from the empty space. The ether even as physical vacuum can be investigated but is not needed to hold or rip the Universe altogether. The AGD cosmological picture which is based on the phase space-time model and involves tangential motions can be used instead of expanding Universe (another interpretation of redshift which is the main feature of it) and of the steady-state one (origination and dissipation of whirls) or can in a sense reconcile both of them, because the observational results can be either interpreted in a different but self-consistent way or preserve the interpretations they have today as formal constructions valid for different scales. The AGD approach preserves all the results of the classical GRT on the planetary system scale and removes the discrepancies between the theory and observations on the galactic scale. Now it requires the specific experiment able to reveal the influence of the anisotropy on the galactic scale.

Chapter 3

Optic-metrical parametric resonance - to the testing of the anisotropic geometrodynamics

3.1 Gravitation waves detection and the general idea of optic-metrical parametric resonance

In the previous chapter we saw that the observations that are the base of the AGD theory belong to the scale that is far beyond the scales that mankind has already mastered in its "laboratory practice", and this laboratory already includes the Solar system as a whole. If in order to test this theory, we want to ask a principally new question which has not yet arisen as a consequence of paradoxical but already achieved observational result, we have to think of new observations on the galactic and higher scales. The signals we definitely obtain from galaxy sources are particles and electromagnetic waves. There is also a hope that there exist the gravitational waves predicted by the GRT. But their experimental detection is a real challenge for experimentalists, since the possible observable effect is terribly small, the needed sensitivity is about 10^{-24}. Nevertheless, J.Weber undertook the experimental search for gravitational waves in the 60s and used massive bars as antennas. Their elastic properties were supposed to provide a resonance with the periodic stretch caused by the gravitation waves produced by some far away catastrophic events like star explosions. It is interesting and instructive to give a list of possible physical effects suggested to be used in the gravitational waves detection since then, and, evidently, the list could not be complete. The classification of the gravitational wave antennas types and methods of observation is given in [Gladyshev (2000)] and includes three groups: ground-based and space-based antennas and astronomical observations.

The ground-based antennas use the methods in which the influence of the gravitational wave upon the registration channel takes place inside a technical device or a special setup located on the Earth surface. From the point of view of action principles, we can speak about the following.

1) Resonance bar antennas [Weber (1960, 1961)]

2) Free masses interferometric antennas [Gladyshev (2000); Gertsenshtein and Pustovoit (1962); Drever (1983)]

3) Gravity radiation pressure [Tamello (1987)]

4) Memory effect [Braginsky and Grischuk (1985); Braginsky and Thorne (1987)]

5) Gravitational wave affecting electromagnetic field in matter [Hizhniakov (1988)]

6) Mössbauer effect [Callagari (1991)]

7) Gravitational wave affecting laser resonators [Iacopini (1979); Danileiko (1984)]

8) Gravitational wave affecting electric capacity [Wagoner *et al.* (1979)]

9) Gravitational wave energy transformation into the electromagnetic field energy [Teissier (1985); Sazhin (1982)]

10) Gravitational wave induction of the electromagnetic field in a superconductor [Jeeva (1995); Huei and Bo (1990)]

11) Combined principle gravitational wave antenna [Kulagin *et al.* (1986)]

12) Gravitational wave affecting elementary particles reactions in the accelerator [Weber (1994)]

The space-based antennas register the variation of the distance between the probe bodies on the scales comparable to the Solar system scale. Their principles of action are the following.

13) Gravitational wave affecting the propagation time of the electromagnetic wave sent to a space satellite [Helling (1983)]

14) Free masses interferometric antennas [Anderson (1987); Jafry *et al.* (1994)]

15) Doppler tracking a space ship [Bertotti (1983)]

Astronomical methods of the gravitational waves registration can be divided into three classes

16) Monitoring the properties of an astrophysical object under the gravitational waves action [Taylor and McCulloch (1980); Smith (1987); Denisov (1989)]

17) Tracing the radiation properties of atoms in the gravitational wave field [Leen *et al.* (1983); Fischer (1994)]

18) Gravitational wave affecting the propagation of the electromagnetic radiation [Fakir (1993); Allen (1989)]

The main experimental efforts are now focused upon the two basic directions: Weber type massive bar antennas and various interferometric antennas. Since Weber's times, the achieved sensitivity of bar antennas grew $10^4 - 10^5$ times, and the expense of the experiments grew correspondingly. In Soviet Union they made a tunnel in Mount Elbrus in order to isolate the experimental chamber from exterior noise. The most spectacular joint NASA-ESA project is the space-based LISA (Laser Interferometer Space Antenna) which will be the first dedicated gravitational-wave detector launched to space presumably in 2017. It will measure gravitational waves by monitoring the fluctuations in the relative distances between three spacecrafts that will fly along an Earth-like heliocentric orbit and be arranged in an equilateral triangle with 5-million-kilometer arms.

While the physicists are still doing their best in order to find a way to detect the gravitational waves in the direct measurement, the astronomers do not care too much about that and are pretty sure that the waves do exist. They have good reasons for that. The Nobel Prize winners R.Hulse and J.Taylor discovered a binary pulsar whose orbital period demonstrated [Taylor and McCulloch (1980)] the decrease coinciding with the theoretical prediction based on the energy loss caused by the gravitational waves emission within the accuracy of 10^{-3}. So, whatever new gravitation theory appears, it has to reproduce this result with none the worse accuracy. However, this does not stop the experimentalists in their efforts.

Of course, the gravitational waves - when registered and measured in one of the mentioned ways - could also be used to test the galaxy scale AGD theory. But what we need on this scale in order to learn more about its geometry, is not the properties of the gravitational waves as they are, but rather the characteristics of the distribution of their sources and the paths the waves follow when propagating through space. If we manage to perform such kind of measurement, and then test and see by what kind of a galaxy scale theory the coming signals are described, this could be an evidence pro or contra AGD. At the same time it would provide a new support to the fundamental relativity ideas concerning the gravitational radiation itself.

The general purposes of this chapter are: i) - to give the theory of the effect called optic-metrical parametric resonance (OMPR); ii) - to give the first observational results obtained in Pushchino radio astronomy observatory of the Russian Academy of Science; and iii) - to explain how the statistical analysis of data could be used in the investigation of the geometrical properties of our galaxy, Milky Way which is, by the way, a spiral one.

The way to these three goals could be a long one, since the theory of the OMPR involves the ideas of quantum optics, regards the behavior of an atom in the multi-component electromagnetic field, exploits the appearance of a non-trivial component in the space maser radiation, all this - in the field of the gravitational wave. And though it would be logical to move step by step and start with the optic-*mechanical* parametric resonance which is the predecessor of the optic-*metrical* parametric resonance, such start would bring us too much aside from the main topic related to gravitation, relativity and the possibility to detect the gravitational waves and use them to test the AGD. This is why the theory of optic-mechanical parametric resonance is moved to the Appendix, there one can find some more technical details. We will begin with the discussion of the physical meaning of optic-mechanical parametric resonance, then skip the theoretical details given in the Appendix, and pass to the analysis of the gravitational wave action on an atom with regard to our goal. After that we will obtain the conditions of the optic-metrical parametric resonance, discuss their physical meaning and then pass to items ii) and iii).

As we know, an atom does not interact with an arbitrary electromagnetic wave, but only with the one called resonant, whose photons have the same or almost the same energy as the difference between the atomic energy levels. The resonant electromagnetic field not only causes absorption and emission of photons by the atoms. It also acts on the atoms mechanically and makes them move. In multi-component fields, the magnitude of the force acting on the atoms changes, and also the very character of this action changes too. As a result, the whole dynamics of the atomic system becomes different. The physical reason of this is that the internal and the translational degrees of freedom of an atom cannot be generally disentangled, but are self-consistent.

Nowadays, the theoretical and experimental research dealing with the diffraction and interference caused by the quantum character of the translational motion of the massive particles has made essential progress. As can be seen from [Bonifacio *et al.* (1997a); Berman (1999); Bonifacio *et al.* (1997b); Moore and Meystre (1998); Berman *et al.* (1995)], the light field can present a diffraction grating for the atoms in resonance, and vice versa, the space distribution of the atoms that interact with the electromagnetic wave in a resonant way can cause the amplification of this wave. The radiation of the atoms located in the light lattice cells can reveal the collective effects. If now the space distribution of the atoms is modulated by some external reason, the amplification of the resonant electromagnetic wave would also be modulated. The last can be detected by the receiver as the modulation of the amplification coefficient of one of the components of the falling radiation spectrum. This is the idea underlying the effect of optic-*mechanical* parametric resonance. The atoms can be made vibrating by an acoustic field in the gas, or we can take ions instead of atoms and use the low-frequency electric field. But it turns out that the reason of atoms motion, the nature of the modulation of atoms distribution does not play any special role, all we need is the periodical change of the atom's velocity.

Let us briefly recollect what is the origin of the light pressure force acting on an atom. It is important that there are two mechanisms of photon emission, there could be spontaneous and stimulated emission. If there were no spontaneous emission, there would be no pressure. The falling and absorbed photon drives the atom, "takes up the trigger". Simultaneously, the atom obtains a certain momentum. Time starts to go. Finally, the trigger comes to an end, the lifetime of the excited state also ends, the atom throws away a photon in an arbitrary direction and obtains a recoil momentum in the arbitrary direction (opposite to that of the emitted photon). When these processes are repeated, no total recoil momentum is accumulated, because of the spontaneous character of the photon emission, whence the momenta obtained with photon absorption are summed. The resulting force produces the light pressure in the weak field.

Now let the field be strong. The photons of the external field have one and the same direction. When an atom is again driven up and the "trigger" is still on its way back to the end, the passing photon knocks atom down, and the atom is stimulated to emit a photon at the same direction as the passing photon had. The recoil momentum again has the exactly opposite direction. And every recoil momentum compensates the momentum obtained before with the absorption of a photon. No momentum is stored, no light pressure takes place. Instead of that, each atom moves periodically to and fro around its equilibrium position following the arrival rate of the incoming photons, i.e. in accordance with the so-called Rabi frequency, $\alpha = \frac{\mu E}{\hbar}$, characterizing the strength of external field (E is the electric stress, μ is the induced dipole moment). Hence, the atomic population oscillates with the same frequency. The mechanical oscillations of different atoms are not correlated, they have different phases and no regular vibrations occur.

The description of the atom dynamics in the resonant field can be found in many books, for example, in [Stenholm (1984)]. It is based on the use of the Bloch's equations for the density matrix components. Figure 3.1

illustrates the above said. The absorption coefficient of the weak probe wave that can be found as a result of the Bloch's equations solution has a bell-shape with the center on the resonance frequency.

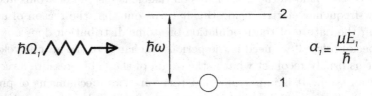

Fig. 3.1 TLA in the strong monochromatic resonant field.

Strictly speaking, since the spontaneous emission has a stochastic character, in some cases it can take place before the atom emits a stimulated photon. This means that there could be some pressure - much smaller than in the weak field case. Thus, paradoxically, the weak field produces stronger pressure than the strong one. It is also important to mention that the falling radiation has a certain line-width, the center of it coincides with the atomic frequency, and all those events of absorption and emission do not take an atom away from the resonance.

Let us now perform a kind of tuning to this vibration frequency. When an atom emits or absorbs a photon, its velocity changes and the Doppler effect causes the shift of its resonance frequency. If alongside with the photons of the strong field with the main frequency, there are present photons of an additional weak field whose frequency is slightly detuned from the main one in due way, Fig. 3.2, then this weak field will interact primarily with those atoms that have just lost or obtained an extra momentum.

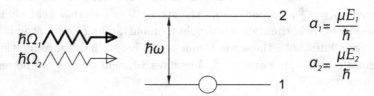

Fig. 3.2 TLA in the bichromatic resonant field.

The line center of the weak field is shifted relative to the line center of the strong field, and this increases the probability of interaction for the atoms moving in the same phase. They do not accumulate any momentum, but the absorption properties of the medium start to change with frequency related to the frequency shift of the additional wave. Effectively, this looks like an appearance of an oscillating force acting on atoms which does not perform any work. Absorption coefficient of the probe wave obtains the non-stationary components at the frequencies located symmetrically close to the resonance frequency. They depend on time and correspond to the change of the absorption to amplification and back.

Similarly, if the atoms vibrate in the direction of the wave vector of the external strong monochromatic field, Fig. 3.3, their velocities depend on time, and it causes the time dependent reconstruction of

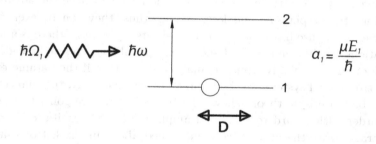

Fig. 3.3   Vibrating TLA in the strong monochromatic resonant field.

the probe wave scattered radiation spectrum, it also obtains the non-stationary component. The question arises, what will happen to the scattered radiation of the weak additional field, Fig. 3.2, regarded as a probe wave for the case Fig. 3.3, when certain conditions of the atom motion are fulfilled? For example, the distuning $\Omega_1 - \Omega_2$ could be set equal to vibration frequency, D, or both of them could be related to Rabi frequency of the strong field, α. All such cases could lead to specific situations called parametric resonances, they can be investigated by means of Bloch's equations solution with the help of special perturbation theory technic based on asymptotic expansion method. The calculations show that the result is the appearance of the signal with large amplitude.

The simultaneous effect of the weak additional field and oscillatory motion of atoms brings to mind the idea of the optic-*metrical* parametric resonance which constitutes the following. Suppose the distribution of atoms presents a space maser which emits the electromagnetic radiation with a narrow spectrum that can be registered and measured. We know a lot of such radio sources now. If the distribution of a space maser atoms and the radiation of this maser are modulated by the periodic gravitational wave, shall we be able to notice this if some conditions of parametric resonance are fulfilled? Obviously, the type of the gravitational wave corresponding to the wave from a catastrophic event, is hardly suitable, because we have to tune our set up, while the star catastrophe in space does not report its parameters beforehand. Therefore, we should look for the sources of the periodic gravitational waves emitted by the short-period binary stars or by rotating pulsars with the known periods. But the amplitudes of such waves are even less than those that we unsuccessfully try to detect for the last half a century. Nevertheless, such gravitational waves have the advantage of being strictly periodic and long lasting, thus, they can be even more suitable for the use in various resonance effects. Of course, there is a lot of limitations on the conditions of such effect possibility, and at first glance, they look discouragingly. But the analysis shows that if the parameters of the astrophysical system are chosen in an appropriate way, the chances are not so bad. Besides, theory shows that the effect we are going to use has zero order with regard to the small amplitude of the gravitational wave, - contrary to all the other effects mentioned above in the list of suitable phenomena. All of them are the first order effects in the powers of the gravitational wave amplitude. So, if certain conditions of the parametric resonance itself and the conditions on the parameters of the astrophysical system are fulfilled, a suitable component of maser spectrum will be periodically absorbed and amplified with the frequency related to the frequency of the gravitational wave source. The extremely low value of the gravitational wave amplitude does not prevent the possibility of parametric resonance. The resonant character of the effect and the nature of the detector, i.e. the atoms of the cosmic maser - quantum particles and not bars or mirrors - make the target signal comparable to the regular one coming from space maser. Therefore, it can be registered and measured by the regular radio telescope which is used to study the radio sources the number of which is large enough. It turned out that the OMPR signal has a specific character - *it is a sufficiently large amplitude periodical variation of the intensity of a certain component in the space maser spectrum.* Despite all the attempts, no other mechanism providing the same effect in a space maser radiation was suggested up to now.

The first encouraging observational results were obtained and reported recently [Siparov and Samodurov (2009a)]. Unfortunately, the rate of the new data arrival is very low, because the method is still poorly known to the astronomical society. When the amount of data is enough for the statistical analysis, such observations of various systems and the comparison of the results could reveal the specific geometrical properties of the Milky Way galaxy including its possible anisotropy as discussed in AGD.

3.1.1 *Space maser as a remote detector of gravitation waves*

In order to understand how to involve the gravitational waves into the practice of spectroscopic experiments described in the Appendix, let us have a look at Figs. 3.4–3.6. Figure 3.4 illustrates the idea of the traditional interferometric method which involves free masses and uses the local

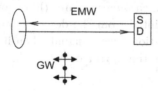

Fig. 3.4   Local detector of gravitational radiation.

setup. One of the masses is the block containing the electromagnetic signal source, S and the receiver/detector, D, the other one is the mirror that reflects the signal back to the detector. When a gravitational wave falls upon this system, the distance between the source/detector and the mirror starts to change, the phase of the signal starts to change too, and the hope is to measure this change. As it was mentioned, the amplitude of the gravitational wave is very small, and the demanded sensitivity is extremely high. This is naturally accompanied by the signal-to-noise problem which is also not so easy to solve. The theoretical foundation of this method is the geodesic declination equation describing the change of the distance between two free-falling bodies in the field of the gravitational wave.

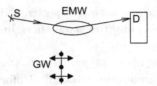

Fig. 3.5   Local detector of gravitational radiation with separated source and detector.

Let us first separate the blocks containing the source and the detector, Fig. 3.5. Then the phase shift appearing because of the gravitation wave that could be discovered in the previous case would be less. If there is an exterior source of the monochromatic radiation, then the falling gravitational wave would change the distance between the mirror and the detector, and there appear the frequency modulation that could be observed and measured in the received signal. The detection problem complicates: earlier we had the fixed phase difference between the wave emitted by the laser and the wave that reached the detector, and waited for its sudden variance with time. This disturbance was supposed to be not long lasting but still recognizable, it disturbs the situation that we organized ourselves by hands with the help of the monochromatic source (laser) of electromagnetic waves. Now the source is disjointed from the detector, and it is harder to recognize the possible gravitational wave signal when it appears. The detecting element in this device is the mirror or a combination of a mirror and the detector.

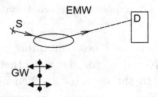

Fig. 3.6   Remote detector of gravitational radiation.

The next step is to pass from the local detector to the remote one, see Fig. 3.6, and let the gravitational wave fall upon the system consisting of a source of the electromagnetic signal and the mirror. However far is the detector receiving the signal, it will be able to analyze it and to discover the phase modulation (if any) in case the carrying signal presents a highly monochromatic one. So, in this kind of experiment we have to build the detector on the Earth and to find a source of a monochromatic radiation and a mirror somewhere in space.

Up to now the reason for the transformation of a local detector with two beams into a remote detector with one beam was unclear. But now we have in mind the OMPR effect, the existence of a possible parametric resonance based on a purely kinematic effect of the atomic behavior, and sufficiently high amplitude of the useful signal in comparison with the signal produced by the carrier, all this makes quite a hint.

There do exist the highly monochromatic sources of the electromagnetic radiation in space, they are space masers working on the meta-stable transitions in hydroxyl, water and some other molecules. These transitions and these states of molecules make them ideal objects to be modeled by the "two-level atom" (TLA) approximation that is commonly used in the theoretical spectroscopy and in optic-mechanical parametric resonance theory. When we try to obtain an atom or a molecule with only two working levels on Earth, we have to overcome lots of problems like high temperature and density of the gas of atoms. In space we have all this for free. The physical processes that make the space masers function are studied already for years and known to be complicated, but for our needs their physics is not so essential.

The usual sources of the gravitational radiation that they have in view when building super sensitive detectors are catastrophic events, but their occurrences are unpredictable, and the space maser observations often cannot be performed permanently for technical reasons. But there are periodic sources of gravitational waves like rotating neutron stars or close binary systems with short periods. These do produce the permanent gravitation radiation which is also monochromatic and can affect the maser cloud and oscillate the distance between its atoms and the distant observer who receives the signal emitted by them. The amplitude of the gravitational waves emitted by the periodical sources is very small, but it is the strict periodicity and not the amplitude that we need now when we think of the parametric resonance. Thus, the "mirror" is the atoms of the space maser itself, which means that we have found both elements of the device that we looked for in space. As to the detector on Earth, we don't even need to construct any new and super sensitive one. The radio telescopes that are used for the study of space masers will do all right, because the signal we expect to register is supposed to have the amplitude comparable to the signal produced by the main wave. This means that the signal-to-noise ratio problem is not so critical.

Finally, we see that the qualitative picture seems promising, but we have to make the quantitative analysis of a lot of details that concern this idea. The gravitation wave can affect

- the structure of atomic levels,
- the electromagnetic wave of the space maser itself, and
- the motion of atoms that form a maser.

The first two effects were separately and independently examined in literature among other methods suggested for the gravitation waves detection (see the list in the previous subsection). We also have to compare all three of them and make sure that they do not cancel out each other, and then study the possibility of a parametric resonance. We have to describe the conditions of the parametric resonance that becomes optic-metrical one and make sure that they do not contradict the corresponding properties of the known astrophysical systems. It would also be good to have a couple of examples of such systems.

3.1.2 *Atomic levels*

Having in view the future study of the possibility of a parametric resonance in the "atom+field" system, one should find out which term in the Bloch's equations describing the dynamics of the TLA will be affected by the possible changes in the structure of the atomic levels. And this is the induced dipole moment, μ, which is present in the definition of Rabi frequency. Our aim now is to evaluate the correction to it due to the interaction between the atom and the gravitational wave [Siparov (2004)].

The state of an atom can be described with the help of Schrödinger equation

$$i\hbar \frac{d}{dt}\Psi = H\Psi \qquad (3.1)$$

The Hamiltonian H here is

$$H(t) = H_0 + H_g(t) + H_{em}(t) \qquad (3.2)$$

where H_0 is the atomic Hamiltonian which is independent of time, $H_g(t)$ is the perturbation of the atomic Hamiltonian due to interaction between the gravitational wave and the atom, $H_{em}(t)$ describes the dipole interaction between the atom and the electromagnetic wave. The last term has the usual form

$$H_{em}(t) = \mu E \tag{3.3}$$

Here μ is, strictly speaking, a matrix element of the induced dipole moment operator, $E = E(t)$ is the electric stress of the electromagnetic wave. The distance between the resonant levels corresponding to the electromagnetic interaction is much larger than the distance between the sublevels that could appear because of the gravitational wave. This means that the dipole moment, μ can be calculated with the help of the perturbation theory using the corrections to the wave functions that can appear because of the gravitational wave action.

Therefore, let us first find the wave functions of the perturbed Hamiltonian

$$i\hbar\frac{d}{dt}\Psi^{(1)} = H_1\Psi = [H_0 + H_g(t)]\Psi^{(1)} \tag{3.4}$$

Fortunately, the expression for the gravitational wave induced perturbation was already found in [Leen *et al.* (1983)], where the shifts induced by space-time curvature arising from gravitational waves were investigated. The effect was studied for low-lying and highly excited states of atomic hydrogen, but here we are interested not in the detailed description and not in the possibilities to register the first-order effect as it was presumed in [Leen *et al.* (1983)], but in the evaluating of the perturbation order, because we want to know how can it affect the parametric resonance. The relativistic calculation performed in [Leen *et al.* (1983)] for the Hamiltonian perturbation gives

$$H_g = mD\left(\frac{GL_g}{4c^3}\right)^{1/2}[\epsilon_{ij}^+ + \epsilon_{ij}^\times]X^i X^j \frac{\cos(K_g r_{gs} - Dt)}{r_{gs}} \tag{3.5}$$

Here D is the gravitational wave frequency, m is the electron mass, G is the gravitation constant, L_g is the full energy flow from a gravitational wave source, c is the light speed, X^i are the coordinate components in the frame corresponding to the atomic center of inertia, K_g is the wave vector of the gravitational wave, r_{gs} is the distance from the atom to the gravitational wave source, $\epsilon_{ij}^+ = \begin{pmatrix} 1 & 0 \\ 0 & -1 \end{pmatrix}, \epsilon_{ij}^\times = \begin{pmatrix} 0 & 1 \\ 1 & 0 \end{pmatrix}$ are the unity polarization tensors with the non-vanishing components in the directions orthogonal to the gravitational wave propagation.

Equation (3.5) may be rewritten as follows

$$H_g = F^{(g)}e^{-iDt} + G^{(g)}e^{iDt} \tag{3.6}$$

$$F^{(g)} = \frac{mD}{2r_S}\left(\frac{GL_g}{4c^3}\right)^{1/2}[\epsilon_{ij}^+ + \epsilon_{ij}^\times]X^iX^je^{iK_g r_S}$$

$$F_{nm}^{(g)} = \int (\Psi_n^{(0)})^*F^{(g)}(\Psi_m^{(0)})dq; \text{ and } F_{nm}^{(g)} = G_{mn}^{(g)*}$$

where $F^{(g)}$ and $G^{(g)}$ are time independent operators. Since $F^{(g)}$ is obviously small, let us use the perturbation theory for the case where the perturbations are periodic functions of time [Landau and Lifshits (1963)]. The first approximation for the wave functions gives

$$\Psi_n^{(1)} = \Psi_n^{(0)} + \sum_k a_{kn}(t)\Psi_k^{(0)} \tag{3.7}$$

where $\Psi_m^{(0)}$ are the wave functions of the unperturbed Hamiltonian, and

$$a_{kn} = -\frac{F_{kn}^{(g)}e^{i(\omega_{kn}-D)t}}{\hbar(\omega_{kn}-D)} - \frac{F_{nk}^{(g)*}e^{i(\omega_{kn}+D)t}}{\hbar(\omega_{kn}+D)} \tag{3.8}$$

Equations (3.7-3.8) describe the wave functions of an atom in the field of the gravitational wave.

The expression for the induced dipole moment is

$$\mu = (1+a_{11}+a_{22})\mu_0 + e\int\sum_{k\geq 2}a_{k1}(\Psi_n^{(0)})^*r(\Psi_2^{(0)})dq + e\int\sum_{k\neq 2}a_{k2}(\Psi_1^{(0)})^*r(\Psi_k^{(0)})dq$$

$$\tag{3.9}$$

where μ_0 is the dipole moment calculated with the wave functions $\Psi_1^{(0)}, \Psi_2^{(0)}$ of the unperturbed Hamiltonian. As follows from eq. (3.8), the expressions for a_{11} and a_{22} are

$$a_{11} = a_{22} = \frac{1}{\hbar D}(F_{11}e^{-iDt} - F_{22}^*e^{iDt}) \tag{3.10}$$

Now let only one pair of levels take part in the electromagnetic interaction, that is let us use the two-level model of an atom. Let the frequency of the external monochromatic field correspond to the 1-2 transition. Then eq. (3.8) means that the two last terms in eq. (3.9) are proportional to the factors oscillating at high frequency; therefore they may be omitted in the further calculation of the dipole moment.

The evaluation of the correction to the dipole moment caused by the gravitational wave gives

$$a_{11} = \frac{1}{\hbar D}F \simeq \frac{1}{\hbar D}\frac{mD}{2r_{gs}}\left(\frac{GL_g}{4c^3}\right)^{1/2} r_a^2 \qquad (3.11)$$

where r_a is the atom radius. To calculate the gravitational energy flow, L_g, falling on the atom, let us use the formula [Amaldi and Pizzella (1979)]

$$L_g = \frac{288GI^2g_e^2D^6}{45c^5} \qquad (3.12)$$

characterizing, for example, the neutron star with the gravitational ellipticity, g_e, and with the moment of inertia, I. Since the expression for the dimensionless gravitational wave amplitude h is given by [Thorne (1987)]

$$h = \frac{GMR_s^2D^2g_e}{c^4r_{gs}} \qquad (3.13)$$

where R_s is the radius of the neutron star, eq. (3.11) gives

$$a_{11} = \frac{1}{\hbar D}12\sqrt{\frac{2}{5}}D^2mr_a^2h = 12\sqrt{\frac{2}{5}}\frac{mr_a^2}{\hbar}Dh \qquad (3.14)$$

The correction to the induced dipole moment is expectedly linear in h, the coefficient is defined by eq. (3.14) in which m and r_a are the atomic mass and radius. Thus, we see that the gravitation wave induced correction to the dipole moment is proportional to $10^{-23}Dh$, where D is the frequency of the gravitational wave and h is its dimensionless amplitude.

3.1.3   *Eikonal*

The most natural way to describe the action of the gravitational wave on the electromagnetic wave is to use the eikonal equation, i.e. consider the phase, ψ of the propagating wave $e^{i\psi}$. The electromagnetic wave eikonal equation in flat space-time can be written as

$$\frac{\partial\psi}{\partial x_i}\frac{\partial\psi}{\partial x^j} = 0 \qquad (3.15)$$

In the simplest case the eikonal is constant

$$\psi = \omega t - k_\alpha x^\alpha \qquad (3.16)$$

where ω and $\mathbf{k}$ are the frequency and the wave vector of the electromagnetic wave.

Rewriting eq. (3.15) for the case when there is a gravitation field and the space-time is curved, one gets [Landau and Lifshitz (1967)]

$$g^{ik}\frac{\partial \psi}{\partial x^i}\frac{\partial \psi}{\partial x^k} = 0 \qquad (3.17)$$

where $g^{ik}(x)$ is the metric tensor. In the case when the gravitational wave propagates in the region that is far from masses, then if the Ox (i.e. x^1) direction corresponds to the direction of the gravitational wave propagation, the linearized metric tensor has the form

$$g^{ik} = \begin{pmatrix} 1 & 0 & 0 & 0 \\ 0 & -1 & 0 & 0 \\ 0 & 0 & -1 + h\cos\frac{D}{c}(x^0 - x^1) & 0 \\ 0 & 0 & 0 & -1 - h\cos\frac{D}{c}(x^0 - x^1) \end{pmatrix} \qquad (3.18)$$

Strictly speaking, this expression corresponds to the gravitational wave in vacuum. It is used here for the same reasons as in the traditional (interferometric) approach, when the laser fields are even larger than that in a cosmic maser and the masses of the parts of the instrument are larger than that of an atom. These reasons take into account the fact that the energy density in the linearized theory of gravitation is of a higher order (see the corresponding discussion in [Isaacson (1968); Srevin *et al.* (2001); Papadopoulos (2002a,b)]).

In eq. (3.18) h is the dimensionless amplitude of the gravitational wave, $h << 1$, and the gravitational wave polarization is considered to be circular (not elliptical) for simplicity. The eikonal equation has the explicit form

$$(\frac{\partial \psi}{\partial x^0})^2 - (\frac{\partial \psi}{\partial x^1})^2 - (\frac{\partial \psi}{\partial x^2})^2 - (\frac{\partial \psi}{\partial x^3})^2 + h[(\frac{\partial \psi}{\partial x^2})^2 - (\frac{\partial \psi}{\partial x^3})^2]\cos\frac{D}{c}(x^0 - x^1) = 0 \qquad (3.19)$$

Since $h << 1$, the eikonal can be expressed as

$$\psi = f_{eik} + h g_{eik} \qquad (3.20)$$

where f_{eik} and g_{eik} are new functions. If we substitute eq. (3.20) into eq. (3.19), neglect all the terms containing h in the powers higher than one and regard separately the equations with and without h, then

$$\left(\frac{\partial f_{eik}}{\partial x^0}\right)^2 - \sum_i \left(\frac{\partial f_{eik}}{\partial x^i}\right)^2 = 0 \tag{3.21}$$

$$2\frac{\partial f_{eik}}{\partial x^0}\frac{\partial g_{eik}}{\partial x^0} - 2\frac{\partial f_{eik}}{\partial x^i}\frac{\partial g_{eik}}{\partial x^i} + \left[\left(\frac{\partial f_{eik}}{\partial x^2}\right)^2 - \left(\frac{\partial f_{eik}}{\partial x^3}\right)^2\right]\cos\frac{D}{c}(x^0 - x^1) = 0 \tag{3.22}$$

Equation (3.21) gives

$$\omega^2 = c^2 k^2 \tag{3.23}$$

which means that the principal term of the electromagnetic wave eikonal in the gravitational wave field corresponds to a flat wave. Substituting this result into eq. (3.22), one gets the equation for the function g_{eik}

$$2\frac{\omega}{c}\frac{\partial g_{eik}}{\partial x^0} + 2k_1\frac{\partial g_{eik}}{\partial x^1} + 2k_2\frac{\partial g_{eik}}{\partial x^2} + 2k_3\frac{\partial g_{eik}}{\partial x^3} + (k_2^2 - k_3^2)\cos\frac{D}{c}(x^0 - x^1) = 0 \tag{3.24}$$

This is a first-order quasi-linear non-homogeneous partial differential equation. Its solution is

$$g_{eik} = -c^2\frac{k_2^2 - k_3^2}{2\omega D}\cos\frac{D}{c}(x^0 - x^1) + \varphi(k_1 x^0 - \frac{\omega}{c}x^1, k_2 x^1 - k_1 x^2, k_3 x^2 - k_2 x^3) \tag{3.25}$$

where φ is an arbitrary function defined by the boundary conditions. This function can be taken zero for simplicity.

Finally, the eikonal for the case when the gravitational wave acts on the electromagnetic wave can be written as

$$\psi = \frac{\omega}{c}x^0 - k_\alpha x^\alpha - hc^2\frac{k_2^2 - k_3^2}{2\omega D}\cos\frac{D}{c}(x^0 - x^1) \tag{3.26}$$

These calculations show that the gravitational wave action upon the electromagnetic wave is equivalent to the phase modulation of the electromagnetic wave, with the frequency of modulation equal to that of the gravitational wave and with the amplitude which is linear in h and is proportional to the ratio of the frequencies of the electromagnetic and gravitational waves. Transforming the phase modulated wave into the superposition of amplitude modulated waves, we get the expression for the stress of the flat electromagnetic wave propagating in the Oy direction

$$E = E_1 \cos(\Omega_1 t - ky) + \frac{1}{2}E_2 \cos[(\Omega_1 - D)t] - \frac{1}{2}E_2 \cos[(\Omega_1 + D)t - 2ky] \quad (3.27)$$

$$E_2 = h\frac{\omega}{D}E_1 \quad (3.28)$$

3.1.4   *Motion of a particle*

The equation they use when dealing with interferometric methods of the gravitational waves detection is the equation of geodesics deviation. But now we cannot use it, because the distance between "two free bodies" is not small enough. The object on which the gravitational wave acts is very far from the detector, that is from the receiver of the signal that this particle emits. The behavior of a particle in the field of the gravitational wave can be described with the help of the geodesics equation. Our aim is to evaluate the amplitude of the particle's periodic displacement orthogonal to the wave vector of the gravitational wave and its influence on the electromagnetic wave modulation.

Let a motionless particle which is located in the coordinate origin transmit periodic electromagnetic signals to the remote detector (emit photons). Then the intervals of time between the subsequently emitted signals, Δt_p, are equal to the intervals of time, Δt_d, between the subsequent signals received by the detector. But if the particle vibrates in parallel to the emitted electromagnetic wave according to a sine law, (D is the frequency of these vibrations, η is the amplitude), then the intervals of time between the subsequently received signals are expressed by the intervals of time between the subsequently emitted signals by

$$\Delta t_p = \Delta t_d + \frac{k\eta}{\omega}\sin(D\Delta t_p) \quad (3.29)$$

This means that the phase suffers a periodic time shift.

Let the location of the atom be at the coordinate origin, Ox-axis is anti-parallel to the direction of the wave vector of the gravitational wave, the Oy-axis is the direction of the wave vector of the electromagnetic wave coinciding with the direction at the Earth (see Fig. 3.7).

Let the maser field be considered absent for a moment. Then the particle in the gravitation field moves along the geodesics

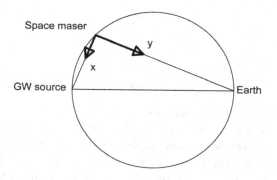

Fig. 3.7   Configuration of the astrophysical system.

$$\frac{d^2 x^i}{ds^2} + \Gamma^i_{kl}\frac{dx^k}{ds}\frac{dx^l}{ds} = 0 \tag{3.30}$$

The Christoffel coefficients Γ^i_{kl} can be expressed with the help of the metric tensor components

$$\Gamma^i_{kl} = \frac{1}{2}g^{im}\left(\frac{\partial g_{mk}}{\partial x^l} + \frac{\partial g_{ml}}{\partial x^k} - \frac{\partial g_{kl}}{\partial x^m}\right) \tag{3.31}$$

Calculating the coefficients Γ^i_{kl} with regard to eq. (3.18) for the gravitational wave metric tensor, one can substitute them all into eq. (3.30), neglect the terms in which the power of h is higher than one, and obtain the following system of equations

$$t'' + \frac{1}{2}h\frac{D}{c^2}(y'^2 - z'^2)\sin\frac{D}{c}(ct - x) = 0 \tag{3.32}$$

$$x'' + \frac{1}{2}h\frac{D}{c}(y'^2 - z'^2)\sin\frac{D}{c}(ct - x) = 0$$

$$y'' + \frac{1}{2}h\frac{D}{c}y'(ct' - x')\sin\frac{D}{c}(ct - x) = 0$$

$$z'' + \frac{1}{2}h\frac{D}{c}z'(ct' - x')\sin\frac{D}{c}(ct - x) = 0$$

Here the common notation $t = x^0/c, x = x^1, y = x^2, z = x^3$ was used again and $w' \equiv \frac{dw}{ds}$. The first two equations of the system eq. (3.32) give

$$ct'' = x'' \tag{3.33}$$

which leads to

$$x = ct + as + b \tag{3.34}$$

where a and b are arbitrary constants.

Let us analyze the meaning of the last expression. At the moment $t = 0$ the atom is in the coordinate origin of the frame which is motionless relative to the gravitational wave source, i.e. $x = 0, s = 0$. That is why $b = 0$. If there were no gravitation field, the atom would never leave its place, therefore, $s = -ct/a, a = 1$. But, since there is a gravitational wave, the value of a in the linear approximation might differ from unity by a value proportional to h. That is $a = 1 + ha_1$ where $a_1 = O(1)$, and eq. (3.34) gives

$$x = ct + (1 + ha_1)s \tag{3.35}$$

Substitute this expression into the third equation of the system eq. (3.32) corresponding to the direction towards the Earth. Neglecting the terms in which the power of h is higher than unity, one gets

$$y'' + h\frac{D}{c}y' \sin[\frac{D}{c}(1 + ha_1)s] = 0 \tag{3.36}$$

The solution of this equation with regard to $h \ll 1$ is

$$y = u_0 s + h\frac{1}{(1 + ha_1)^2}\frac{c}{D}\sin[\frac{D}{c}(1 + ha_1)s] + d \tag{3.37}$$

where u_0 and d are the integration constants. Since $s = \frac{x-ct}{1+ha_1}$, and at $t = 0$, the particle is at the coordinates origin

$$y = \frac{u_0(x - ct)}{1 + ha_1} - h\frac{1}{(1 + ha_1)^2}\frac{c}{D}\sin\frac{D}{c}(ct - x) \tag{3.38}$$

There is a sine dependence of the space coordinate of the atom parallel to the wave vector of the electromagnetic wave on the time coordinate of the atom. The amplitude η of the corresponding oscillations of the particle in the gravitational wave field is

$$\eta = h\frac{1}{(1+ha_1)^2}\frac{c}{D} \simeq h\frac{c}{D} \tag{3.39}$$

Let us compare these results with eq. (3.29). The particle that emits a signal is influenced by the gravitational wave in such a way that there is a periodical phase shift in the signals received by the detector. The frequency of the shift is equal to the gravitational wave frequency, and its amplitude is equal to

$$k\eta = \frac{\omega}{c}h\frac{c}{D} = h\frac{\omega}{D} \tag{3.40}$$

The comparison of eqs.(3.38), (3.40) with eqs.(3.29), (3.26), (3.27) shows that both frequencies and amplitudes of the gravitational wave action on the maser electromagnetic field and on the mechanical vibration of the atom are the same.

3.2 OMPR in space maser

Equation (3.14) shows that the gravitational wave action on the reconstruction of the atomic levels that could lead to the change in the dipole interaction is negligible in comparison with the other two effects, defined by eqs. (3.27) and (3.40). Therefore, no additional terms are needed in the expression for the dipole moment in Bloch's equations for the density matrix components. The corrections that should be present in the Bloch's equations describe the deformation of the strong electromagnetic wave emitted by a maser and the oscillatory motion of atoms. Several examples of the treatment of the similar situations are given in the Appendix.

The method we are going to apply (and modify with regard to the presence of the parametric resonance) is the so-called asymptotic expansion method, (see also the Appendix). The only demand this method has is the presence of a small parameter, but the value of this parameter and even the order of it is arbitrary. This arbitrariness starts to play an important role when the mathematical solution is used for the interpretation of a physical problem. Estimating the value of the small parameter, we have to consider many details and parameters of a concrete astrophysical system that are usually known only approximately, and this could subsequently lead to disappointments in the observational testing. But this is the cost we have to pay when passing to a principally new approach to the problem of the GW detection.

Bloch's equations for the components of the density matrix describing the dynamics of a two-level atom in the strong field of electromagnetic radiation have the form [Stenholm (1984)]

$$\frac{d}{dt}\rho_{22} = -\gamma\rho_{22} + 2i\frac{\mu E_1}{\hbar}\cos(\Omega_1 t - ky)(\rho_{21} - \rho_{12}) \qquad (3.41)$$

$$(\frac{\partial}{\partial t} + V\frac{\partial}{\partial y})\rho_{12} = -(\gamma_{12} + i\omega)\rho_{12} - 2i\frac{\mu E_1}{\hbar}\cos(\Omega_1 t - ky)(\rho_{22} - \rho_{11})$$

$$\rho_{22} + \rho_{11} = 1; \rho_{12} = \overline{\rho_{21}}$$

Here ρ_{22} and ρ_{11} are the populations of the levels, ρ_{12}, ρ_{21} are the polarization terms, γ is the decay rate, γ_{12} is the transversal decay rate (when the lower level is the ground one, $\gamma_{12} = \gamma/2$); $\alpha_1 = \frac{\mu E_1}{\hbar}$ is the Rabi parameter of the strong electromagnetic field; k is the wave vector. In our case eq. (3.41) must be rewritten with regard to the results of the previous section. The weak gravitational perturbation of the monochromatic electromagnetic wave produces the phase modulation that can be described in terms of the amplitude modulation according to eq. (3.27)

$$E(t) = E_1\cos(\Omega_1 t - ky) + \frac{1}{2}E_2\cos((\Omega_1 - D)t) \qquad (3.42)$$

$$-\frac{1}{2}E_2\cos[(\Omega_1 + D)t - 2ky]$$

$$E_2 = h\frac{\omega}{D}E_1$$

In these terms it means that the electromagnetic wave acquires the weak sidebands. The first term corresponds to the regular maser radiation which is spectroscopically strong. Here $\Omega_1 = \Omega \cong \omega$ is the frequency of the maser field which is resonant to the atomic transition frequency. The second and third (additional) components appear due to the gravitational wave action and are much weaker than the first one. It is important to notice that though the terms corresponding to the weak electromagnetic waves are small, they are not vanishingly small. In fact, the ratio of amplitudes of the weak and strong waves is of the same order as the ratio of the decay rate and the Rabi frequency of the strong wave. This demand indicates what atomic transitions and what conditions might take part in the effect.

The wave vector of the gravitational wave is considered to be orthogonal to the wave vector of the electromagnetic wave. In view of the results of the previous section, the gravitational wave action on the atom motion can be interpreted as the appearance of a periodical component V_{osc} in the component of the atomic velocity parallel to the wave vector of the electromagnetic wave. According to eq. (3.38)

$$V_{osc} = -V_1 \cos Dt \qquad (3.43)$$
$$V_1 = hc$$

In this case the velocity of an atom becomes

$$V = V_0 - V_1 \cos Dt \qquad (3.44)$$

and this must be accounted for in the expression $\frac{\partial}{\partial t} + V\frac{\partial}{\partial y}$. Here it is important that this field can be described classically. The lower level of the atom is considered to be the ground one. Then the Bloch's equations become

$$\frac{d}{dt}\rho_{22} = -\gamma\rho_{22} + 2i[\alpha_1 \cos(\Omega_1 t - k_1 y) + \frac{1}{2}\alpha_2 \cos((\Omega_1 - D)t) \qquad (3.45)$$

$$-\frac{1}{2}\alpha_2 \cos[(\Omega_1 + D)t - 2ky]](\rho_{21} - \rho_{12})$$

$$(\frac{\partial}{\partial t} + V\frac{\partial}{\partial y})\rho_{12} = -(\gamma_{12} + i\omega)\rho_{12} - 2i[\alpha_1 \cos(\Omega_1 t - k_1 y) + \frac{1}{2}\alpha_2 \cos(\Omega_1 t - Dt)$$

$$-\frac{1}{2}\alpha_2 \cos[(\Omega_1 + D)t - 2ky]](\rho_{22} - \rho_{11})$$

$$\rho_{22} + \rho_{11} = 1$$

Here $\alpha_2 = \frac{\mu E_2 h\omega}{D\hbar}$ is Rabi parameter of the additional wave; the wave vectors of both waves components can be taken equal to k. The usual way to deal with the problem like that expressed by eq. (3.45) is to neglect the swiftly oscillating terms like $e^{i(\Omega_1 t - ky)}$, but preserve the slow oscillating ones whose frequencies are much less than that of the atomic transition. This is achieved by the use of the rotating wave approximation (RWA). The RWA substitutions give

$$\rho_{21} = R_{21}e^{i(\Omega_1 t - ky)} \qquad (3.46)$$
$$\rho_{12} = R_{12}e^{-i(\Omega_1 t - ky)}$$
$$\rho_{22} = 2^{-1/2}\widetilde{\rho_{22}}$$

Rewriting eqs. (3.45), one gets

$$\frac{d}{dt}2^{-1/2}\widetilde{\rho_{22}} = -2^{-1/2}\gamma\widetilde{\rho_{22}} + 2i[\alpha_1\frac{1}{2}(R_{21} - R_{12}) \tag{3.47}$$

$$+\frac{1}{2}\alpha_2\frac{1}{2}(R_{21}e^{i(Dt-ky)} - R_{12}e^{-i(Dt-ky)})$$

$$-\frac{1}{2}\alpha_2\frac{1}{2}(R_{21}e^{-i(Dt-ky)} - R_{12}e^{i(Dt-ky)})]$$

$$\frac{\partial}{\partial t}R_{12} = -\{\gamma_{12} + i\omega - i\Omega_1 + ik[V_0 + V_1\sin(Dt - ky)]\}R_{12}$$

$$-2i[\alpha_1\frac{1}{2} + \frac{1}{2}\alpha_2\frac{1}{2}e^{i(Dt-ky)} - \frac{1}{2}\alpha_2\frac{1}{2}e^{-i(Dt-ky)}](2^{1/2}\widetilde{\rho_{22}} - 1)$$

The goal now is to obtain the expression for the absorption coefficient of the additional electromagnetic wave corresponding to the gravitational wave. As it is known [Stenholm (1984)], absorption coefficient is proportional to the imaginary part $Im(R_{21})$ of the polarization term. Let us introduce the notation:

$\alpha_1\sqrt{2}t \equiv \tau$ - normalized time measured in Rabi periods;

$\frac{\gamma}{\sqrt{2}\alpha_1} \equiv \Gamma\varepsilon; \varepsilon$ - small parameter characterizing the strength of the field, $\Gamma = O(1)$;

$\frac{\alpha_2}{\alpha_1} \equiv a\varepsilon; a = O(1)$ describes the presence of the weak additional wave;

$\frac{\omega-\Omega_1}{\sqrt{2}\alpha_1} \equiv \delta$ - normalized detuning measured in Rabi frequencies;

$\frac{k}{\sqrt{2}\alpha_1} \equiv \kappa$ - normalized wave vector of the electromagnetic wave;

$\frac{D}{\alpha_1\sqrt{2}} \equiv \delta_d$ - normalized frequency of atom oscillations measured in Rabi frequencies; at the same time, δ_d is exactly the value of the normalized distuning between the main wave (maser radiation) and the effective additional weak waves related to the gravitational wave action on the electromagnetic wave;

$V_1 = v_1\varepsilon; v_1 = O(1)$ - oscillating velocity parameter

Now divide eqs. (3.47) by α_1 and by $\alpha_1\sqrt{2}$. Then they take the form

$$\frac{d}{d\tau}\widetilde{\rho_{22}} = -\Gamma\varepsilon\widetilde{\rho_{22}} - iR_{12} - a\varepsilon\sin(Dt - ky)R_{12} + iR_{21} \tag{3.48}$$

$$-a\varepsilon\sin(Dt - ky)R_{21}$$

$$\frac{\partial}{\partial\tau}R_{12} = -i\widetilde{\rho_{22}} + a\varepsilon\sin(Dt - ky)\widetilde{\rho_{22}} - i(\delta + \kappa V_0)R_{12} - \frac{1}{2}\Gamma\varepsilon R_{12}$$

$$-i\kappa\varepsilon v_1\sin(Dt + ky)R_{12} + i2^{-1/2} - a2^{-1/2}\varepsilon\sin(Dt - ky)$$

$$\frac{\partial}{\partial\tau}R_{21} = i\widetilde{\rho_{22}} + a\varepsilon\sin(Dt - ky)\widetilde{\rho_{22}} + i(\delta + \kappa V_0)R_{21} - \frac{1}{2}\Gamma\varepsilon R_{21}$$

$$+i\kappa\varepsilon v_1\sin(Dt + ky)R_{21} - i2^{-1/2} - a2^{-1/2}\varepsilon\sin(Dt - ky)$$

Let us choose the origin in such a way that ky is equal to zero. Now rewrite the last equation in the matrix form

$$\frac{\partial}{\partial \tau}\mathbf{W} = (\mathbf{Q}_0 + \varepsilon \mathbf{Q}_1(\tau))\mathbf{W} + \mathbf{C}(\tau) \tag{3.49}$$

where

$$\mathbf{Q}_0 = i \begin{pmatrix} 0 & -1 & 1 \\ -1 & -\sigma & 0 \\ 1 & 0 & \sigma \end{pmatrix}; \mathbf{W} = \begin{pmatrix} \tilde{\rho}_{22} \\ R_{12} \\ R_{21} \end{pmatrix}; \tag{3.50}$$

$$\mathbf{C} = i2^{-1/2} \begin{pmatrix} 0 \\ 1 + i\varepsilon a \sin \delta_d \tau \\ -1 + i\varepsilon a \sin \delta_d \tau \end{pmatrix}; \tag{3.51}$$

$$\mathbf{Q}_1 = \begin{pmatrix} -\Gamma & -a \sin \delta_d \tau & -a \sin \delta_d \tau \\ a \sin \delta_d \tau & -\frac{1}{2}\Gamma - i\kappa v_1 \sin(\delta_d \tau) & 0 \\ a \sin \delta_d \tau & 0 & -\frac{1}{2}\Gamma + i\kappa v_1 \sin(\delta_d \tau) \end{pmatrix}.$$

$$\sigma = \delta + \kappa v; v = V_0$$

In these matrices, σ accounts for the detuning, δ, and for Doppler shift, κv_0, and these parameters enter matrix $\mathbf{Q}_0$; the perturbations stemming from the additional waves distuned by δ_d from the main frequency enter the non-diagonal components of matrix $\mathbf{Q}_1$ and the non-homogeneous term $\mathbf{C}$; and the perturbations stemming from the atom vibrations that also have frequency, δ_d enter the diagonal terms of matrix $\mathbf{Q}_1$. The method of solution of the systems of this type with regard to the parametric resonance is given in more detail in the Appendix.

The parametric resonance condition is: the normalized detuning of the strong wave from the frequency of the atomic transition with regard to the Doppler shift is equal to the distuning between the strong and the weak waves

$$F = \delta_d \tag{3.52}$$

where $F = \sqrt{\sigma^2 + 2}$. In the initial physical notation, this means

$$(\omega - \Omega_1 + kv_0)^2 + 4\alpha_1^2 = D^2 \tag{3.53}$$

So, if we consider that the resonance takes place, i.e. $\omega - \Omega_1 + kv_0 \sim 0$, then the frequency condition for the parametric resonance is

$$D = 2\alpha_1 \tag{3.54}$$

that is the frequency of the gravitational wave must be equal to the doubled Rabi frequency of the maser radiation which is proportional to the intensity of the maser. The intensity grows with distance from the center of a maser towards Earth, that is the region where the parametric resonance condition is fulfilled could be present not only on the border closest to the Earth, but in the depth of maser cloud as well. Thus, in order to choose an astrophysical system suitable for observations, we have to compare frequency (of the source of the gravitational waves) and intensity (of the maser) expressed in the frequency units.

The further calculation of vector $\mathbf{W}(\tau)$ goes along the lines described in Appendix and ends with the integration

$$\mathbf{W}(\tau) = \mathbf{U}\exp(id\tau)\exp(\varepsilon\mathbf{K}\tau)\int_0^\tau \exp(-\varepsilon\mathbf{K}x)\exp(-idx)\mathbf{U}^{-1}\mathbf{C}(x)dx \tag{3.55}$$

where $\mathbf{U}, \mathbf{d}$ and $\mathbf{K}$ are the auxiliary matrices that are needed to obtain the

solution of our problem with the help of perturbation theory methods. Let us give the expressions for these matrices and mention their properties that were used in the process of solution. The idea of the solution is the same as in more simple problems discussed in the Appendix.

Matrix $\mathbf{U}$ is defined in such a way that it provides $U^{-1}Q_0 U \equiv E = diag\{0, iF, -iF\}$, that is

$$U = \frac{1}{F}\begin{pmatrix} \sigma & 1 & 1 \\ -1 & -\frac{1}{F+\sigma} & \frac{1}{F-\sigma} \\ -1 & \frac{1}{F-\sigma} & -\frac{1}{F+\sigma} \end{pmatrix}$$

and $F = \sqrt{\sigma^2 + 2} \Rightarrow F^2 - \sigma^2 = 2; \frac{F-\sigma}{2} = \frac{1}{F+\sigma}$. In order to follow the tuning at the parametric resonance eq. (3.52), the tuning parameter $\nu = O(1)$ was introduced in such a way that

$$F = \delta_d + \varepsilon\nu$$

and the following designations were used for the exponents

$$e^{iF\tau} = A_1\beta; \beta = \exp(i\varepsilon\nu\tau)$$

Matrix $\mathbf{K}$ is constructed in the following way

$$K = \begin{pmatrix} 1 & 0 & 0 \\ 0 & \beta & 0 \\ 0 & 0 & \beta^{-1} \end{pmatrix} \exp\{-\mathbf{E}\tau\} \mathbf{M}_1(\tau) \exp\{\mathbf{E}\tau\} \begin{pmatrix} 1 & 0 & 0 \\ 0 & \beta^{-1} & 0 \\ 0 & 0 & \beta \end{pmatrix}$$

where $\mathbf{M}_1(\tau) = \mathbf{U}^{-1}\mathbf{Q}_1(\tau)\mathbf{U}$. Therefore,

$$K = \frac{1}{F^2} \begin{pmatrix} -\Gamma\left(\sigma^2 + 1\right) & i(H - G) & -i(H - G) \\ -i(H + G) & -B + i\nu & 0 \\ i(H + G) & 0 & -B - i\nu \end{pmatrix}$$

where the designations are $B \equiv \frac{1}{2}\Gamma\left(F^2 + 1\right); H \equiv \frac{1}{2}\left(2a\sigma^2 - 4a\right); G = \kappa v_1 F$ and

$$\det K = -\frac{\Gamma\sigma^2 B^2 + \Gamma\sigma^2\nu^2 + \Gamma B^2 + \Gamma\nu^2 - 2H^2 B + 2G^2 B}{F^6}$$

In order to calculate matrix $\exp(-idx)\mathbf{U}^{-1}\mathbf{C}(x)$, that stands in the integral in eq. (3.55), we need matrix

$$\exp(idx) = \begin{pmatrix} 1 & 0 & 0 \\ 0 & A_1 & 0 \\ 0 & 0 & A_1^{-1} \end{pmatrix}$$

and

$$\begin{aligned}
&\exp(-idx)\mathbf{U}^{-1}\mathbf{C}(x) \\
&= \begin{pmatrix} 1 & 0 & 0 \\ 0 & A_1^{-1} & 0 \\ 0 & 0 & A_1 \end{pmatrix} \frac{1}{F} \begin{pmatrix} \sigma & -1 & -1 \\ 1 & -\frac{1}{F+\sigma} & \frac{1}{F-\sigma} \\ 1 & \frac{1}{F-\sigma} & -\frac{1}{F+\sigma} \end{pmatrix} i2^{-1/2} \begin{pmatrix} 0 \\ 1 + \varepsilon a(A_1 - A_1^{-1}) \\ -1 + \varepsilon a(A_1 - A_1^{-1}) \end{pmatrix} \\
&= \frac{i}{\sqrt{2}} \begin{pmatrix} 0 \\ -A_1 \\ A_1 \end{pmatrix}
\end{aligned}$$

Then

$$\mathbf{W}(\tau) = \mathbf{U}\exp(id\tau)\exp(\varepsilon\mathbf{K}\tau)\int_0^\tau \exp(-\varepsilon\mathbf{K}x)\exp(-idx)\mathbf{U}^{-1}\mathbf{C}(x)dx$$

$$= \mathbf{U}\exp(id\tau)\exp(\varepsilon\mathbf{K}\tau)\cdot$$

$$\cdot[\exp(-\varepsilon\mathbf{K}\tau)\frac{1}{-\varepsilon K + i\delta_d}\frac{i}{\sqrt{2}}e^{i\delta_d\tau}\begin{pmatrix}0\\-1\\1\end{pmatrix} + O(\varepsilon)]$$

$$= \mathbf{U}\exp(id\tau)\left[\frac{\exp(i\delta_d\tau)}{\delta_d\sqrt{2}}\begin{pmatrix}0\\-1\\1\end{pmatrix}\right]$$

Thus, we see that the principal term of the asymptotic expansion is proportional to ε^0. We are interested in the second component of the obtained result

$$\{\mathbf{W}\}_2 = \left\{\mathbf{U}\exp(id\tau)\frac{\exp(i\delta_d\tau)}{\delta_d\sqrt{2}}\begin{pmatrix}0\\-1\\1\end{pmatrix}\right\}_2 = \frac{i}{\delta_d\sqrt{2}}e^{2i\delta_d\tau}$$

whose imaginary part, $Im\{\mathbf{W}\}_2 = ImR_{12}$ is responsible for absorption and amplification. Therefore,

$$ImR_{12} = \frac{1}{\delta_d\sqrt{2}}\cos 2\delta_d\tau = \frac{E\mu}{D\hbar}\cos 2Dt \tag{3.56}$$

The physical meaning of the obtained result is the following.

The mathematical method used to find the solution of the Bloch's equations eq. (3.45) presumed the presence of the parametric resonance the condition of which is eq. (3.52). If it is not fulfilled, the method is not valid, and the solution would not contain a periodic term obtained here. Thus, the first physical result is: if there is no gravitational radiation or if there is a gravitational radiation, but it does not suffice the OMPR conditions, no characteristic term like that of expression eq. (3.56) appears, no prediction for the experimental search of a periodical feature can be made, and there is nothing of this kind to look for in observations, because no periodic component can appear in maser signal.

Second, if there is an astrophysical system such that the OMPR conditions are fulfilled, then the mathematical model can be applied to its description, the principal term of the asymptotic expansion that is used to calculate the absorption coefficient of the additional wave will contain a non-stationary periodic component, and there is a hope to observe the corresponding feature on the maser spectrum. Notice, that the order of the signal amplitude in eq. (3.56) is ε^0, that is there is no dependence on the amplitude of the gravitational wave. The meaning of this result is that in the frequency region close to atomic transition there will be frequency at which the absorption coefficient would change its sign and correspond in turn to absorption and amplification of the signal at this frequency. And the frequency of this sign change corresponds to the doubled frequency of the gravitational wave. This change also means that the energy balance is not disturbed and the average level of the signal that is received at those frequencies is constant. Notice also, that in other words, the parametric resonance causes only the redistribution of energy between the main wave and its effective sidebands that appear because of the interaction with the gravitational wave.

The most important result is the prediction of the periodic change of intensity for a component of maser signal. Whatever structure the stationary and well observed maser signal has, the appearance of a non-stationary component either observed with the help of special instrumentation like gated detector or revealed by the special data processing would indicate at the existence of a parametric resonance in the system. In case under discussion, it is the optic-metrical parametric resonance dealing with the influence of the periodic gravitational waves on space maser. Since the amplitude of ImR_{12} in eq. (3.56) is proportional to ε^0 and not to ε^1 or ε^{-1}, the signal cannot become too small or too large.

The conclusion is that if the non-stationary signal discussed above is observed, and if its frequency can be attributed to the corresponding source of the gravitational waves, then it would be a direct evidence of the gravitational wave existence. The effect can be used in two ways: either take a known source of the gravitational radiation and search for suitable masers to find a non-stationary component in the signal, or take a space maser with a non-stationary component in the spectrum and look for a suitable source of the gravitational radiation. These investigations present a clear perspective for the development of a new branch of astronomy, though the road to it is not simple.

3.3   Astrophysical systems

As it was already mentioned, the character of problems with the gravitational waves detection by means of the OMPR effect shifts from technical one like sensitivity increase, signal/noise ratio increase etc., characteristic to the usual methods, to the search of an astrophysical system suitable for observations. The OMPR conditions essential for the solution of the OMPR problem are

$$\frac{\gamma}{\alpha_1} = \Gamma \varepsilon; \Gamma = O(1); \varepsilon << 1 \tag{3.57}$$

$$\frac{\alpha_2}{\alpha_1} = \frac{\omega h}{8D} = b\varepsilon; b = O(1); \varepsilon << 1 \tag{3.58}$$

$$\frac{kv_1}{\alpha_1} = \frac{\omega h}{\alpha_1} = \kappa \varepsilon; \kappa = O(1); \varepsilon << 1 \tag{3.59}$$

$$(\omega - \Omega + kv_0)^2 + 4\alpha_1^2 = D^2 + O(\varepsilon) \Rightarrow D \sim 2\alpha_1 \tag{3.60}$$

In a more qualitative physical terms that are needed when searching for a suitable astrophysical system, they can be formulated as

i) strong field - the electromagnetic field produced by the space maser must be spectroscopically strong

$$\alpha >> \gamma \tag{3.61}$$

ii) frequency condition - GW frequency is related to maser intensity

$$D \sim \alpha \tag{3.62}$$

iii) amplitude condition - there is still a specific limitation on the GW amplitude

$$h\frac{\omega}{D} \sim \frac{\gamma}{\alpha} \tag{3.63}$$

where h is the dimensionless amplitude of the GW. This last condition means (see also Appendix) that the amplitudes of an additional wave and the amplitude of mechanic oscillations of an atom being small still shouldn't be too small.

The analysis of the suitable astrophysical systems must include the GW sources and their parameters, the space masers or radio sources and the geometrical configuration and distances. But this is not all. As it was mentioned in the beginning of the previous section, this analysis cannot give the definite answer to the question which system will suit, because of the insufficiently accurate values of the parameters of the system. Therefore, in order to find a system sufficing the amplitude condition iii), some luck would be of help.

The theory of the gravitation waves emission [Landau and Lifshitz (1967)] gives the following expression for the loss of the energy flow, $\mathcal{E}$ due to the gravitational waves emission

$$-\frac{d\mathcal{E}}{dt} = \frac{G}{45c^5} \frac{d^3 \mathcal{D}_{\alpha\beta}^2}{dt^3} \tag{3.64}$$

where $\mathcal{D}_{\alpha\beta} = \int \rho(3x^\alpha x^\beta - \delta_{\alpha\beta} x_\gamma^2) d^3 x$ is the quadrupole moment of mass distribution. The numerical value of this loss is extremely small even for a double star system and is about 10^{-12} part of its total energy. The energy loss suggests that the distance, $\mathcal{R}$, between stars with masses m_1, m_2 decreases with the speed $\frac{d\mathcal{R}}{dt} = -\frac{64G^3 m_1 m_2 (m_1 + m_2)}{5c^5 \mathcal{R}^3}$, and this causes the decrease of the orbital period. The last was measured in [Hulse and Taylor (1975)] and [Taylor and McCulloch (1980)], and the obtained value coincided with the theoretical one within 10^{-3} accuracy. For a single rotating body, the GW radiation can be caused by the asymmetry of the body which is described by the so-called gravitational ellipticity, g_e. The correlation between the distance from the GW source, r_g, the GW source parameters and the GW amplitude is given by expression [Thorne (1987)]

$$h = \frac{GM\mathcal{R}^2 D^2 g_e}{c^4 r_g} \tag{3.65}$$

where M is the mass of a source, $\mathcal{R}$ is its characteristic radius, g_e is its gravitational ellipticity.

3.3.1 GW sources

One of the possible candidates able to take part in the OMPR effect as a GW source is a neutron star (a pulsar). The typical values of mass and radius of a pulsar are: $M = 10^{30}$kg, $\mathcal{R} = 10^4$m. The gravitational ellipticity, g_e, is a parameter which is not so easy to estimate. In [Thorne (1987)] one can find $g_e = 10^{-3}$, while in earlier paper [Zimmermann (1978)] $g_e = 10^{-6}$. There is a theoretical lower limit for the rotation period of a pulsar [Friedmann (1922)], which is equal to $0.5 - 2$ ms depending on the equation of state. In [Andersson *et al.* (1999)] it was also shown that if a pulsar has a period shorter than a certain critical value, then within a year this period will increase to the critical value. For the simple model of pulsar regarded in [Andersson *et al.* (1999)] the critical value of the period is about 20 ms which is consistent with the estimated initial spin of the Crab pulsar. Thus, there is the upper limit for the frequency of the GW produced by a pulsar, and it is equal to $(5 - 20) * 10^2$ s^{-1}. Nowadays, hundreds of pulsars are known. Some of them have planets [Konacki and Wolszezan (2003); Miller and Hamilton (2001)] whose masses and dimensions are several times larger than those of Jupiter. Presumably, such planets could have thick and vast atmospheres.

Another possible periodical GW source is a binary star system. In view of the possible observations of the change of the electromagnetic signal, the binaries with the shortest orbital periods suffice best. Therefore, these binaries consist of close components. These binaries could be: helium star - white dwarf/neutron star, double white dwarfs, white dwarf - neutron star/black hole, double neutron star, double black hole. The orbital period of such binaries also has a characteristic value. When the separation between the components of a binary is small enough, the mass transfer process begins. The mass transfer does not constrain the GW radiation, and the binaries with the mass transfer process going on could be of some additional interest, since the change in their periods due to GW emission presents the data that could be related to the results of observations discussed here. The short-period mass transferring objects of this kind are known as AM CVn systems and ultra-compact X-ray binaries (UCXB's). The comprehensive reviews of these binaries are presented in [Nelemans (2003); Verbunt (1997); Verbunt and Nelemans (2001)]. For example, there are such binaries as RX J1914.4+2456, V407 Vul [Cropper (1998)] with the period of 9.5 min, KUV 01584-0939, ES Cet [Warner and Woudt (2002)] with the period of 10.3 min and XTE J1807-294 [Markwardt *et al.* (2003)] with the period of 40.1 min. The binary with the shortest period now known is RX J0806.3+1527 [Israel (2002)] which has a period equal to 5.3 min and is about 100 pc away from the Earth. Thus, the binary stars provide the GW frequencies $D \sim 10^{-4} - 10^{-3}$ s^{-1}. The orbital radius of a binary can be evaluated with the help of Kepler law, $\mathcal{R}^3 = \frac{GMT^2}{4\pi^2}$.

3.3.2 Space masers

The number of known space masers (radio sources) is more than a thousand. Their radiations correspond to various transitions in atoms and molecules found in space. The following types of the cosmic masers could be distinguished:

Star forming region type. As it was noted in [Kaplan and Pikel'ner (1979); Varshalovich (1982); Minier (2002)], some space masers originate from the clouds that are going to transform into protostars. If there is a suitable GW source (see eq. (3.62)), such masers could be of interest.

Circumstellar type. The gas density in space near the star could provide the conditions to start the masering process. Two cases of interest are possible: a) this star is a GW source itself (a pulsar or a binary) and b) there is a distant GW source acting on this maser. According to [Kaplan and Pikel'ner (1979)], the gas density value which is needed to start a strong masering process in molecules like OH, H_2O and SiO is $n \geq 10^7 - 10^9$ cm^{-3}. To evaluate the strength of such maser field, one should consider the pumping mechanism. If the pumping is due only to the resonant radiation, then the decay rate, γ, corresponds to the natural bandwidth with regard to the influence of the strong field. If the atomic collisions take part in pumping, then the value of the decay rate should be corrected with account to the collisions frequency, i.e. with account to the gas density.

Interstellar type. There are masering regions in space that are not related to stars. Sometimes they could be related to the $\lambda = 21$ cm hydrogen line. When in space far away from stars, it is the radiation excitation that dominates over the collision excitation [Kaplan and Pikel'ner (1979)]. Notice, that Einstein coefficient, A, characterizing the spontaneous decay rate, γ, could be very small. For example, for $\lambda = 21$ cm hydrogen transition, it is equal to $2.85 \cdot 10^{-15}$ s^{-1}, so the value of this order could enter the OMPR condition.

Since the frequency condition of the OMPR eq. (3.62) suggests $D \sim \alpha$, and $\alpha = \frac{\mu E}{\hbar}$ is the Rabi frequency of the "resonant field+atom" system, one should consider the electric stresses E of the cosmic EMW sources, that is their intensities. The intensities of cosmic EMW sources is usually measured in terms of their brightness temperature T_b. For the isotropic distribution of the maser intensity the correlation between the electric stress and brightness temperature is given by [Slysh (2003)]

$$E = 8\pi\nu_0\sqrt{\frac{kT_b\delta\nu}{c^3}} \qquad (3.66)$$

where $\nu_0 = \frac{\omega}{2\pi}$ is the transition frequency (this notation should not be mixed up with $\nu = O(1)$ used for the tuning parameter in the previous section), $\delta\nu = \frac{\gamma}{2\pi}$ is the bandwidth, $k = 1.38 \cdot 10^{-23}$ J/K is Boltzmann constant.

In the absence of external field, the natural bandwidth is equal to Einstein coefficient, $\gamma = A = \frac{32\pi^3\nu_0^3\mu^2}{3\hbar c^3}$. But in the presence of the spectroscopically strong field (such maser is called saturated) the time that an atom spends on the upper level is reduced by the factor $\frac{kT_b}{\pi\hbar\nu_0}$ [Slysh (2003)], and the bandwidth accordingly increases by the same factor, $\delta\nu = \frac{\gamma}{2\pi}\frac{kT_b}{\pi\hbar\nu_0}$.

To obtain the lower limit of the brightness temperature of a maser providing the possibility of the OMPR in case when the pumping is radiative and the influence of the collisions is neglected, $\delta\nu = \frac{A}{2\pi}\frac{kT_b}{\pi\hbar\nu_0}$, consider $E = \frac{\hbar\alpha}{\mu}$, $\alpha \sim D$ and $\nu_0 = \frac{\omega}{2\pi}$ and substitute these into eq. (3.66)

$$T_b = \frac{\sqrt{3\pi}}{8}\frac{\hbar^2 c^3}{k}\frac{D}{\omega^2\mu^2} \sim 10^{-7}\frac{D}{\omega^2\mu^2} \qquad (3.67)$$

If $\omega \sim 10^9$ s^{-1} and $\mu \sim 3 \cdot 10^{-19}$ CGS$_q$·cm, then the minimal needed brightness temperature of a maser is $T_b \sim 10^{14}$ K for the GW observations dealing with a pulsar ($D \sim 10^2$ s^{-1}), and $T_b \sim 10^8$ K for the GW observations dealing with a binary ($D \sim 10^{-4}$ s^{-1}). If one thinks of a space laser (see below) and takes $\omega \sim 10^{15}$ s^{-1} and $\mu \sim 3 \cdot 10^{-19}$ CGS$_q$·cm, then the minimal needed brightness temperature of a laser is $T_b \sim 10^2$ K for the GW observations dealing with a pulsar ($D \sim 10^2$ s^{-1}), and $T_b \sim 10^{-4}$ K for the GW observations dealing with a binary ($D \sim 10^{-4}$ s^{-1}).

The usual values of brightness temperatures measured for the spots in cosmic masers (condensations) are known to be 10^9 K for methanol masers, 10^{12} K for OH-masers and 10^{15} K for H_2O-masers. Therefore, the existing masers could in principle provide the intensities of the EMW needed for the OMPR. If the collisions take part in pumping, then γ increases, and the brightness temperature that is needed to provide the same electric stress is lower.

In a saturated maser the intensity grows linearly with distance from the center to the border. It means that if the intensity of a maser is higher than that needed for the OMPR, there could exist a region in the depth of maser where the OMPR condition is satisfied, but at the same time there appears additional limitations on maser parameters [Siparov (2001a)]. Then it is this region that could be the source of a non-stationary signal which is characteristic for the OMPR.

3.3.3 *Distances*

Equation eq. (3.63) suggests to account for the following: i) the decay rate, γ, which is defined by the atomic transition and the density of atoms; ii) the frequency of the atomic transition, i.e. the frequency of the radio source, ω; iii) the intensity of the maser field expressed by its Rabi frequency, α; iv) the frequency of the GW source, D; v) the GW amplitude, h, that depends on the distance between the GW source and maser, r_g, and is defined by eq. (3.65).

Let us find the correlation between the GW frequency, D, and the corresponding values of μ, ω, T_b, and γ. Considering $D \sim \alpha$, and

$$\alpha = \frac{\mu E}{\hbar}; E = 8\pi\nu_0\sqrt{\frac{kT_b\delta\nu}{c^3}}; \delta\nu = \frac{\gamma}{2\pi}\frac{2kT_b}{2\pi\hbar\nu_0} \qquad (3.68)$$

we get that $D = f(\mu, \omega, T_b, \gamma)$, and eq. (3.68) gives

$$D^2 = \frac{16k^2}{\pi c^3\hbar^3}\gamma\mu^2\omega T_b^2 \qquad (3.69)$$

Let us substitute the consequence of the conditions eq. (3.62) and eq. (3.63), i.e. $h = \frac{\gamma}{\omega}$, into eq. (3.65) together with eq. (3.69)

$$r_g = \frac{16Gk^2}{\pi c^7\hbar^3} \cdot M\mathcal{R}^2 g_e \cdot \mu^2\omega^2 T_b^2 \qquad (3.70)$$

The first factor is the combination of constants, the second factor characterizes the GW source, the third factor characterizes the EMW source. In the further evaluations, we will use

$$r_g \sim 10^{-31} \cdot M\mathcal{R}^2 g_e \cdot \mu^2\omega^2 T_b^2 [\text{cm}] \qquad (3.71)$$

and $M \sim 10^{33}$ g, $\mathcal{R} \sim 10^{10}$ cm, $g_e \sim 10^0$ - for a binary, $M \sim 10^{33}$ g, $\mathcal{R} \sim 10^5$ cm, $g_e \sim 10^{-3}$ - for a pulsar, $\mu \sim 3 \cdot 10^{-19}$ CGS$_q\cdot$ cm, $\omega \sim 10^9$ s^{-1} - for a cosmic maser and $\mu \sim 3 \cdot 10^{-19}$ CGS$_q\cdot$cm, $\omega \sim 10^{15}$ s^{-1} - for a cosmic laser. Then the results of the evaluation fit the following Table 3.1.

Table 3.1   The dependence of the distance to the GW-source providing the OMPR effect on brightness temperature.

GW-source	Maser	Laser
Binary	$r_g \sim 10^3 T_b^2 [cm]$	$r_g \sim 10^{15} T_b^2 [cm]$
Pulsar	$r_g \sim 10^{-10} T_b^2 [cm]$	$r_g \sim 10^2 T_b^2 [cm]$

Table 3.2   Brightness temperatures that provide OMPR for circumstellar EMW sources.

GW-source	Circumstellar maser	Circumstellar laser
Binary	$T_b \sim 10^6 K$	$T_b \sim 10^0 K$
Pulsar	$T_b \sim 10^{12} K$	$T_b \sim 10^{11} K$

Table 3.3   Brightness temperatures that provide OMPR for interstellar EMW sources.

GW-source	Interstellar maser	Interstellar laser
Binary	$T_b \sim 10^8 K$	$T_b \sim 10^2 K$
Pulsar	$T_b \sim 10^{14} K$	$T_b \sim 10^{13} K$

These correlations mean that for a given brightness temperature of EMW source, a GW-source which is able to provide the astrophysical situation suitable for the OMPR observations must be at the corresponding distance. Let us evaluate the needed brightness temperatures of the EMW sources for the "circumstellar" and "interstellar" distances between the EMW source and the GW source. Taking $r_S \sim 10^{14}$ cm (about $10^0 - 10^1$ astronomical units) for the circumstellar EMW source in the vicinity of a GW source, we find the needed temperatures given in Table 3.2.

Taking $r_S \sim 10^{18}$ cm (about $10^0 - 10^1$ pc) for the case when the GW source is at the interstellar distance from the EMW source, we get the values given in Table 3.3.

Comparing these results with the brightness temperatures mentioned in the end of the previous subsection, we obtain the following results:

- a cosmic maser located far away from the GW source can be affected in an OMPR-observable manner both by pulsars and binaries, while the circumstellar type of a maser with a GW source as a central body cannot be used for the OMPR based observations; the last is an answer to the natural doubts concerning the possibility of the existence of a stable maser in the vicinity of an active pulsar: even if such a maser exists, it would not be possible to use it.

The obtained result means that the initial intention to allocate the remote detector based on the OMPR effect closer to the GW source does not lead to strong restrictions, that is the number of masers suitable for observations could be large enough. It also means that the chosen maser could be simultaneously affected by several GW sources. But this cannot destroy the effect. It means that the observer who uses the radio telescope can find several non-stationary components with different frequencies for one and the same maser signal, that is the radio telescope in search for gravitational radiation would work similarly to a home radio receiver.

It comes to mind that in this case the masers in the laboratories on the Earth could probably be used for the same purpose. But in this case the intensity of a maser should be rather low to provide the Rabi frequency of the order of $(10^{-3} - 10^2)$ Hz (compare with the Rabi frequency of the common laboratory laser which is about 10^9 Hz). And this leads to the signal/noise problems well known in traditional approach. Besides, the atomic decay rates for the laboratory masers are essentially higher than those characteristic to the space masers, and the TLA model can be applied and give sensible predictions only in special cases.

A space laser seems suitable in all cases, but the relatively large value of γ leads to the break of the amplitude condition eq. (3.63) when the distance becomes large and h decreases according to eq. (3.65). Therefore,

- a cosmic laser could reveal the action of the GW only in case when located in the vicinity of the GW source.

It seems possible that a laser effect could appear not only in a star shell (like that in η Carinae, see [Johansson and Letokhov (2002)]), but also in the atmosphere of a giant planet near a pulsar. But this is a question addressed to astrophysicists. Notice, that the solution of the OMPR problem for a laser needs some modifications (see below), because the pumping can take place, and the lower level could be not only the ground one.

The considerations given above make clear that the variety of possibilities to observe the OMPR is rather wide. The masers of all types - star forming, circumstellar, interstellar - can be affected by the GW from the far off pulsars or binary stars. A cosmic laser can also be used for the GW observation with the help of the OMPR if it is close to a pulsar or to a double star.

3.3.4   *Examples*

Let us give several examples [Siparov (2008b)] of the astrophysical systems that could be suitable for the GW observations based on the OMPR method. All the mentioned masers are H_2O ones which means that their brightness temperatures suffice the obtained condition (at least for the condensations). The distances between the GW sources (pulsars and double star) and the corresponding masers are tens of parsecs, d is the estimated distance from the Earth. The data was taken from [The ATNF Pulsar Database (2005); Palagi+ (1993)].

The values of distances could vary in different sources, sometimes essentially. For example, in the original paper [Israel (2002)], the distance to the white dwarf binary RXJ0806.3+1527 was estimated as just $d \geq 100$ pc, while now it is considered to be essentially larger.

3.3.4.1   *On the OMPR problem for a cosmic laser*

In [Johansson and Letokhov (2002)] they found a *laser* effect ($\omega \sim 10^{14}$ s^{-1}) in the cloud surrounding the hot η Carinae. They discussed the laser action appearing in a quantum transition between the excited states in $FeII$, its higher levels being optically pumped by the intense $HLy\alpha$ radiation. Here we are not interested in the detailed mechanism of this effect (it is described in [Johansson and Letokhov (2002)]). The important thing is that the pair of regular levels (no metastability) can be found lasing in space and, thus, can also take part in the OMPR. This means that the approach given above to describe the maser involved situation should be modified to account for the fact that now the lower atomic level is not the ground one. On the other hand, it involves more levels than two, and the theoretical model of an atom becomes less applicable. Nevertheless, we can try to apply the developed approach to this situation also, at least, to get a qualitative idea of what could happen. In [Siparov (2005)] the corresponding system of Bloch's equations was solved for the additionally simplified case when there was only one additional weak wave

Table 3.4   Examples of the astrophysical systems suitable for the search of OMPR effect.

1	Name	RaJ	DecJ	d (pc)	GW-frequency (Hz)
Pulsar	J1908+0734	19:08:17.01	+07^{0}34'14.36"	580	4.7091472
Maser-1	IRC+10365	18:34:59.0	+10^{0}23'00.0"	500	
Maser-2	RT AQL	19:35:36.0	+11^{0}36'18.0"	530	

2	Name	RaJ	DecJ	d (pc)	GW-frequency (Hz)
Pulsar	J0205+6449	02:05:37.92	+64^{0}49'42.8"	3200	15.22386
Maser-1	IRAS00117+6412	00:11:44.6	+64^{0}12'04.0"	3170	
Maser-2	W3(1)	02:21:40.8	+61^{0}53'26.0"	3180	

3	Name	RaJ	DecJ	d (pc)	GW-frequency (Hz)
Pulsar-1	B1133+16	11:36:03.2477	+15^{0}51'04.48"	360	0.841812
Pulsar-2	J1022+1001	10:22:58.006	+10^{0}01'52.8"	300	60.77945
Maser	AF Leo	11:25:16.4	+15^{0}25'22"	270	

4	Name	RaJ	DecJ	d (pc)	GW-frequency (Hz)
Pulsar	B0656+14	06:59:48.134	+14^{0}14'21.5"	290	2.59813686
Maser	U ORI	05:52:51.0	+20^{0}10'06.0"	280	

5	Name	RaJ	DecJ	d (pc)	GW-frequency (Hz)
Pulsar	J0538+2817	05:38:25.0632	+28^{0}17'9.07"	1770	6.98527635
Maser	HH4	05:37:21.8	+23^{0}49'24.0"	1700	

6	Name	RaJ	DecJ	d (pc)	GW-frequency(Hz)
Pulsar	B0031-07	00:34:08.86	-07^{0}21'53.4"	720	1.06050049872
Maser	U CET	02:31:19.6	-13^{0}22'02.0"	660	

7	Name	RaJ	DecJ	d (pc)	GW-frequency (Hz)
Binary	RXJ0806.3+1527	08:06.3	+15^{0}27	100	0.0031153
Maser	RT Vir	13:00:06.1	+05^{0}27'14"	120	

$$\frac{d}{dt}(\rho_{22} - \rho_{11}) = -\gamma(\rho_{22} - \rho_{11}) + 4i[\alpha_1 \cos(\Omega_1 t - ky) + \qquad (3.72)$$
$$\alpha_2 \cos(\Omega_2 t - ky)](\rho_{21} - \rho_{12}) + \Lambda$$
$$\frac{d}{dt}\rho_{12} = -(\gamma_{12} + i\omega)\rho_{12} - 2i[\alpha_1 \cos(\Omega_1 t - ky) +$$
$$\alpha_2 \cos(\Omega_2 t - ky)](\rho_{22} - \rho_{11})$$
$$\rho_{12} = \overline{\rho_{21}}$$
$$\frac{d}{dt} = \frac{\partial}{\partial t} + V\frac{\partial}{\partial y}; V = V_0 - V_1 \cos Dt$$
$$V_1 = hc$$

Here Λ describes the incoherent pumping of any nature. The rotating wave approximation suggests to use the substitutions

$$\rho_{21} = R_{21}e^{i(\Omega_1 t - ky)} \tag{3.73}$$
$$\rho_{12} = R_{12}e^{-i(\Omega_1 t - ky)}$$
$$R = 2^{-1/2}(\rho_{22} - \rho_{11})$$

Then we get

$$\frac{d}{d\tau}R = -\varepsilon R - iR_{12} + iR_{21} - iR_{12}a\varepsilon e^{i\delta_d \tau} + iR_{21}a\varepsilon e^{-i\delta_d \tau} + \lambda \tag{3.74}$$

$$\frac{\partial}{\partial\tau}R_{12} = -[\Gamma\varepsilon + i\delta + i\kappa V_0 - i\kappa v_1\varepsilon\cos(\delta_d\tau)]R_{12} - i[1 + a\varepsilon e^{-i\delta_d\tau}]R$$

where $\lambda = \frac{\Lambda}{2\alpha_1}$, or in the matrix form

$$\frac{\partial}{\partial\tau}\mathbf{W} = (\mathbf{Q}_0 + \varepsilon\mathbf{Q}_1(\tau))\mathbf{W} + \mathbf{C} \tag{3.75}$$

where

$$\mathbf{W} = \begin{pmatrix} R \\ R_{12} \\ R_{21} \end{pmatrix}; \mathbf{Q}_0 = i\begin{pmatrix} 0 & -1 & 1 \\ -1 & -\sigma & 0 \\ 1 & 0 & \sigma \end{pmatrix}; \mathbf{C} = \begin{pmatrix} \lambda \\ 0 \\ 0 \end{pmatrix}; \tag{3.76}$$

$$\mathbf{Q}_1 = \begin{pmatrix} -1 & -iae^{i\delta_d\tau} & iae^{-i\delta_d\tau} \\ -iae^{-i\delta_d\tau} & -\Gamma + i\kappa v_1\cos(\delta_d\tau) & 0 \\ iae^{i\delta_d\tau} & 0 & -\Gamma - i\kappa v_1\cos(\delta_d\tau) \end{pmatrix}; \sigma = \delta + \kappa V_0$$

Equation (3.75) can be solved by the same asymptotic expansion method with $\varepsilon = \frac{\gamma}{2^{1/2}\alpha_1}$ as a small parameter characterizing the strength of the field. The non-stationary component of the signal that is proportional to $Im R_{21}$ appears to be

$$Im R_{21} = -\frac{\lambda\sigma H\sqrt{B^2 + \nu^2}}{\varepsilon F\{(\sigma^2 + 2\Gamma)(B^2 + \nu^2) + 2BH^2\}} \cdot \tag{3.77}$$
$$\cdot\cos(Dt + \arctan\frac{\nu}{B})$$

$$H = F(\frac{a}{F - \sigma} + 2\kappa v_1) = 2^{1/2}\frac{\omega D}{\alpha_1\gamma}\frac{V_1}{c}(1 - h\frac{\alpha_1^2}{D\omega}\frac{c^2}{V_0 V_1})$$

$$B = 1 + \Gamma(\sigma^2 + 1) = 1 + \frac{\gamma_{12}}{\gamma}(1 + \frac{1}{2\alpha_1^2}(D + \frac{\omega V_0}{c})^2)$$

According to the conditions of OMPR eq. (3.63), $\varepsilon \sim \frac{\hbar\omega}{D}$, and since all the other parameters in eq. (3.74) for the amplitude are of the order of unity, one can see that the non-stationary component appears to be large because of ε in the denominator. The result seems attractive, but the applicability of the method here requires justification. This result illustrates the possibility to involve into consideration other transitions than those that are common for cosmic masers, for example, like that mentioned in [Johansson and Letokhov (2002)].

3.4 Observations and interpretations

3.4.1 *Radio sources observation methods*

The monochromatic cosmic radio radiation corresponding to certain transitions in the H_2O, OH and other molecules is produced by the space masers [Bochkarev (1992); Zel'dovich and Novikov (1988); Elitzur (1992); Cook (1977)] - the clouds with the diameters of $10^0 - 10^3 a.u.$ Such masers radiation spectra occupy the bands not more than 10^2 km/s wide and contain the series of narrow details corresponding to the spots with dimensions of about $0.1 a.u.$ The parts of the maser cloud can have various densities and can move relative to one another. The origin of the pumping may vary for various masers and the spectra and the intensities of their details may also vary with time. The scales of these time variations vary from years to minutes.

The cycles of the flux change with periods of 5-15 years were monitored for many years (1979-2010) with the 22 m radio telescope RT-22 at Pushchino Radio Astronomy Observatory (Astro Space Center of the Lebedev Institute of Physics, Russian Academy of Science) [Berulis *et al.* (1998)]. The instrumentation included a 1.35 cm wavelength receiver with a helium-cooled field-effect transistor amplifier and a 128-channel filter-bank spectrometer with a resolution of 0.1k m/s. Since December 2005, the 2048-channel autocorrelator with a total bandwidth of 168.5 km/s and a velocity resolution of 0.082 km/s was used for the spectral analysis of the signal.

The flare-like activity that lasts for months is known practically since the moment of H_2O maser first discovery in 1969. The flux density variations in the spectra details lasting for days were reported in [Rowland and Cohen (1986)], in some of them the flux density doubled in 2.4 days. The so-called intraday variability or ultra-rapid variations were first noticed in megamasers [Argon *et al.* (1994)], when the source in the galaxy IC 10 doubled its flux in an hour. In [Samodurov and Logvinenko (2001); Richards (2005); Samodurov (2006)] it was discovered that the radiation of space masers belonging to our galaxy can have ultra-rapid fluctuations (tens of minutes) also.

The investigations of the ultra-rapid (hour and less) variations confront specific problems:

- adjustment of the telescope: when the weather is fine, the diagram can drift for $\pm 1'$ angular minute due to the change of the relative positions of the Sun and the telescope; when the weather is not fine, the flux density strongly depends on the weather conditions especially on the absolute humidity, the flux can suffer 1.5-3 times decrease;

- strong flux density dependence on the elevation of the source above horizon due to radiation absorption in the atmosphere of the Earth; sometimes this dependence disagree with the theoretical model, and the source which is at $20°$ degrees above horizon gives a flux twice as large as the source at 5-$7°$ degrees;

- the problems with stable calibration (lack of intense calibrating sources);

- the fluctuations of the source flux density on the non-homogeneities of the interstellar plasma with the characteristic time of 0.5-1 hour, etc.

Hence, even the best telescopes under ideal weather conditions usually have the calibration accuracy about 5%.

The principal feature of the mentioned problems is the changes in the spectrum as a whole. In order to manage this difficulty, the self-calibration method was used: only the variations of the spectral details' weights relative to the total flux were measured and analyzed.

Thus, only the following causes for the details flux variations are left:

- the linear polarization of some spectral features results in flux variations of these features on a time scale of several hours. The method to overcome this is to store data on an extended interval of unchanged projected position angles of the polarization for several consecutive days. Then, by comparing the obtained time series, the identical diurnal trends in the fluxes can be eliminated;

- some sources have a complex structure, i.e. include several maser spots within less than $2'$ angular minutes, whereas the beam width of the used telescope is $2.6'$ angular minutes. The telescope pointing must be carefully checked, and such sources should be discarded.

The errors for the sources with line flux $50 Jy$ and exposition 10 minutes were $1.5 - 2\%$ for σ and 10% for 5σ. The errors for the sources with line flux $5000Jy$ and exposition 2.5 minutes were 0.1% for σ and 0.5% for 5σ. Everything larger the 5σ threshold gives a reliable evidence.

The goal of such investigation is to register and study the individual spectral features that possess an intrinsic variability. The physics of masers suggests to regard the internal processes in a maser itself as the cause of this variability. Such processes could present some specific inner changes in the maser condensations seen as the spots of radiation; or they can represent the competition for pumping between close condensations; or they can be some kind of swirling-type phenomena; or maybe they are the result of the total flux change because of the superposition of the projections of the moving condensations on the line of sight. In any of these cases one should take into account the space dimensions of a condensation and of the maser itself and relate them to the duration of a fluctuation.

In view of the theory given in the previous sections of this chapter, one might suggest that if a space maser or, strictly speaking, if one of its condensations suffices the astrophysical conditions needed for the OMPR, the corresponding detail on the maser spectrum would reveal the *periodic* variation with the period corresponding to a periodic source of the gravitational wave. Because of the need for the tuning (OMPR conditions) and of the specific features known from observations (see above), this behavior must have the following characteristics:

- (1) space maser radiation reveals a periodic variation;
  (2) if the spectrum is complicated, the periodic variations should be observed not for the whole spectrum but for only one detail of it;
  (3) the period of variations should be such as to get noticed during the reasonable period of observations, that is neither too short to be completely averaged out, nor too long to be mixed with some other phenomena in space masers and their surroundings;
  (4) the periodic variations in spectral details could be not seen every time because of the variety of non-periodical fluctuations of various time-scales that could break the parametric resonance conditions.

The last item makes the search for the OMPR signal in space maser even less determined than it already was because of inaccurate estimation of parameters. The periodic sources of the gravitation radiation mentioned in the theory which was developed in the previous sections included pulsars and binary stars. Both of those have known periods, but the periods of pulsars are usually parts of seconds, and, therefore, they require the instrumentation whose performance provide the possibility to deal with spectra obtained and stored for such periods of time. The close binaries whose periods are known to be minutes or tens of minutes suffice much better in view of the available instrumentation. Currently, the number of the known such short-period binaries is not more than a dozen. But the longer period has a binary, the more problematic would be the observation of the effect in view of the reasons given in (4). Thus, from the point of view of the search for the OMPR effect, in order to find the periodicity in a signal, the observations of various maser sources should be performed and the obtained data processed. It is also clear that no specially designed observations are needed to discover the periodic signal (if any). It can be tracked in any data that can be used for the search of the variability of the signal by means of the special processing.

3.4.2 *Ultra-rapid variability and signal processing*

In the period 2002-2009 the search for the ultra-rapid variability was performed for 42 radio sources.

In 32 of them, namely, in IRAS 16293-2422, 18316-0602, 18335-0713, 1855+0136, 21144+5430, Agfl 5142, G111.26-00.77, G 9.62+0.20, G 35.2-1.74, G 43.8-0.1, GGD4, Mon R2, NGC 2071, NGC 7538, NML Cyg, ON 1,R Crt, RCW 142, RR Aql, RS Vir, RT Aql, S 128, S 140, S 255, S 269, Sgr B2, W 3(2), W 43M3, W 44C, W 75N, W 75S, W B652, no rapid variability was found.

In 5 of them, namely, in Lk Hα 234, Ori A, VY CMa , W3OH, W 51 Main, there were detected various kinds of fluctuations on time scales less than an hour, probably related to the polarization of the details.

In 4 of them, namely, in Cep A, W 31A, W 33B, W 49N, there was found ultra-rapid variability that did not include polarization.

In 2 of them, namely, in Cep A and, RT Vir there were reasons to suggest that the periodical variations in the components of the spectra were detected.

In Fig. 3.8 there is an example of the time sequence of spectra obtained for the source W 49N on February 07, 2008 [Siparov and Samodurov (2009a,b)] and one can see how it looks.

In Fig. 3.9 there is an example of the sequence of spectra obtained for the source Sep A on June 24, 2008. Notice the change with time of the detail $(-14.38 - -11.63)$ corresponding to the shoulder peak to the left of the main peak.

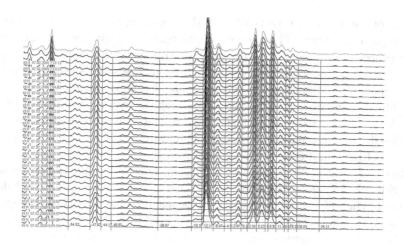

Fig. 3.8   W49N spectra, RT-22, Pushchino RAO, February 07, 2008.

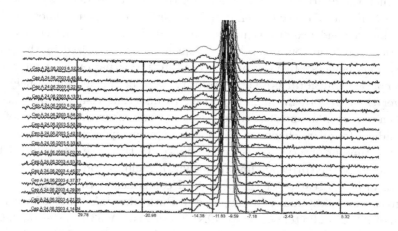

Fig. 3.9   Sep A spectra, RT-22, Pushchino RAO, June 24, 2008.

The registration of the space maser signal is performed in the following way. The radio telescope is tuned at a given frequency and its antenna follows the chosen maser for several hours. During a certain exposure time (from 3 to 8 minutes for different sessions and sources), the signal including all the adjacent frequencies is stored, then the spectrum is constructed – the dependence of the flow intensity (Jy*km/s) at a given frequency on the value of the frequency (actually, on the detuning of the given frequency from the main one). Then the exposure regime is switched on again and the procedure repeats N times until the observation conditions permit. As a result, a set of spectra corresponding to the signals received in the subsequent (approximately equal) intervals of time is obtained.

The data processing starts with obtaining the average value of the flow, I_0 , for every spectrum as a whole. Then the interval of frequencies (velocities) is divided into parts containing different details of the spectra (vertical lines on Figs.), and the time dependent averages, I_i , for every part are calculated. Then the differences, $I_{id} = I_i - I_0$, are found, and the processing provides the dependences $I_0(n), I_i(n), I_{id}(n)$ for $n = 1, 2, ..., N$. The calculation of the differential flows $I_{id}(n)$ makes it possible to exclude the effect of the variability of the source as a whole (and also the effects of the variability of the instrument properties, atmosphere, interstellar medium) on the behavior of the details' variations. If the equidistant time points are needed, the spline approximation can be used.

Then the Fourier transformations for the time dependencies of the flow as a whole and of every chosen detail are performed. As it is known, when the time dependence of a signal is complicated (stochastic processes), the Fourier transform contains a set of frequencies characterized by the peaks of various heights. In view of our problem, the situation is of interest if the Fourier transform presents a solitary peak or a peak whose height is larger than the others. It is important to underline that the presence of such a plot even in one observation session would mean the existence of the periodical (regular) change of the flow for a given spectrum detail at least for the time of observation (several hours). In this case one could define the values of the corresponding frequency, ν_j, and period T_j.

In order to search for the periodical components, the following radio sources were additionally investigated in Pushchino RAO in 2008 with the help of RT-22 radio telescope: W 49N (9 sessions), W 33B (3 sessions), Cep A (9 sessions), S 128 (1 session), W 3(2) (1 session), W 44C (1 session), W 75N (3 sessions), RT Vir (1 session). Together with these measurements performed specially, the measurements performed in 2005-2007 in search of ultra-rapid variations of the flow for the same sources were also processed and analyzed (20 session more). The time dependencies of the flow for the various spectral details vary for different sources and dates.

In Fig. 3.10 there is an example of a characteristic Fourier spectra obtained for W 33B radio source, which doesn't contain any visible regularity, that is, no specific periods are characteristic for the time fluctuations in these sources.

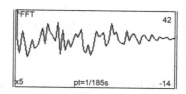

Fig. 3.10   Example of Fourier spectra: W 33B.

But if it is possible to suggest a suitable astrophysical system, whose properties when measured independently suffice the OMPR conditions, the existence of the periodical components in the variations of the details intensities could be interpreted as the evidence of the possible gravitational waves action on a space maser.

3.4.3   *Search for the periodic components in space maser signals*

3.4.3.1   *Radio source Cep A*

In Fig. 3.11 there are the examples of the data samples [Siparov and Samodurov (2009a,b)] for the fluctuations observed for the spectrum detail that demonstrated almost monochromatic periodic intensity changes with the same period at different dates.

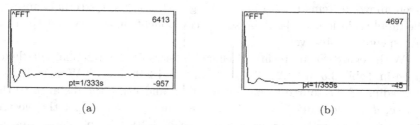

Fig. 3.11   Fourier spectra. (a) Cep A, RT-22, Pushchino RAO, February 07, 2008. (b) Cep A, RT-22, Pushchino RAO, June 24, 2008.

Table 3.5 contains the characteristics of the spectrum detail in the ra-

Table 3.5   Cep A: Spectrum detail characteristics

Source	Date	Detail (km/s)	T_j min	I_d $(Jy \cdot km/s)$
Cep A	February 7, 2008	$-11.5 \div -7.7$	22.2	1631
Cep A	June 6, 2008	$-11.6 \div -7.2$	23.7	2775

dio source Cep A. The detail here is the central peak of the spectrum. The fluxes were calculated with regard to the antenna coefficient of the telescope. Cep A radio source is located at the distance $d = (0.70 \pm 0.04)$ kpc from the Earth and has the coordinates RaJ 22h 54min 19.2s and Dec $61°45'44.1''$. As we see from Table 3.5, the corresponding periods of the observed intensity change were 22.2 min and 23.7 min. The difference in these two values could be explained by the different exposure times for two sessions (4 min and 7 min respectively). On the other dates, this peak of the Fourier spectrum can also be seen on the plots, but alongside with it there are also present the others that correspond to other terms of the Fourier expansion.

In order to interpret the obtained results, one should notice that it is not so easy to suggest a mechanism that would make a detail on a space maser radiation spectrum variate periodically, even if this process takes only several hours and the number of periods is not more than a dozen, and even if such event took place only once. However complicated are the physical properties of the masering system itself, however long and full of adventures could be the radiation path to the Earth, however complicated might be the electric and electronic details of the instrument, the origin of periodicity for only one - concretely this or that frequency component on the spectrum - is hard to suggest. On the other hand, the predictions given above in items (1)-(4) and dealing with a certain external cause for the rapid variations of the radio source component suggest exactly the type of behavior we observe.

With regard to the technical characteristics of the antenna, the fluxes given in Table 3.5 are $1.21 \cdot 10^{-18}$ W/m^2 and $2.06 \cdot 10^{-18}$ W/m^2. Taking the radius of the source to be about 10^{13}m and using the distance to the source, d and formula $S = \varepsilon_0 c E^2$ for the energy flow, we get the electric stress, E on maser border proportional to $E \sim 10^{-2}$ V/m. The frequency condition of the OMPR includes $\alpha_1 = \frac{\mu E}{\hbar}$, where μ is the induced dipole moment and $\hbar = 1,05 \cdot 10^{-34}$ J·s is Planck's constant. The period of the gravitational source, i.e. the orbital period of the needed binary, coincides with the period of the gravitational wave, T_{gw}. According to the theory given above, the frequency of the observed flux change ν_j corresponds to $2D$, therefore, the period of the gravitation wave T_{gw} is equal to $T_{gw} = 2T_j$. Taking the value of T_j from Table 3.5, one can see that the effective induced moment, μ, providing the OMPR condition in our case must be essentially less than the dipole moment of the water molecule and have the order of 10^{-35} Cl·m. This is in the qualitative agreement with the known fact that the transition in the H_2O molecule providing maser generation in a space maser is the rotational one.

The known compact binaries with the corresponding periods are the double pulsars XTE J1807-294 with the period of 40.1 min [Markwardt *et al.* (2003)], XTE J1751-305 with the period of 42.4 min [Markwardt (2002)] and XTE J0920-314 with the period of 44 min [Markwardt (2002)]. Besides, there is the double star V429 Car 10:41:17.5 - 59:40:37 which has a period of 44.64 min according to the measurements performed by Hipparcos [Hipparcos (1997)]. Two last objects though having periods closer to the measured values do not satisfy the condition for the geometrical configuration. As one can see from Fig. 3.7, this condition has two limiting cases: if a) the distance between the Earth and the maser is essentially less than the distance the Earth and the GW-source, or b) the distance between the GW source and the maser is essentially less than the distance between the Earth and the GW-source, then the direction from the Earth at the maser must be almost perpendicular to the line connecting the Earth and the GW-source. This is needed in order to provide the vibrations of the maser atoms in the direction at the Earth.

The first two mentioned binaries more or less suffice the geometrical conditions but their periods are a couple of minutes less than the measured ones. One can see that there is a 1.5 minutes difference in two measurements. So, the future measurements could appear to be closer to the period of the binaries. If it is not so, the search for other binaries is needed.

3.4.3.2   *Radio sources RT Vir and W Hya*

Another search was initiated by the famous binary RXJ0806.3+1527 which has the shortest known period equal to 321 s and is mentioned in Table 3.4. With time, it turned out that the estimated distance to it is much larger than it was assumed earlier. This suggests that all the radio sources located at the "cap" of the sphere, Fig. 3.12 are suitable for the search of the OMPR effect. The variety of masers located on a cap could serve as a kind of antenna set directed at a point where there could be the source of gravitational radiation. If there appear a periodic signal of the same

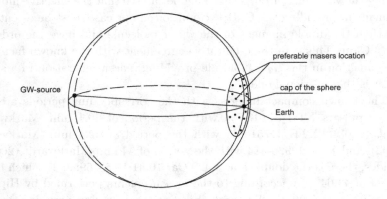

Fig. 3.12   The sky region preferable for the search of radio sources.

frequency in several different masers, such scheme would help to define the position of the GW-source. Up to now no signal corresponding to the frequency of gravitational waves emitted by RXJ0806.3+1527 was found. Instead, in the data collected on April 23 and 24, 2009 and on June 25, 2009 in the signal of the radio source RT Vir there was found a periodic component with the period of approximately 84 minutes. Up to now, there was performed 15 sessions of observations, and in 5 of them there are reasons to suspect the presence of the periodic component. The flux spectrum and the Fourier spectrum of the data collected for RT Vir on January 29, 2010 are given in Fig. 3.13 and in Fig. 3.14.

In the data obtained after processing the signal received from the radio source W Hya, there was found another example of the signal with the same period. In one of the 5 observation sessions there appeared the same period as registered before for RT Vir. The flux spectrum and the Fourier spectrum of the data collected for W Hya on March 05, 2011 are given in Fig. 3.15 and Fig. 3.16.

Both periods correspond to the point No. 120 of the total number of points for which the calculation of the Fourier coefficient was performed. This means that the period is 83 minutes. Table 3.2 contains the charac-

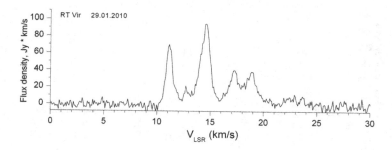

Fig. 3.13 RT Vir radio source flux spectrum, RT-22, Pushchino RAO, January 29, 2010.

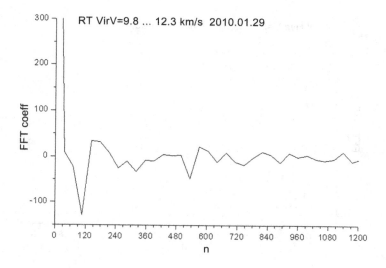

Fig. 3.14 RT Vir radio source Fourier spectrum for the detail, RT-22, Pushchino RAO, January 29, 2010.

teristics of the spectra details of the radio sources.

The coordinates of the radio source RT Vir and the distance to it are the following: RaJ 13 : 00 : 06.1, DecJ $+05°27'14''$, $d \sim 120 - 220$pc. The coordinates of the radio source W Hya and the distance to it are the following: RaJ 13 : 46 : 12, DecJ $-28°7'9''$, $d \sim 100$pc. Unfortunately, the estimations for the distances are not precise. Both these masers, RT Vir

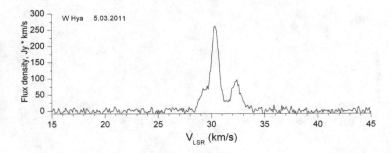

Fig. 3.15   W Hya radio source flux spectrum, RT-22, Pushchino RAO, January 29, 2010.

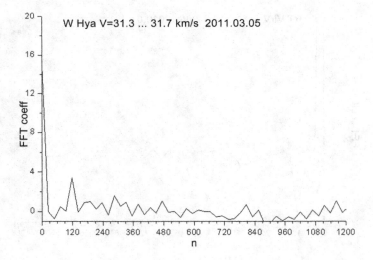

Fig. 3.16   W Hya radio source Fourier spectrum for the detail, RT-22, Pushchino RAO.

(H_2O-maser) and W Hya (OH-maser), belong to the circumstellar type and surround rather old stars. Their fluxes are weak, and their dimensions are of the order of 10^2 of astronomical units. No source of the gravitation waves with the period corresponding to the value mentioned above is known, but the coincidence of the periods attracts attention. Let us mention once again that only the components of the spectrum and not the whole of it possessed the noticed periodic character, they were found in different sources and on

Table 3.6   RT Vir and W Hya: spectra details characteristics

Source	Date	Detail (km/s)	T_j min	I_d ($Jy \cdot$km/s)
RT Vir	January 29, 2010	$9.8 - 12.3$	83.5	70
W Hya	March 3, 2011	$31.3 - 31.7$	83.5	40

different dates.

As it was already mentioned, the main (and not the only one) problem with these observations is the maser signal variability the nature of which is not clear. It can break the conditions of the OMPR and prevent the observation of the periodic component. It means that in order to get more reliable statistics, the observations should be continued and the independent observation of this effect would be helpful. Besides, the cap opposite to one presented in Fig. 3.12 should also be investigated. The same should be performed for the regions corresponding to other known short-period binaries.

3.5   On the search for the space-time anisotropy in Milky Way observations

In this section we finally come to the discussion of the possibility to perform a specific test of the AGD approach and find out if there is any space anisotropy caused by the motion of gravitation sources on a galaxy scale. As we discussed earlier, this anisotropy causes the metric tensor velocity dependence. Up to now in this chapter, we didn't touch it, because, first, we had to give the details concerning the theory and observations of the OMPR effect that deals with the gravitational waves detection. Now we are going to explain [Siparov and Brinzei (2008b); Siparov (2008b, 2007)] how the effect of optic-metrical parametric resonance can be used for the exposure of the geometrical properties of the real physical space or, better say, how it can be used for the choice of the most appropriate model to describe the real physical space.

As we saw above ([Siparov (2004, 2006b)]), the distances between the GW sources (pulsars or doubles) and space masers are not small but are of interstellar scale. Our galaxy, Milky Way, is a typical spiral galaxy, which means that there is an obvious specified direction coinciding with the axis of rotation and a series of locally specified directions coinciding with the ordered (convective) circular motion of stars in any given place.

According to the approach developed in the previous chapter, all this must be reflected in the expression of metric. One and the same GW source could affect several masers, and one and the same maser could be affected by several GW sources. The orientations of all these systems relative to the galactic plane could be different. What could be the influence of space-time anisotropy caused by the motion of masses on the OMPR effect? In which way could the orientation act on it, and will the OMPR be possible in the anisotropic space at all? The answer must include the description of the gravitation waves in the anisotropic space, the analysis of the possibility of the OMPR effect and the changes in the OMPR conditions (if any). The comparison between the results of such theory and the subsequent observations and measurements would simultaneously provide the evidence for the AGD and the data for the description of the geometrical properties of physical space on the galaxy scale for our galaxy.

3.5.1 *Mathematical formalism and basic equations*

Let us again regard $M = \mathbf{R}^4$ which is a differentiable 4-dimensional manifold of class C^∞, TM its tangent bundle, and $(x, y) = (x^i, y^i)$; $i = 0, ..., 3$ the coordinates in a local frame [Brinzei and Siparov (2007)]. We call *locally Minkowskian* a metric with the property that there exists a system of local coordinates on TM in which it does not depend on the positional variables, x^i, but may depend on the directional variables, $y^i = \dfrac{\partial x^i}{\partial t}$ (t is a parameter), $i = 0 - 3$. Let us regard a metric tensor $g_{ij}(x, y) = \gamma_{ij}(y) + \varepsilon_{ij}(x, y)$, $\forall (x, y) \in TM$, where $\gamma = \gamma(y)$ is a Finsler-locally Minkowskian undeformed metric tensor on M and $\varepsilon = \varepsilon(x, y)$ is a small anisotropic deformation of γ. In anisotropic spaces, the tangent spaces $T_x M$, $x \in M$ are generally curved. In these models, TM turns to be an 8D Riemannian manifold. As before, the Christoffel symbols, $\Gamma^i_{\ jk}$, calculated with regard to $g(x, y)$, depend on y, that is $\Gamma^i_{\ jk} = \dfrac{1}{2}\gamma^{ih}\left(\dfrac{\partial \varepsilon_{hj}}{\partial x^k} + \dfrac{\partial \varepsilon_{hk}}{\partial x^j} - \dfrac{\partial \varepsilon_{jk}}{\partial x^h}\right)$.

3.5.1.1 *Einstein equations*

For our linearized model, the Einstein equations for the empty space [Miron and Anastasiei (1994, 1987); Balan and Brinzei (2005)] appear to be :

$$R_{ij} - \frac{1}{2}R\gamma_{ij} = 0 \qquad (3.78)$$

$$(\delta^i_s \delta^l_j - \gamma^{il}\gamma_{sj})\frac{\partial \Gamma^s_{li}}{\partial y^b} = 0 \qquad (3.79)$$

The first set of equations eq. (3.78) involves only the x-derivatives of the deformation, ε, while the second ones, eq. (3.79), contain mixed derivatives of ε of the second order. In order to integrate eq. (3.78), we apply the same procedure as in the classical Riemannian case and look for the solutions satisfying the harmonic (Lorentz) gauge conditions $\gamma^{ij}\Gamma^h{}_{ij} = 0$, which are actually

$$\frac{\partial \varepsilon^i_j}{\partial x^i} - \frac{1}{2}\frac{\partial \varepsilon}{\partial x^j} = 0. \qquad (3.80)$$

Consequently, eq. (3.78) becomes

$$\Box \varepsilon_{ij} = 0; \qquad (3.81)$$

which demonstrates explicitly the existence of the GW in the anisotropic space. We look for a wave solution in the form

$$\varepsilon_{jh} = Re(a_{jh}(y)e^{iK_m(y)x^m}), \qquad (3.82)$$

where i denotes the imaginary unit. Strictly speaking, we should take the perturbation as a series each term of which corresponds to a wave. But for simplicity, we regard only one term of this series. Both the amplitude $a_{jh}(y)$ and the wave vector $K_m(y)$ of the GW are no longer isotropic, but may depend on direction. Substituting eq. (3.82) into eq. (3.81), we see that either ε itself is zero, or we must have $\gamma^{hl}K_hK_l = 0$. By eq. (3.78), and eq. (3.80), we infer that in the anisotropic space $a_{jh}(y)$ and $K_m(y)$ should obey the algebraic system

$$\begin{cases} \gamma^{hl}K_hK_l = 0 \\ a^i{}_j K_i = \frac{1}{2}a^i{}_i K_j \end{cases} \qquad (3.83)$$

Thus, the wave solutions eq. (3.82) obeying eq. (3.80) of the Einstein equations eqs. (3.78, 3.79) must suffice:

$$\begin{cases} \gamma^{hl}K_hK_l = 0 \\ a^i{}_j K_i = \frac{1}{2}a^i{}_i K_j \\ \frac{\partial}{\partial y^b}(\frac{1}{2}a^i{}_i K_j) = \overset{0}{C}{}^i{}_{lb}(2a^l{}_j K_i - a^l{}_i K_j), \end{cases} \qquad (3.84)$$

where the third equation comes from the "mixed Einstein equation" eq.(3.79), $\overset{0}{C}{}^{i}{}_{lb} = \frac{1}{2}\gamma^{ih}\frac{\partial\gamma_{hl}}{\partial y^b}$. We also see that the amplitude $a^{i}{}_{j}$ and the wave vector K_i now depend on each other.

Such solutions will behave tensorially under coordinate transformations $\tilde{x}^i = \Lambda^i_j x^j$ ($\Lambda^i_j \in \mathbf{R}$) on the manifold $M = \mathbf{R}^4$ (which include homogeneous Lorentz transformations, [Carroll (1997)], in the Minkowski case).

3.5.1.2 *Eikonal equation*

In case when the space is anisotropic, the first approximation of the eikonal, $\psi(x^i, y^i)$, corresponding to some wave (e.g. to the EMW), can be written as $\psi = \psi_0 + \frac{\delta\psi}{\delta x^i}dx^i + \frac{\partial\psi}{\partial y^a}\delta y^a$. Then, eq. (3.17) becomes:

$$g^{ij}\frac{\delta\psi}{\delta x^i}\frac{\delta\psi}{\delta x^j} = 0, \tag{3.85}$$

where the "adapted derivative" $\frac{\delta}{\delta x^i}$ is characteristic for Finsler and Lagrange geometries, [Miron and Anastasiei (1987, 1994)] and insures the tensorial character of $\frac{\delta\psi}{\delta x^i}$. Under our assumptions on the metric structure and on coordinate transformations, this equation becomes simply:

$$g^{ij}\frac{\partial\psi}{\partial x^i}\frac{\partial\psi}{\partial x^j} = 0. \tag{3.86}$$

Let us look for the eikonal in the form $\psi(x,y) = \widehat{f}(x,y) + h\widehat{g}(x,y)$; ($h \ll 1, h^2 \simeq 0$) that satisfies eq. (3.86). With $g^{ij} = \gamma^{ij} - h\tilde{a}^{ij}\cos(K_m x^m)$, for our model, we get

$$\gamma^{ij}\widehat{f}_{,i}\widehat{f}_{,j} + h(2g^{ij}\widehat{f}_{,i}\widehat{g}_{,j} - \tilde{a}^{ij}\cos(K_m x^m)\widehat{f}_{,i}\widehat{f}_{,j}) = 0. \tag{3.87}$$

In search for the solutions sufficing

$$\begin{cases} \gamma^{ij}(y)\widehat{f}_{,i}\widehat{f}_{,j} = 0 \\ 2\gamma^{ij}\widehat{f}_{,i}\widehat{g}_{,j} - \tilde{a}^{ij}\cos(K_m x^m)\widehat{f}_{,i}\widehat{f}_{,j} = 0 \end{cases}, \tag{3.88}$$

we get the following class of solutions:

$$\widehat{f} = k_i(y)x^i + \Phi_1(y) \tag{3.89}$$

$$\gamma^{ij}k_i k_j = 0, \tag{3.90}$$

where $\Phi_1 = \Phi_1(y)$ is arbitrary. Here k_i is the wave 4-vector of the EMW, $k_0 = -\frac{\omega}{c}$, and K_m is the wave 4-vector of the GW. Considering $k^i = \eta^{ij}k_j$, we get

$$\widehat{g} = \frac{1}{2}\frac{\tilde{a}_{ij}k^{i}k^{j}}{K_{i}k^{i}} \sin\left(K_{i}x^{i}\right) + \Phi_{2}(y, x). \tag{3.91}$$

If we choose, for simplicity, $\Phi_1 = \Phi_2 = 0$, we obtain the eikonal

$$\psi = k_i(y)x^i + h\tilde{A}(y)\sin\left(K_i x^i\right), \quad \gamma^{ij}k_ik_j = 0, \ k_i = k_i(y) \tag{3.92}$$

where $\tilde{A}(y) = \dfrac{1}{2}\dfrac{\tilde{a}_{ij}k^{i}k^{j}}{K_{i}k^{i}} = \dfrac{1}{2}\dfrac{\tilde{a}^{ij}k_{i}k_{j}}{\gamma^{ij}K_{i}k_{j}}$. In anisotropic spaces, the components, k_i, generally also depend on the directional variables y^i.

3.5.1.3 *Generalized geodesics*

The Finslerian function, F, corresponding to the deformed locally Minkowskian metric, namely, $F^2 = (\gamma_{hl}(y) + \varepsilon_{hl}(x, y))y^h y^l$, leads to the Euler-Lagrange equations

$$\frac{\partial F^2}{\partial x^i} - \frac{d}{ds}(\frac{\partial F^2}{\partial y^i}) = 0, \tag{3.93}$$

that are equivalent to

$$g_{ij}^*\frac{dy^j}{ds} + \frac{1}{2}(\frac{\partial^2 F^2}{\partial y^i \partial x^j}y^j - \frac{\partial F^2}{\partial x^i}) = 0, \tag{3.94}$$

where s is the arclength $s = \int\limits_{0}^{t}F(x(\tau), y(\tau))d\tau$, and $g_{ij}^* = \dfrac{1}{2}\dfrac{\partial^2 F^2}{\partial y^i \partial y^j}$. Performing the computations, we get, in linear approximation,

$$\frac{dy^i}{ds} + \gamma^{it}(\Gamma_{tls} + \frac{1}{2}\frac{\partial \varepsilon_{sl}}{\partial x^j \partial y^t}y^j)y^s y^l = 0 \tag{3.95}$$

This equation for the simplest case, $\gamma^{ik}(y) = \eta^{ik} = diag\{1, -1, -1, -1\}$, was already used in the previous chapter and we saw that it has a physical meaning. For the locally Minkowskian space with small anisotropic deformation, the force potentials consist of two terms. The second term in brackets, originating from the anisotropy of the metric deformation, is associated with the velocity, it provides an analogue to the second term in the expression for the Lorentz force in electrodynamics and illustrates the ideas that were initially discussed in the end of [Siparov (2006a, 2007)].

If $\varepsilon_{ij}(x, y) = h\tilde{a}_{ij}(y)\cos(K_m(y)x^m)$, then, performing the derivations, we obtain the geodesic equations:

$$\frac{dy^i}{ds} + hA^i(y)\sin(K_m x^m) + hB^i_p(y)x^p \cos(K_m x^m) = 0 \qquad (3.96)$$

where the tensorial coefficients

$$A^i = -\frac{1}{2}\gamma^{it}y^l y^s [y^j \frac{\partial(K_j \tilde{a}_{sl})}{\partial y^t} + (K_s \tilde{a}_{tl} + K_l \tilde{a}_{ts} - K_t \tilde{a}_{ls})], \qquad (3.97)$$

$$\dot{B}^i_p = -\frac{1}{2}\gamma^{it}y^l y^s y^j K_j \tilde{a}_{sl}\frac{\partial K_p}{\partial y^t} = -\frac{1}{2}\gamma^{it}K\tilde{a}\frac{\partial K_p}{\partial y^t}$$

depend only on the directional variables y^i. Here $\tilde{a} \equiv \tilde{a}_{nm}y^n y^m$ and $K \equiv K_i y^i$. In particular, if K_i are constant, then $B^i_p = 0$, $i = 0 \div 3$ and the equations of geodesics simplify. Solving eq. (3.96), one can find that unit-speed geodesics of the perturbed metric $g_{ij}(x,y) = \eta_{ij}(x,y) + h\tilde{a}_{ij}(y)\cos(K_m(y)x^m)$ are described by

$$x^i(s) = \alpha^i s + \beta^i - \frac{h}{2}\eta^{it}\frac{\partial}{\partial y^t}\left(\frac{\tilde{a}}{K}\right)\sin(K_m x^m) - \frac{hx^p}{2}\eta^{it}\frac{\tilde{a}}{K}\frac{\partial K_p}{\partial y^t}\cos(K_m x^m),$$
$$(3.98)$$

where α^i and β^i depend on the initial conditions. In particular, if K_m are constant, geodesics of the perturbed metric obey

$$x^i(s) = \alpha^i s + \beta^i - \frac{h}{2}\gamma^{it}\frac{\partial}{\partial y^t}\left(\frac{\tilde{a}}{K}\right)\sin(K_m x^m). \qquad (3.99)$$

From eq. (3.98), we get that along geodesics $hx^i(s) \simeq h(\alpha^i s + \beta^i)$ and $hy^i(s) = h\frac{dx^i}{ds} \simeq h\alpha^i$.

3.5.2 *Weak anisotropic perturbation of the flat Minkowski metric*

In order to find explicitly the way in which the anisotropy could reveal itself in observations, we will search for those solutions which are as close to the solutions discussed in the previous Sections as possible. Let the initial metric be the flat Minkowskian one $\eta = diag(1, -1, -1, -1)$; then the system eq. (3.84) becomes

$$\begin{cases} \eta^{hl} K_h K_l = 0 \\ a^i{}_j K_i = \frac{1}{2}a^i{}_i K_j \\ (\frac{1}{2}a^i{}_i K_j)_{.b} = 0. \end{cases} \qquad (3.100)$$

Let us choose $K_0 = -K_1 = \dfrac{D}{c}$, where $D, c \in \mathbf{R}$, $K_2 = K_3 = 0$, and $a_2^2 = -a(y)$, $a_3^3 = a(y)$, $a^i{}_j = 0$ for all other (i, j). Then eq. (3.100) is identically satisfied and the perturbed Minkowski metric in the weakly anisotropic case can be expectedly presented as

$$
g_{ij} = \begin{pmatrix}
1 & 0 & 0 & 0 \\
0 & -1 & 0 & 0 \\
0 & 0 & -1 + a(y)\cos(\frac{D}{c}(x^0 - x^1)) & 0 \\
0 & 0 & 0 & -1 - a(y)\cos(\frac{D}{c}(x^0 - x^1))
\end{pmatrix},
$$

(3.101)

where $a(y)$ is an arbitrary scalar 0-homogeneous function, small enough such that $a^2 \simeq 0$. When $a(y)$ is a constant, this metric reduces to the perturbed Minkowski metric for the isotropic empty space and the results of the previous sections are valid.

3.5.2.1 *Eikonal*

Let $a(y)$ in the eq. (3.101) be equal to $a(y) = h\tilde{a}(y^i)$, where $\tilde{a}(y^i)$ is an arbitrary scalar 0-homogeneous function, $h^2 \simeq 0$. Then, $\tilde{A} = \dfrac{1}{2}\dfrac{\tilde{a}^{ij}k_i k_j}{\eta^{ij}K_i k_j} = \dfrac{1}{2}\dfrac{c^2\tilde{a}(y^i)(k_2^2 - k_3^2)}{D(ck_1 - \omega)}$, and the eikonal eq. (3.92) takes the form

$$
\psi = -\frac{\omega}{c}x^0 + k_1 x^1 + k_2 x^2 + k_3 x^3 + \frac{h}{2}\frac{c^2\tilde{a}(y^i)(k_2^2 - k_3^2)}{D(ck_1 - \omega)}\sin\left(K_i x^i\right). \quad (3.102)
$$

Example:

Let $v \in X(M)$ be an arbitrary vector field and $\tilde{a} = \dfrac{K_i y^i}{\eta_{ij}v^i y^j}$, where K is the wave vector. Thus, $\tilde{a}$ is globally defined. In the given local frame, in which $K_0 = \dfrac{D}{c}$, $K_1 = -\dfrac{D}{c}$, $K_2 = 0, K_3 = 0$ (which is chosen such that the GW propagates antiparallel to the Ox axis), $\tilde{a}$ is equal to $\dfrac{\frac{D}{c}(y^0 - y^1)}{\eta_{ij}v^i y^j}$. In a simple case, $v^i = 0, i = 1, 2, 3$ and $v^0 = -\dfrac{D}{c}$, we get

$$
\tilde{a} = \frac{y^0 - y^1}{y^3}
$$

hence,

$$
\psi = -\frac{\omega}{c}x^0 + k_1 x^1 + k_2 x^2 + k_3 x^3 + \frac{h}{2}\frac{c^2(y^0 - y^1)(k_2^2 - k_3^2)}{Dy^3(ck_1 - \omega)}\sin\left(K_i x^i\right).
$$

(3.103)

3.5.2.2 *Geodesics*

In the case of a small anisotropic perturbation of the Minkowski metric $diag(1, -1, -1, -1)$, geodesics are described by eq. (3.99):

$$x^i(s) = \alpha^i s + \beta^i - \frac{h}{2} \eta^{it} \frac{\partial}{\partial y^t} \left(\frac{\tilde{a}}{K} \right) \sin(K_m x^m).$$

In particular, for $\tilde{a} = \tilde{a}(\frac{y^0 - y^1}{\phi(y^2, y^3)})$ and $K_0 = \frac{D}{c}$, $K_1 = -\frac{D}{c}$, $K_2 = K_3 = 0$, $D, c \in \mathbf{R}$, we substitute it into eq. (3.99) and get $x^0 - x^1 = \alpha s + \beta$; $y^0 - y^1 = \alpha \in \mathbf{R}$. Particularly, the Cartesian Oy coordinate is

$$x^2 = \nu + u_0 \left\{ \frac{x^0 - x^1 - \beta}{\alpha} + \frac{hc}{2D\alpha} \frac{\partial \tilde{a}}{\partial y^2} \sin(\frac{D}{c}(x^0 - x^1)) \right\}, \qquad (3.104)$$

where $u_0, \nu \in \mathbf{R}$.

Examples:

1) For $a = h = const$, where $\tilde{a} = 1$, we get the expression obtained in the previous section and in [Siparov (2004)].

2) If $a = h \frac{y^0 - y^1}{y^3}$ as earlier, we get $a = \frac{h\alpha}{y^3} = \frac{1 + ha_1}{y^3}$ along geodesics, and, with $x^i(0) = 0$,

$$x^2 = u_0 \frac{x^0 - x^1}{1 + ha_1} + u_0 \frac{hc}{y^3 D} \sin(\frac{D}{c}(x^0 - x^1)). \qquad (3.105)$$

Then the Oy-component of the atom velocity, that is the component directed at the Earth, will contain a term of the form

$$y^2 \sim u_0 \frac{hc}{y^3} \cos(\frac{D}{c}(x^0 - x^1))$$

and the amplitude factor in front of the cosine depends on the velocity component, y^3, orthogonal to Ox and Oy axes. Therefore, one can see in which way the orientation of the (GW-source)-maser-Earth system may affect the conditions of the OMPR for the model regarded in this example.

3.5.3   *Modification of the OMPR conditions*

The physical interpretation of the obtained solutions leading to the modifications in the OMPR effect is the following. One can see that the anisotropy does not destroy the solution of the OMPR problem based on the Bloch's equations. For a simple anisotropic deformation of the Minkowski metric, we get the dependence of eqs. (3.103, 3.105) on the directional variable orthogonal to Ox and Oy, i.e. to the plane containing the Earth, the space maser and the GW source.

Geodesics describe the trajectory of the particle, and the sample eq. (3.105) means that the amplitude of the oscillations of the space maser atom velocity component oriented at the Earth, y^2, depends on y^3. This means that when the system "Earth-space maser-GW source" is located close to the periphery of the galaxy, the orientation of this system (see Fig. 3.17) might affect the OMPR conditions. For the last example given above,

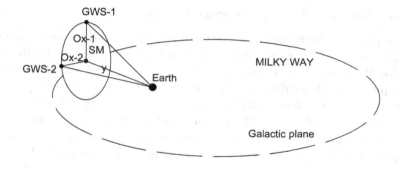

Fig. 3.17 Configuration of the astrophysical system for the galaxy geometrical properties investigation.

second and third of the OMPR conditions eq. (3.57) must be modified and take the form

$$\frac{\alpha_2}{\alpha_1} = \frac{\omega h}{8Dy^3} = b\varepsilon; b = O(1); \varepsilon << 1 \qquad (3.106)$$

$$\frac{kv_1}{\alpha_1} = \frac{\omega hc}{\alpha_1 y^3} = \kappa\varepsilon; \kappa = O(1); \varepsilon << 1 \qquad (3.107)$$

that illustrates the qualitative analysis given in [Siparov (2007)]. This means that the experimental investigation of the astrophysical systems with various orientations might provide the information on the quantitative characteristics of the geometrical anisotropy of our galaxy.

Thus, it has been shown that in the anisotropic case Einstein equations for the empty space still have wave solutions (gravitational waves), but now they become direction dependent and their amplitudes, $a_{jk}(y)$, and wave numbers, $K_j(y)$, can become coupled. After the corresponding generalizations of the equations for the eikonal and for the geodesics, they were used to find how the OMPR conditions change in the anisotropic space. It turned out that the orientation of the astrophysical system that takes part in the OMPR could cause the changes in the observable effect, thus, giving one the possibility to experimentally investigate the space-time geometrical properties on the galactic scale.

3.5.4 *Investigations of the space-time properties*

Let us present a superficial analysis [Siparov (2008b)] of the possible results of the OMPR based experiment that could be performed in future and discuss their meaning. If with time it is possible to get the GW signals from pulsars, such sources would become kind of beacons with frequencies known to the eight decimal digits. The more close perspective is to verify the applicability of the AGD approach and to obtain some geometrical characteristics of our galaxy. The investigation of the geometrical properties of space-time in our galaxy is based on the possibility to use one GW source for several masers or to use one space maser as a receiver of signals from several GW sources with regard to Fig. 3.17. It is important that the orientations of the systems correspond both to the galaxy plane and to the planes orthogonal to it.

In order to be able to use the suggested observations on the galaxy scale, sufficient amount of data must be stored. The observations should be performed and analyzed in the following way. The GW sources should be chosen in various places of the galaxy and several masers should be found around each of them. Then the fulfillment of the OMPR conditions eq. (3.57) must be checked for every system and the search for the OMPR effect should be performed with the help of radio telescopes. The same should also be done for the systems consisting of one maser and several GW sources acting upon it. All the systems fall into two groups: the first includes the systems that are closer to the inner part of the galaxy (IPG), the second includes the systems that are closer to the periphery part of the galaxy (PPG). The observations analysis may give only nine possible outcomes listed below. They also form two groups: the first deals with the situations when we cannot track any anisotropy in our galaxy, the second can point at the existence of anisotropy and provide data that can be used to measure it. The first group of possible outcomes contain:

1: OMPR signal is found in all the suitable systems with various orientations corresponding both to the IPG and to the PPG and is in accord with predictions made on the base of eq. (3.57).

Consequences: another specific test of AGD is needed, GW astronomy is possible.

2: No OMPR signal for all the systems corresponding to the IPG and to the PPG.

Interpretation and consequences: a) OMPR theory doesn't work; b) there are no gravitation waves ⇒ Einstein equation for the empty space doesn't have the structure of the wave equation ⇒ something is wrong with the classical GRT. In case b) there is a need to turn back to the choice of scalar in the variation principle, maybe conformal gravity based on Weyl tensors would help. There also appear problems with the interpretation of [Taylor and McCulloch (1980)] results for the decrease of orbital period. Notice that until all this is fixed, the introduction of dark matter would be hasty.

3,4: OMPR signal is either present for all the systems corresponding to the IPG and absent for all the systems corresponding to the PPG or vice versa.

Interpretation and consequences: the outcomes point at the different properties of geometrodynamics in the IPG and PPG. Still, no direct evidence of the influence of anisotropy is provided. GW astronomy is possible in the region where the OMPR effect is observed.

The rest five situations realize when instead of equations eqs. (3.57), the corresponding equations of the type eqs. (3.106,3.107) should be used to describe the conditions of OMPR. That is the behavior of some systems follows the predictions based on the regular OMPR conditions, while other systems - presumably with the alternative orientation - do not show the effect. This would point at the anisotropy discussed in the previous section.

5: OMPR signal corresponding to expectations is present only for the systems with certain orientation allocated both in the IPG and in the PPG.

Interpretation and consequences: confirmation of the validity of the AGD approach, GW astronomy is possible with regard to the anisotropy of phase space-time.

6,7: OMPR signal is either present in the systems with certain orientation corresponding to the IPG and absent in all systems corresponding to the PPG, or vice versa.

Interpretation and consequences: the outcomes point at the presence of space-time anisotropy in the regions where the OMPR effect is present. In these regions the GW astronomy is possible with regard to the anisotropy of phase space-time. In the regions where there is no OMPR effect, some other properties of geometrodynamics are involved.

8,9: OMPR signal is either present for all the systems corresponding to the IPG and present only in the systems with certain orientation corresponding to the PPG, or vice versa.

Interpretation and consequences: the outcomes point at the presence of space-time anisotropy in the regions where the OMPR effect is present only for certain orientations of the astrophysical systems. In these regions the GW astronomy is possible with regard to the anisotropy of phase space-time. In the regions where the effect is present for all orientations, the influence of anisotropy becomes vanishingly small.

In Table 3.4, there are the examples of the pairs of masers corresponding to the GW sources located both in the IPG and in the PPG. If it happens that the situations 5-9 are realized in observations, then the systematic observations interpreted with the help of the expression like eq. (3.82) could provide the coefficients for the perturbation of the locally Minkowskian metric

$$g_{ij}(x,y) = \gamma_{ij}(y) + \varepsilon_{kj}(x,y) \qquad (3.108)$$

Thus, if the phase space-time anisotropy is observable on the galactic scale in our galaxy, then its quantitative characteristics could be obtained in the OMPR based observations.

Appendix A

Optic-mechanical parametric resonance

A.1 Brief review

A.1.1 *Mono-chromatic excitation*

Atom is a complex system, and in the theoretical analysis the simplified models of atom with the limited number of levels are used. The simplest model is, of course, a two-level atom (TLA). Nevertheless, such model preserves many principal features of the real atom and is widely used in scientific research.

The study of the interaction between a single TLA and intense electromagnetic field appeared to be very fruitful for the quantum optics and laser spectroscopy. In one of the earliest papers, Autler and Townes [Autler and Townes (1955)] described the atom excitation by the strong radio frequency field, and the absorption and fluorescence were measured for the weak probe radiation which corresponded to the transition from one of the excited atomic energy levels to another one. In spectroscopy, when they speak of the strong field, they mean such a field that makes the stimulated transitions dominate over the spontaneous ones. It is convenient to characterize the strength of the field with its Rabi parameter, $\alpha = \frac{E\mu}{\hbar}$ (E is the electric stress of the field, $\hbar$ is Planck constant, μ is the induced dipole moment of the atom) which is also called Rabi frequency because its measurement units are those of frequency. Besides, it is this value that characterizes the frequency of the population oscillation between upper and lower levels when the stimulated emission dominates and the atom reaches a stationary absorbing-emitting state. The interest in these questions grew up when the stability and intensity of laser radiation made it possible to investigate the optical region. Mollow [Mollow (1969)] calculated the fluorescence spectrum of the two-level atom driven by the quasi-resonant monochromatic laser field whose Rabi frequency was essentially larger than the decay rate of the excited state. Later this spectrum was observed in the experiments of several groups of researchers. It consists of the steep coherent component with the frequency of the laser field and the three-peak non-coherent component whose lines correspond to the laser frequency and two sidebands shifted from the central peak by Rabi frequency. This structure of the fluorescence spectrum in the strong field is known as 'Mollow triplet', though, to be fair, one should mention Burshtein's paper [Burshtein (1962)] in which this structure was obtained seven years earlier. In analogue to the splitting of atomic spectral lines in the constant electric field, this splitting was called dynamic Stark effect. If a laser is tuned exactly at the transition frequency, the absorption spectrum in the Autler-Tauns effect contains two stationary sidebands symmetrically shifted from the resonance frequency. More detailed reviews can be found in [Letohov and Chebotaev (1990); Stenholm (1984)].

A.1.2   *Poly-chromatic excitation*

When the excitation is performed with the help of the poly-chromatic (multi-component) field, the physical situation complicates, and the absorption and fluorescence spectra complicate too. Their views essentially depend on the number of the excitation field components, on their Rabi parameters, on their detuning from the atomic transition frequency. Bonch-Bruevich, Vartanian and Chigir' were the first [Bonch-Bruevich *et al.* (1979)] to use the strong probe radiation for the investigation of the two-level atom in the strong excitation field. A series of shallow resonances was discovered at the frequencies $\omega_i = \omega_0 \pm \frac{\alpha_R}{n}$, i.e. at the detunings from the atomic transition frequency equal to the multiples of Rabi frequency. That is why they are called 'subharmonic resonances', and their appearance is considered to be due to the n-photon absorption between the atomic levels. In experiments performed by Lunis, Jelezko and Orrit [Lounis *et al.* (1997)], the subharmonic resonances in the admixture molecule in solid were observed for the first time. The subharmonic spectroscopy has applications for the lasers working in the inversionless regime.

Poly-chromatic fields are also used for laser cooling of atoms. An atom moving in the field of standing wave can be regarded as an atom moving in the field of two waves propagating towards each other. Then the velocity dependence of the force acting on this atom must have several sharp peaks that Kirola and Stenholm [Kyrola and Stenholm (1977)] called dopplerons, they were subsequently experimentally discovered by Bigelow and Prentis [Bigelow and Prentiss (1990)]. In another case, an ion harmonically oscillating in a trap can be cooled by the laser radiation [Wineland and Dehlmelt (1975)]. It was also shown that the cooling is most effective when the excitation laser frequency is detuned down from the transition frequency by the frequency of the ion oscillations.

A.1.3 Bichromatic fields

Zhu, Lezama Wu and Mossberg [Zhu *et al.* (1989)] measured the fluorescence spectrum of the two-level atom when the excitation radiation consisted of the two components that had equal amplitudes and were symmetrically detuned from the transition frequency. They found out that the spectrum consists of the central component with the frequency of the atomic transition and of the series of sidebands that were equidistant from the central one with the step equal to the detuning frequency. Surprisingly, the displacements of the sidebands did not depend on Rabi frequency which still played an important role and defined the total number and relative intensity of the sidebands in a rather complicated manner. This was completely different from the measurements obtained earlier for Mollow triplet when the displacements of the sidebands linearly depended on Rabi frequency, and their total number and intensities did not depend on it. In a series of papers [Freedhoff and Ficek (1997); Ficek *et al.* (1999)] the interaction between the bichromatic field and atoms was regarded. In the first of them they investigated the effect of the excitation of the two-level atom by the two strong waves that had the same frequency but different intensities. They calculated the fluorescence spectrum and Autler-Townes absorption spectrum of such a "dressed atom". It was found that the spectrum of such double dressed atom consists of a staircase sequence of doublets. These sequences appeared in the fluorescence spectrum instead of the sidebands of the Mollow triplet and the Autler-Townes absorption spectrum also becomes the sequence of doublets. In the papers [Rudolph *et al.* (1998)], they analyzed the fluorescence spectrum of the two-level atom excited by the two laser fields of different frequencies and amplitudes. They regarded a strong field resonant to the transition and a weak one that was detuned from the main field and from its subharmonics by Rabi frequency. The spectra were essentially different from Mollow triplet characteristic for the monochromatic excitation field, they consisted of the triplets located at the harmonics and subharmonics of Rabi frequency of the strong field. It was discovered that such a multi-peak spectrum structure originates from the splitting of the energy levels produced by the strong and weak fields. It was suggested that the splitting appears when the weak field interacts with the levels in a multi-photon process. In addition to the level splitting, it was found that the weak field shifts the resonances - the effect that could be related to the shift in the dynamic Stark effect. When the weak field is tuned at the shifted resonance, there is no fluorescence at the atomic frequency. These results were obtained numerically and explained in terms of double dressed atoms.

A.1.4 *Mechanical action on atoms*

Mechanical action of the electromagnetic field on atoms in the gaseous medium is a well-known phenomenon studied both theoretically and experimentally. It is not only interesting from the point of view of the fundamental research but also has practical applications such as monochromatization of velocities in the atomic ensembles by means of laser radiation, focusing and collimation of the atomic beams, slowing down the atomic beams and cooling the atoms to the temperatures essentially less than liquid helium temperature. In particular, it was possible to achieve 0.1-0.01 K for a thermal atomic beam and to cool the localized ions down to the hundredth of Kelvin.

A remarkable feature of the resonant light pressure physics that plays an important role is the intermixing of mechanical and optical phenomena. The importance is due to the possibility to perform a frequency and velocity selection in the mechanical action (pressure) of the monochromatic laser radiation on atoms and ions. The frequency selectivity that appears while acting on atoms by the monochromatic radiation is possible because an atom effectively scatters photons of the monochromatic radiation only if they are resonant to the one of the atomic transitions. Here follows the possibility to use the light pressure for the selective action only upon definite atoms, for example, on the atoms with the same isotopic composition.

At the same time, one can notice that due to Doppler effect, such resonant interaction depends on the magnitude and direction of the atomic velocity. And here appears the possibility for the velocity selection with the help of the monochromatic light pressure on atoms, that is the dependence of the intensity of photon scattering on the magnitude and direction of the atomic velocity. Really, the light pressure on a moving atom is essential only if the Doppler shift, kv is compensated by the corresponding detuning of the light source frequency, $\Omega = \omega - kv$. Here Ω is the light frequency, ω is the atomic transition frequency, k is the wave vector, v is the atom velocity. That is why the pressure of the monochromatic light acts only on atoms that have a certain velocity. The demand for the intensities of the sources appears to be moderate: the intensity of 0.1 W/cm^2 is enough to saturate the atomic optical transitions.

There are two most important groups of problems related to the resonance interaction between the electromagnetic field and atoms. The first of them deals with the laser spectroscopy of ultra-high resolution and with the construction of the precision-built frequency standards. The possibility to change atoms' velocities by means of the resonant light action led to a principally new idea of deep cooling of the atomic particles [Wineland and Dehlmelt (1975); Dehlmelt (1976)]. The theoretical study showed that with the help of radiational cooling the lowest temperature that can be achieved is defined by the natural width of the atomic transition [Javanainen (1980)]. For typical optical transitions the order of minimal temperature is about $10^{-3} - 10^{-4}$K. The study of the radiation cooling of atomic particles is one of the main directions in the problem of resonant light pressure [Stenholm (1978a)]. The theoretical analysis of the possibility to obtain the cold atoms by the slowing down of the atomic beams with the help of the oncoming radiation can also be found in [Minogin *et al.* (1981)]. The theoretical problems of the radiation cooling of the atomic ions located in electromagnetic traps were regarded in [Javanainen (1980, 1981)].

The second group of problems deals with the localization of the cold atoms in the electromagnetic (light) field [Letokhov and Pavlik (1976)]. The use of the non-resonant light field for the stable localization even of the ultra-cold atoms calls for high intensity of laser radiation and leads to experimental difficulties. In case of the resonant radiation, the demands for intensity are lower. The theoretical study of this problem showed that the stable localization of the cold atoms is possible when the atoms are in the coherent superposition of the ground and excited states. Studying the change of the atomic momentum in the field of the plane running wave, one can see that the stimulated transitions do not change the average momentum. The radiational force is defined only by the spontaneous transitions rate, γ. They speak about the force of the spontaneous light pressure, and it does not exceed $f_{sp} = \hbar k \gamma / 2$ [Ashkin (1970)]. Since this force is small, the translational motion changes slowly, and the atomic dynamics is rather simple. The possibilities of the laser action increase, when the non-homogeneous fields like focused beams and their combinations or oncoming and standing waves are used. The induced processes in such fields play an important role, and the force of the light pressure depends on the gradients of the field. The stimulated light pressure could surpass f_{sp} already for comparatively weak swept laser. This is especially characteristic for the field of standing waves when the gradient can reach the highest values. Turning back to the forces arising when the electromagnetic field interacts with atoms, one can mention the following possibilities to create a potential well for the cold atoms: the use of the gradient force produced by the several (orthogonal) intersecting laser beams; the joint use of the gradient force and of the light pressure force in the field of the two light beams coming from the opposite directions; the use of the light pressure force acting on an atom in the light field with central symmetry. These effects make it possible to store the cold atoms (including single ones) in traps. This is important for the precision experiments in atomic physics and spectroscopy. For example, the cold localized atoms can be used as the frequency standards. If there are many cold atoms accumulated in a trap, this system is interesting from the point of view of atoms scattering at ultra low temperatures. Phase transitions like Bose-condensation are also possible in such ensembles.

Thus, we see that though there is a lot of applications of the resonant interaction between the electromagnetic field and atoms, though there are multiple hints at the significance of such parameter as Rabi frequency, in the above there was still no idea how to use the possible parametric resonance, i.e. the coincidence of certain parameters of the system other than the energy of a photon and the difference between energies corresponding to the atomic levels. Usually, they speak of the parametric resonances [Naife (1984)] when a weak perturbation acting upon a system has a frequency close to the difference between the proper frequencies of the unperturbed system. In [Fradkin (1983)] the parametric resonances were the situations when an intersection or a quasi-intersection of the quasi-levels of the TLA located in the modulated field took place. As it follows from [Kazakov (1990)], when a bichromatic field with a weak additional wave acts on a TLA, these definitions of the parametric resonances coincide. In papers [Kazakov (1992, 1993a)] the influence of the parametric resonance on absorption spectra in the gases of two- and three-level atoms were studied. There appears a possibility to use the parametric resonance effects in sub-Doppler spectroscopy. In [Kazakov (1993b)] it was shown that the presence of the parametric resonance can lead to the (periodic) amplification of the probe wave when there is no population inversion.

All these issues are rather far from the main topic of this book. They are given here in order to show what kind of phenomenon is going to be discussed below and what is its place in the well-developed field of theoretical spectroscopy to which it naturally belongs. Its application to the field of relativistic physics can be grounded better, if there is an understanding how it works in laboratory where it could be studied in detail.

A.2 Force acting on a two-level atom

In this section we use the TLA model and regard the situation when an atom suffers the action of a specially chosen bichromatic field of laser radiation. We study the mechanical action of such field on an atom for the case when the conditions of parametric resonance are fulfilled. Here a bichromatic field is understood as a field consisting of two slightly distuned monochromatic waves resonant to the atomic transition and running in the same direction. Then the result of parametric resonance could be the appearance of a specific feature in the atom dynamics.

Let the atom be placed in the field of resonant wave ($\Omega_1 \approx \omega$ where Ω_1 is the wave frequency). The field is strong, i.e. the electric component of the field, E_1 is such that Rabi parameter, $\alpha_1 = \frac{E_1 \mu}{\hbar}$ is essentially larger than the decay rate, γ (the inverse lifetime of the excited state of an atom). This wave will be called the main one. In the presence of a strong field, the populations of the TLA levels would periodically change with the frequency defined by the Rabi frequency. The ordinary laboratory instrumentation for the optical spectroscopic measurements provides the value of Rabi parameter of the order $10^8 - 10^{10}$ s^{-1}, that is Rabi frequency belongs to the radio band of the wavelengths.

Let us now have an additional monochromatic wave that acts on an atom and has the frequency, Ω_2 close to the resonant frequency, therefore, the field is bichromatic. We demand this frequency to suffice

$$\Omega_1 - \Omega_2 = 2\alpha_1 \tag{A.1}$$

Now we demand that the amplitude of the additional wave is such that its Rabi frequency, $\alpha_2 = \frac{E_2 \mu}{\hbar}$ is of the same order as the atomic decay rate, γ. Then this wave is a small perturbation in comparison to the main one. At the same time its action during the lifetime of the excited state of an atom leads to the distortion of the atom's dynamics and causes an essential deformation of the probe wave absorption spectrum and the reconstruction of the quasi-energy states [Kazakov (1990, 1992, 1993a)]. What would be

the mechanical action of such bichromatic light field on an atom when the
additional mode is weak and the parametric resonance condition eq. (A.1)
is fulfilled?

A.2.1   *Dynamics of an atom in the bichromatic field*

Let us describe the physical problem. There are atoms modelled by the
approximate but reasonable TLA model. The populations of the upper
and lower levels are ρ_{22} and ρ_{11}, and in the free state, $\rho_{11} = 1$. The lifetime
of the upper level is the inverse of the decay rate, and γ, and γ_{12} are the
longitudinal and transversal decay rates of the atom (since in our model
level 1 is the ground level, which corresponds to $\rho_{22} + \rho_{11} = 1$, we have $\gamma_{12} = \gamma/2$). Now we put the atom into a strong resonant field with frequency Ω_1
accompanied by a weak field with frequency Ω_2, both frequencies are close
to the atomic transition frequency, ω. The wave vectors of these components
in our case can be considered equal because of the space averaging, and
$k_1 = k_2 = k$. If there were only one strong wave, Ω_1, the population starts
to occupy upper and lower level in turn, and these oscillations will be defined
by Rabi frequency, α_1. But in case there are two waves, the situation can
change.

The problem of the TLA dynamics in the bichromatic field with a weak
additional mode cam be regarded as a perturbation theory problem. The
corresponding equations [Stenholm (1984)] have the form

$$\frac{d}{dt}\rho_{22} = -\gamma\rho_{22} + 2i[\alpha_1\cos(\Omega_1 t - k_1 y) \tag{A.2}$$

$$+\alpha_2\cos(\Omega_2 t - k_2 y)](\rho_{21} - \rho_{12})$$

$$\frac{d}{dt}\rho_{12} = -(\gamma_{12} + i\omega)\rho_{12} - 2i[\alpha_1\cos(\Omega_1 t - k_1 y)$$

$$+\alpha_2\cos(\Omega_2 t - k_2 y)](\rho_{22} - \rho_{11}) \tag{A.3}$$

$$\rho_{22} + \rho_{11} = 1$$

$$\frac{d}{dt} = \frac{\partial}{\partial t} + v\frac{\partial}{\partial y} \tag{A.4}$$

Our goal is to find the expression for the force acting on an atom,
but the force changes the atom velocity and could take the atom out of
the resonant interaction with the electromagnetic field. Therefore, strictly
speaking, eq. (A.2) should be supplemented by the equation describing the
atomic motion. But as it will be shown below, the velocity change caused
by the force is negligibly small in comparison with the Doppler contour

change that is needed for the break of resonance. That is why here and below the velocity will be considered a parameter of the problem.

In order to solve eq. (A.2), let us use the substitutions of the so-called rotating wave approximation justified in [Stenholm (1984)]

$$\rho_{21} = R_{21}e^{i(\Omega_1 t - ky)}$$
$$\rho_{12} = R_{12}e^{-i(\Omega_1 t - ky)}$$
$$\rho_{22} = 2^{-1/2}\widetilde{\rho_{22}}$$

and introduce the following notations

$$2^{1/2}\alpha_1 t = \tau; \quad \frac{\Omega_2 - \Omega_1}{\alpha_1} = 2^{1/2}\delta_d; \quad \frac{\gamma}{\alpha_1} = 2^{1/2}\varepsilon; \quad \frac{\gamma_{12}}{\alpha_1} = 2^{1/2}\Gamma\varepsilon; \quad \frac{k}{\alpha_1} = 2^{1/2}\kappa;$$
$$\frac{\omega - \Omega_1}{\alpha_1} = 2^{1/2}\delta; \quad \frac{\alpha_2}{\alpha_1} = a\varepsilon$$

Here τ is the dimensionless time counted in the number of Rabi periods of the main wave; δ_d is the dimensionless distuning between the main and additional waves counted in Rabi frequencies; ε is the dimensionless small parameter characterizing the spectroscopic strength of the field; Γ is the dimensionless transversal decay rate; δ is the dimensionless detuning between the main wave and the atomic transition frequencies; κ remains dimensional and is defined in such a way that the combination κv is dimensionless; a is the dimensionless amplitude of the additional (weak) wave normalized by the amplitude of the main wave. The $2^{1/2}$ factor is not obligatory and is introduced here to make the form of the equations fit the form already used in [Kazakov and Siparov (1997)].

Then we come to the following system of equations

$$\frac{\partial}{\partial \tau}\mathbf{W} = (\mathbf{Q}_0 + \varepsilon\mathbf{Q}_1(\tau))\mathbf{W} + \mathbf{C}(\tau) \tag{A.5}$$

where vectors and matrices are

$$\mathbf{W} = \begin{pmatrix} \widetilde{\rho_{22}} \\ R_{12} \\ R_{21} \end{pmatrix}; \quad \mathbf{Q}_0 = i\begin{pmatrix} 0 & -1 & 1 \\ -1 & -\sigma & 0 \\ 1 & 0 & \sigma \end{pmatrix}; \quad \mathbf{C} = i\sqrt{2}\begin{pmatrix} 0 \\ 1 + \varepsilon a e^{-i\delta_d\tau} \\ -1 - \varepsilon a e^{i\delta_d\tau} \end{pmatrix};$$

$$\mathbf{Q}_1 = \begin{pmatrix} -1 & -iae^{i\delta_d\tau} & iae^{-i\delta_d\tau} \\ -iae^{-i\delta_d\tau} & -\Gamma & 0 \\ iae^{i\delta_d\tau} & 0 & -\Gamma \end{pmatrix}; \quad \sigma = \delta + \kappa v$$

and σ characterizes the dimensionless detuning of the main wave from the atomic transition with account to Doppler shift. Let us specify the mathematical assumptions that characterize the problem. We are going to construct the stationary solution of the system described by eq. (A.5), when $\varepsilon << 1$, and the other parameters such as a, δ_d, Γ, satisfy the condition $a, \delta_d, \Gamma \sim O(1)$. Now we designate $J = (\sigma^2 + 2)^{1/2}$, and one can check that the eigenvalues of matrix $\mathbf{Q}_0$ are $0, iJ$ and $-iJ$. In order to construct the solution of a system of equations that looks like eq. (A.5), they usually exploit the perturbation theory and find the expansion of the solution in the powers of small parameter. But we see that matrix $\mathbf{Q}_1$ contains the harmonics whose frequencies could coincide (or be asymptotically close) with the difference of matrix $\mathbf{Q}_0$ eigenvalues. This could lead to parametric resonances [Kazakov (1993b)], and the standard procedure of the perturbation approach demands modification. Constructing the solution of eq. (A.5), we will use the many scales method - similarly to what was done in [Kazakov (1992)] and obtain the solution as a principal order of the asymptotic expansion.

A.2.2 *Stationary dynamics of atom*

Let us first find the solution of the homogeneous problem related to eq. (A.5):

$$\frac{\partial}{\partial \tau} \mathbf{N}(\tau) = (\mathbf{Q}_0 + \varepsilon \mathbf{Q}_1(\tau)) \mathbf{N}(\tau) \qquad (A.6)$$
$$\mathbf{N}(0) = \mathbf{I}$$

where $\mathbf{I}$ is a unit matrix. Let us introduce the auxiliary matrix

$$\mathbf{U} = \frac{1}{J} \begin{pmatrix} \sigma & 1 & 1 \\ -1 & -\frac{1}{J+\sigma} & \frac{1}{J-\sigma} \\ -1 & \frac{1}{J-\sigma} & -\frac{1}{J+\sigma} \end{pmatrix}$$

with the property $\mathbf{U}^{-1} \mathbf{Q}_0 \mathbf{U} \equiv \mathbf{E} = diag\{0, iJ, -iJ\}$. Notice, that eq. (A.1) together with the definition of matrix $\mathbf{Q}_0$ eigenvalue, J, give $J = 2\alpha_1$, where α_1 is the Rabi frequency of a TLA moving with velocity v in the field of the main wave. If $\mathbf{N}(\tau) = \mathbf{U} \exp(\mathbf{E}\tau) \mathbf{L}(\tau)$, then

$$\frac{\partial}{\partial \tau} \mathbf{L} = \varepsilon \mathbf{G}(\tau) \mathbf{L} \qquad (A.7)$$

where $\mathbf{G}(\tau) = \exp(-\mathbf{E}\tau)\mathbf{U}^{-1}\mathbf{Q}_1(\tau)\mathbf{U}\exp(\mathbf{E}\tau)$. Let us introduce the so-called slow time $\tau_1 = \varepsilon\tau$, and look for the solution of equation

$$(\frac{\partial}{\partial\tau} + \varepsilon\frac{\partial}{\partial\tau_1})\mathbf{L}(\tau, \tau_1) = \varepsilon\mathbf{G}(\tau)\mathbf{L}(\tau, \tau_1) \tag{A.8}$$

in the form

$$\mathbf{L}(\tau, \tau_1) = \mathbf{L}_0(\tau, \tau_1) + \varepsilon\mathbf{L}_1(\tau, \tau_1) + ... \tag{A.9}$$

as it is prescribed by the general procedure [Naife (1984)]. Substituting eq. (A.9) into eq. (A.8), we obtain the recurrent system of equations for matrices $\mathbf{L}_n(\tau, \tau_1)$. The first two of these equations are

$$\frac{\partial}{\partial\tau}\mathbf{L}_0(\tau, \tau_1) = 0 \tag{A.10}$$

$$\frac{\partial}{\partial\tau}\mathbf{L}_1(\tau, \tau_1) = \mathbf{G}(\tau)\mathbf{L}_0(\tau, \tau_1) - \frac{\partial}{\partial\tau_1}\mathbf{L}_0(\tau, \tau_1)$$

The first of equations in eq. (A.10) gives $\mathbf{L}_0(\tau, \tau_1) = \mathbf{L}_0(\tau_1)$. With regard to the general rules of many scales method, the second equation gives

$$\frac{\partial}{\partial\tau_1}\mathbf{L}_0(\tau_1) = <\mathbf{G}(\tau)> \mathbf{L}_0(\tau_1) \tag{A.11}$$

where $< ... >$ means the removing of all the high frequency harmonics from the matrix $\mathbf{G}(\tau)$. According to eq. (A.8), in matrix $\mathbf{G}(\tau)$, the harmonics $\pm(J \pm \delta_d), \pm(2J \pm \delta_d)$ may be present. It can be shown that the harmonics $\pm(2J \pm \delta_d)$ are absent. Thus, we have to study the possible parametric resonances that correspond to the conditions

$$J = \delta_d + \varepsilon\nu; \nu = O(1) \tag{A.12}$$

and

$$J = -\delta_d + \varepsilon\nu; \nu = O(1) \tag{A.13}$$

where parameter ν describes the accuracy of tuning. As we will see, it doesn't matter which sign of δ_d corresponds to the parametric resonance condition. Further, we will consider eq. (A.12). Substituting eq. (A.12) into eq. (A.8) and changing $\varepsilon\tau = \tau_1$, we come to the equation

$$\frac{\partial}{\partial\tau}\mathbf{L}_0(\tau_1) = \mathbf{G}_1(\tau)\mathbf{L}_0(\tau_1);$$

$$\mathbf{G}_1(\tau) = \begin{pmatrix} -\sigma^2 - 2\Gamma & iaJ\frac{\exp(i\nu\tau_1)}{J-\sigma} & -iaJ\frac{\exp(-i\nu\tau_1)}{J-\sigma} \\ iaJ\frac{\exp(-i\nu\tau_1)}{J-\sigma} & -1 - \Gamma(1+\sigma^2) & 0 \\ -iaJ\frac{\exp(i\nu\tau_1)}{J-\sigma} & 0 & -1 - \Gamma(1+\sigma^2) \end{pmatrix}$$

This system of equations can be transformed into the system with constant coefficients in the following way. Let us designate $\mathbf{Z}(\tau_1) = diag\{1, \exp(-i\nu\tau_1), \exp(i\nu\tau_1)\}$ and let $\mathbf{L}_0(\tau_1) = \mathbf{Z}(\tau_1)\mathbf{S}(\tau_1)$, then for $\mathbf{S}(\tau_1)$ we have

$$\frac{\partial}{\partial\tau_1}\mathbf{S}(\tau_1) = \mathbf{K}\mathbf{S}(\tau_1); \tag{A.14}$$

$$\mathbf{K} = \begin{pmatrix} -\sigma^2 - 2\Gamma & \frac{iaJ}{J-\sigma} & -\frac{iaJ}{J-\sigma} \\ \frac{iaJ}{J-\sigma} & -1 - \Gamma(1+\sigma^2) + i\nu & 0 \\ -\frac{iaJ}{J-\sigma} & 0 & -1 - \Gamma(1+\sigma^2) - i\nu \end{pmatrix}$$

Solving eq. (A.14), we find

$$\mathbf{L}_0(\tau, \tau_1) = \mathbf{Z}(\tau_1)\exp(\mathbf{K}\tau_1)$$

Turning back to matrix $\mathbf{N}(\tau)$, we get

$$\mathbf{N}(\tau) = \mathbf{U}\exp(\mathbf{E}\tau)\mathbf{Z}'(\tau)\exp(\varepsilon\mathbf{K}\tau)\mathbf{U}^{-1}$$

where $\mathbf{Z}'(\tau) = diag\{1, \exp(-i\varepsilon\nu\tau), \exp(i\varepsilon\nu\tau)\}$. The definition of parameter ν gives $\exp(\mathbf{E}\tau)\mathbf{Z}'(\tau) = \exp(i\mathbf{d}\tau)$, where matrix $\mathbf{d} = diag\{0, \delta_d, -\delta_d\}$. Therefore,

$$\mathbf{N}(\tau) = \mathbf{U}\exp(i\mathbf{d}\tau)\exp(\varepsilon\mathbf{K}\tau)\mathbf{U}^{-1} \tag{A.15}$$

Expression eq. (A.15) presents the main term of the asymptotic expansion of the solution of the homogeneous equation eq. (A.6) in the powers of ε. The solution of the non-homogeneous equation eq. (A.5) can be built with the help of eq. (A.15) as

$$\mathbf{W}(\tau) = \mathbf{N}(\tau)\mathbf{W}(0) + \mathbf{N}(\tau)\int_0^\tau [\mathbf{N}(x)]^{-1}\mathbf{C}(x)dx$$

An important circumstance should be underlined. Matrix $\mathbf{K}$ has three eigenvalues, and their real parts are strictly negative. That is why we can throw away the first term on the right-hand side of the last equation when constructing the stationary solution. This leads to

$$\mathbf{W}(\tau) = \mathbf{U}\exp(id\tau)\exp(\varepsilon\mathbf{K}\tau)\int_0^\tau \exp(-\varepsilon\mathbf{K}x)\exp(-idx)\mathbf{U}^{-1}\mathbf{C}(x)dx$$

(A.16)

In the integral on the right-hand side of eq. (A.16) there are the harmonics with frequencies $-\varepsilon K_n \pm i\delta_d$, $-\varepsilon K_n$, where K_n are the eigenvalues of matrix $\mathbf{K}$. Then in the principal term of the asymptotic expansion we have

$$\exp(-idx)\mathbf{U}^{-1}\mathbf{C}(x) = \mathbf{C}_0 + \mathbf{C}_1\exp(i\delta_d x) + \mathbf{C}_2\exp(2i\delta_d x)$$
$$+\mathbf{C}_{-1}\exp(-i\delta_d x) + \mathbf{C}_{-2}\exp(-2i\delta_d x) \text{ (A.17)}$$

where

$$\mathbf{C}_0 = \varepsilon\widetilde{C_0} = \frac{i\varepsilon a}{\sqrt{2}J(J-\sigma)}\begin{pmatrix}0\\-1\\1\end{pmatrix}; \mathbf{C}_1 = \frac{i}{\sqrt{2}}\begin{pmatrix}0\\0\\1\end{pmatrix}; \mathbf{C}_{-1} = -\frac{i}{\sqrt{2}}\begin{pmatrix}0\\1\\0\end{pmatrix};$$

$$\mathbf{C}_2 = \frac{i\varepsilon a}{\sqrt{2}J(J+\sigma)}\begin{pmatrix}0\\0\\1\end{pmatrix}; \mathbf{C}_{-2} = -\frac{i\varepsilon a}{\sqrt{2}J(J+\sigma)}\begin{pmatrix}0\\1\\0\end{pmatrix}$$

Now substitute eq. (A.17) with regard to explicit form of its terms into the right-hand side of eq. (A.16). Throwing away the small terms in the asymptotes and preserving only the main ones, we get for the stationary dynamics of vector $\mathbf{W} = \begin{pmatrix}\widetilde{\rho_{22}}\\R_{12}\\R_{21}\end{pmatrix}$ the following expression

$$\mathbf{W}_c(\tau) = \mathbf{U}\{i\delta_d^{-1}\mathbf{C}_{-1} - i\delta_d^{-1}\mathbf{C}_1 - \exp(id\tau)\mathbf{K}^{-1}\widetilde{\mathbf{C}_0}\}$$

(A.18)

It is important to underline that if $\nu \to \infty$, i.e. if we are shifting away from the parametric resonance, then matrix $\mathbf{K}$ tends to the diagonal matrix. In this limit the integral on the right-hand side of eq. (A.16) can be calculated, and we get

$$\mathbf{W}_c \to \mathbf{U}(-\varepsilon\varsigma - i\mathbf{E})\mathbf{U}^{-1}\langle\mathbf{C}\rangle$$

(A.19)

where $\varsigma = diag\{1, \Gamma, \Gamma\}$ and $\langle \mathbf{C} \rangle$ is the averaged value of vector $\mathbf{C}(\tau)$. The result given by expression eq. (A.19) coincides with the usual result for the stationary dynamics of atom in the strong monochromatic field, i.e. when only one resonant strong wave is acting on an atom. It leads to the well-known expressions for the light pressure force acting on an atom in the monochromatic resonant field. But when deriving eq. (A.18), we presumed that $\nu = O(1)$ and threw away the terms of order epsilon. Therefore, the expression eq. (A.19) and its consequences cannot be derived from eq. (A.18) with the help of limit transition $\nu \to \infty$. Far from the parametric resonance, the action of the weak wave is revealed only in the next term of the asymptotic expansion in the powers of epsilon.

A.2.3   *Light action on an atom*

Let us calculate the light pressure of the weakly modulated bichromatic field on an atom in conditions close to the parametric resonance with the help of the usual expression

$$F = - \sum_k k \sum_{\alpha, \beta} Im\{\mu_{\alpha\beta} E_k \rho_{\alpha\beta}\} \tag{A.20}$$

where α and β enumerate the components of the corresponding matrices. In our case the external field consists of two components. But when calculating the value of the force in the main term of the asymptotic expansion, it is enough to account only for the influence of the main wave. Thus,

$$F = -k\mu_{12}E_1 Im\{\rho_{12}\} = -k\mu_{12}E_1 Im\{\mathbf{SU}[i\delta_d^{-1}\mathbf{C}_{-1} - i\delta_d^{-1}\mathbf{C}_1 \\ - \exp(id\tau)\mathbf{K}^{-1}\widetilde{\mathbf{C}_0}]\}$$

where $\mathbf{S} = (0, 1, 0)$ - is a row vector. Matrix $\mathbf{U}$ has real elements as well as vectors $i\mathbf{C}_{-1}, i\mathbf{C}_1$, and they don't enter the imaginary value of expression in brackets. Thus, we get

$$F = k\mu_{12}E_1 Im\{\mathbf{SU}\exp(id\tau)\mathbf{K}^{-1}\widetilde{\mathbf{C}_0}\}$$

Substituting the explicit expressions and taking into account that $\mu_{12}E_1 = \alpha_1\hbar$, one gets the expression for the force acting on an atom in the parametric resonance [Kazakov and Siparov (1997)]

$$F = \hbar k \alpha_1 H(a, \sigma, \nu) \cos[\delta_d \tau + \psi(\nu)] \tag{A.21}$$

where

$$H(a, \sigma, \nu) = \frac{a(\sigma^2 + 2\Gamma)\sqrt{1 + \Gamma(1 + \sigma^2)^2 + \nu^2}}{\sqrt{2}J^5(J - \sigma)\det\{\mathbf{K}\}} \cdot$$
$$\cdot \left[1 + \frac{J - \sigma}{\delta_d} + \frac{2\sigma(1 - \Gamma)}{\delta_d(\sigma^2 + 2\Gamma)}\right];$$

$$\det\{\mathbf{K}\} = -\nu^2(\sigma^2 + 2\Gamma) - \sigma^2 - \frac{2a^2J^2}{(J - \sigma)^2} - 2\Gamma^3(1 + \sigma^2)^2$$
$$-2\Gamma\left[1 + \sigma^2(1 + \sigma^2) + a^2J^2\frac{1 + \sigma^2}{(J - \sigma)^2}\right]$$
$$-\Gamma^2(1 + \sigma^2)[4 + \sigma^2(1 + \sigma^2)];$$

$$\psi(\nu) = \arccos\frac{1 + \Gamma(1 + \sigma^2)}{\sqrt{1 + \Gamma(1 + \sigma^2)^2 + \nu^2}} + \pi$$

Notice, that if $a, \sigma, \nu = O(1)$, then $H(a, \sigma, \nu) = O(1)$. Now $a = \frac{\alpha_2\sqrt{2}}{\gamma}$ is a saturation parameter for the additional wave, and σ is the ratio of the detuning of the main wave to Rabi frequency with regard to Doppler shift.

Let us discuss the meaning and the characteristic features of this result.

First, the expression for the force eq. (A.21) does not contain constant component. Thus, in the main term of the asymptotic expansion, the force contains only the component oscillating with δ_d (or, in the initial terms, with $\Omega_2 - \Omega_1$). The constant term can be present only in the next term of the asymptotes proportional to ε. The calculation of this term presents a much more complicated problem. This is in accord with the known expression for the light pressure force acting on an atom in the monochromatic resonant field

$$F = \hbar k\gamma\xi(v, \Omega_1, E_1) \tag{A.22}$$

where $\xi = O(1)$. Since in our notations $\gamma/\alpha_1 = \varepsilon\sqrt{2}$, we see that expression eq. (A.22) corresponds to the next (smaller) term of the asymptotic expansion. Thus, according to eq. (A.21), the amplitude of the oscillating force in the vicinity of the parametric resonance is essentially larger than the value of the constant force in the field of the monochromatic wave.

Second, the frequency, δ_d of the oscillating force does not depend on ν, i.e. on the parameter that describes the tuning on the parametric resonance and is related to the atomic velocity. Hence, the force acting on various atoms in the vicinity of parametric resonance oscillates with one and the same frequency. Parameter ν describes only the phase, $\psi(\nu)$ of the force oscillations. It seems natural to call the phenomenon of the appearance of the high amplitude oscillating force that acts on an atom in the specially composed strong resonant field with the special choice of parameters the direct optic-mechaniclal parametric resonance.

A.2.4 *Groups of atoms in optical-mechanical parametric resonance*

Let us describe the possibility of the parametric resonance for the groups of atoms. With regard to the definition, $J = (\sigma^2 + 2)^{1/2}$, the condition eq. (A.12) of the parametric resonance can be written as

$$\sqrt{\sigma^2 + 2} = |\delta_d| + \varepsilon\nu; \nu = O(1) \qquad (A.23)$$

From the physical point of view, this gives the relation between the detuning, δ of the main wave from the atomic transition and the distuning, δ_d between the main and additional waves. In this equation σ depends on velocities because of Doppler effect, and in order to find the atoms sufficing the condition given by eq. (A.23), the distuning between the main and additional waves must suffice the additional threshold condition

$$|\delta_d| \geq \sqrt{2} \qquad (A.24)$$

It should be underlined that the condition eq. (A.24) is understood in the asymptotic sense.

If $|\delta_d| < \sqrt{2}$, i.e. $\sqrt{2} - |\delta_d| = O(1) > 0$, then there is no atoms in the parametric resonance. Let us regard the case when $\sqrt{2} - |\delta_d| = \varepsilon\eta, \eta = O(1)$ (and η can be negative as well). This situation corresponds to the parametric resonance close to the threshold. Then the condition eq. (A.23) can be rewritten as follows

$$\sigma = \pm 2^{3/4}\varepsilon^{1/2}(\nu - \eta)^{1/2} \qquad (A.25)$$

Here $\nu > \eta$. Thus, the atoms with the velocities defined by

$$v = \frac{\Omega_1 - \omega}{k} \pm 2^{3/4}\varepsilon^{1/2}(\nu - \eta)^{1/2}\kappa^{-1}$$

take part in the parametric resonance. This group of atoms is about $\varepsilon^{1/2}$ wide in the units of $\kappa^{-1} = \alpha_1 c\sqrt{2}/\omega$ where c is the light speed.

If $|\delta_d| > \sqrt{2}$, then eq. (A.23) gives

$$\sigma = \pm \left[\sqrt{\delta_d^2 + 2} + \frac{\varepsilon \nu \delta_d}{\delta_d^2 - 2} \right]^{1/2}$$

or

$$v = \frac{\Omega_1 - \omega}{k} \pm \kappa^{-1} \left[\sqrt{\delta_d^2 + 2} + \frac{\varepsilon \nu \delta_d}{\delta_d^2 - 2} \right]^{1/2} \qquad (A.26)$$

In this case the parametric resonance takes place beyond the threshold and there are two groups of atoms in the parametric resonance corresponding to the choice of sign in eq. (A.26). The centers of these groups of atoms have the velocities $\frac{\Omega_1 - \omega}{k} \pm \kappa^{-1}(\delta_d^2 + 2)^{1/2}$. Parameter ν can take any value, $\nu = O(1)$. The widths of these groups of atoms is of the order of ε in the units of κ^{-1}. When $|\delta_d|$ decreases, the velocities of these groups "move" towards each other and become equal when $|\delta_d|$ starts to differ from $\sqrt{2}$ by ε and less, then the two groups of atoms become one single group.

A.2.5 *Velocity change due to the force action*

Let us discuss some details related to the derivation of expression for the force eq. (A.21) which is given above and for which ν was presumed to be independent of time. As it has already been mentioned, this parameter which describes the tuning of the atomic velocity at the parametric resonance can change under the action of force eq. (A.21) according to the mechanics equations, and the atom might drop out of the parametric resonance. As it follows from eq. (A.21), the atom velocity, v (and, thus, ν also) oscillates with frequency δ_d. Let us evaluate the possible changes and assume the following values for the parameters that we can control in a laboratory measurement: $\Omega_1 \sim \omega \sim 10^{15}\text{Hz}$, $\Omega_1 - \Omega_2 \sim 10^9\text{Hz}$, $\alpha_1 \sim 10^9\text{Hz}$, $\gamma \sim 10^7\text{Hz}$ which means that $\varepsilon \sim 10^{-2}$. Then we choose a saturation parameter, a, for an additional wave in such a way that $H(a, \sigma, \nu) = 1$. Then we have

$$\frac{dv}{d\tau} = \frac{\hbar k \alpha_1 \cos(\delta_d \tau)}{M}$$

where M is the atomic mass. We have to evaluate the change of ν for half of the period of the force oscillations when the force, F does not change sign. Then $\Delta \tau \sim 1/\alpha_1$ and

$$\Delta v = \frac{\hbar k}{M} \sim 3 \text{ cm/s} \tag{A.27}$$

Let us perform the evaluations near the threshold and assume $\Omega_1 \sim \omega$, that is the center of the velocity distribution of the group of atoms corresponds to $v = 0$. Then according to eq. (A.25) we get for the tuning parameter, ν

$$\nu = \frac{(kv)^2}{2^{3/2}\varepsilon} \tag{A.28}$$

and

$$\Delta \nu \sim 2\kappa^2 \frac{v}{\varepsilon} \Delta v \tag{A.29}$$

In order to evaluate the velocity, v for $\nu = 1$, let us use the relation eq. (A.28), and find $v \sim 3 \cdot 10^3$ cm/s. Substituting this value into eq. (A.29), we find $\Delta \nu \sim 10^{-3}$. Therefore, we see that the change of the tuning parameter which could be due to the oscillations is really negligibly small as we suggested earlier.

According to the given estimations, the width of the velocity distribution of the group of atoms that take part in the parametric resonance near the threshold is about $\Delta v \sim 10^2$m/s, the center of it is considered to have $v = 0$ and $\varepsilon \sim 10^{-2}$. The average heat velocity of an atom at temperature of about $T_r = 300$ K is $v_r \sim 10^3$ m/s. Then, at $T = 10^{-2}T_r = 3$ K we have $v_T < \Delta v$, and practically all the atoms take part in the parametric resonance.

A.2.6 *Main result*

The physical meaning of the phenomenon known as light pressure on an atom was discussed in the beginning of the last chapter. The main result of this section is the derivation of the expression for the force [Kazakov and Siparov (1997)] acting on an atom in the condition of parametric resonance $\sqrt{\sigma^2 + 2} = \delta_d$ which gives the relation between the detuning, δ of the main wave from the atomic transition and the distuning, δ_d between the main and additional waves. For example, the exact tuning corresponds to $\Omega_2 - \Omega_1 = 2\alpha_1$. If the OMPR condition is fulfilled, the force is

$$F = \hbar k \alpha_1 H(a, \sigma, \nu) \cos[(\Omega_2 - \Omega_1)t + \psi(\nu)] + O(\varepsilon); \varepsilon = \frac{\gamma}{\alpha_1 2^{1/2}} \tag{A.30}$$

where $\hbar$ is Planck's constant, and H is a certain combination of the problem parameters which has the order of $O(1)$. As we see, the force depends on time and oscillates with frequency equal to $(\Omega_2 - \Omega_1)$ while its amplitude is essentially larger than the value of the constant force acting on an atom in the field of a strong monochromatic wave. Thus, the relatively small perturbation of the external field in case of the parametric resonance leads to the essential changes in the value of the force acting on an atom, and alongside with the small acceleration in the direction of the wave vector, the TLA suffers an intensive shaking along this vector with the doubled Rabi frequency.

A.3 Probe wave absorption

The obtained result makes one suggest that a kind of an inverse parametric resonance is also possible. For example, if an external force of an arbitrary nature acts on a TLA in such a way that it starts mechanically vibrating along some axis with a certain frequency, then the intensity (i.e. Rabi frequency) of a strong monochromatic wave resonant to the atomic transition and parallel to this axis may be chosen such that a certain parametric resonance takes place again. This effect could be experimentally traced by the observation of the probe wave absorption spectrum on which some specific features could appear. This resembles the Brillouin scattering - a powerful method of the atomic structure and atomic dynamics research. Now there is a possibility to discover the following new and appealing feature: Rabi frequency depends on the intensity of the radiation, therefore, it would be possible to control the location of the sidebands that could appear because of the parametric resonance and place them onto the frequency region suitable for observations and measurements.

The use of the oncoming probe wave [Stenholm (1984); Letohov and Chebotaev (1990)] is one of the most effective experimental methods in the study of the atomic structure details. The descriptions of the corresponding set ups can also be found in [Stenholm (1984); Letohov and Chebotaev (1990)]. They can be used to measure the probe wave absorption spectrum in case of the parametric resonance and compare it with the theoretical one which we are going to obtain.

The influence of various characteristics of the electromagnetic field on the atomic systems behavior is discussed in literature more often than vice versa. Notice, for example, the mechanical action of the running or standing resonant wave on the moving atom regarded in [Minogin and Letohov (1986); Kazantsev *et al.* (1991)], in a sense this effect is reverse to what we are going to discuss below. When a two-level ion suffers the simultaneous action of the resonant laser field and of a low frequency electric field [Minogin and Letohov (1986)], the result of the light pressure is the heating or cooling of the localized particles ensemble.

The powerful method of the study of the atomic behavior is the use of a weak probe wave that can scan the interesting region of frequencies. If an atom is excited by the strong monochromatic wave and the probe wave is parallel (or anti-parallel) to the strong beam, then the atomic transition line is saturated, and there will be a dip on the probe wave absorption spectrum. The details of this effect are used for the study of atom dynamics. In [Boyd and Sargent III (1988)] the dynamic Stark effect was discussed. It was found that it causes the shifting of the atomic energy levels, and this causes the new resonances in the non-linear optical susceptibility describing the probe wave absorption. The comparison of these calculations with the experiment was performed in [Gruneisen *et al.* (1988)]. The interaction of the new energy levels produced by the dynamic Stark effect can also be seen in the resonance fluorescence [Liaptsev (1994)]. All these processes can affect the performance of various devices and instruments exploiting the non-linear response of the atomic systems. If we take an ion instead of an atom and correlate the low frequency of the electric field action with the Rabi frequency of the laser field, the parametric resonance can appear and be measured.

What we have to do now is to write down the Bloch's equations for the components of the density matrix with regard to the low frequency mechanical atomic vibrations caused by the arbitrary source [Siparov (1997)]. Then we use the rotating wave approximation and come to the equation that can be solved by the perturbation theory methods for the case of the parametric resonance.

A.3.1   *Problem and its solution*

As before, we describe the laser field classically and use the density matrix $\rho(v, y)$ to describe the atomic dynamics. Then the Bloch's equations take the form

$$\frac{d}{dt}(\rho_{22} - \rho_{11}) = -\gamma(\rho_{22} - \rho_{11}) + 4i[\alpha_1 \cos(\Omega_1 t - k_1 y)$$

$$+\alpha_p \cos(\Omega_p t - k_p y)](\rho_{21} - \rho_{12}) + \Lambda \qquad (A.31)$$

$$\frac{d}{dt}\rho_{12} = -(\gamma_{12} + i\omega)\rho_{12} - 2i[\alpha_1 \cos(\Omega_1 t - k_1 y)$$

$$+\alpha_p \cos(\Omega_p t - k_p y)](\rho_{22} - \rho_{11}) \qquad (A.32)$$

Here k_p is the wave vector of the probe propagating in the direction opposite to that of the main wave, α_p is its amplitude in the frequency units, Λ is a saturation parameter. Let us regard the situation when the atom's velocity, v oscillates with (low) frequency, D and amplitude, V_1 due to some periodic external action.

The full time derivative can be written as

$$\frac{d}{dt} = \frac{\partial}{\partial t} + V\frac{\partial}{\partial y}; V = v + V_1 \cos Dt$$

where v is the atom's heat velocity. If the scale characterizing the atom-field interaction in a laboratory set up is of the order 10^{-2}m, then we can take $k_1 = k_p = k$. We will also demand that the amplitude, V_1 of the velocity change due to the atomic oscillations is small in comparison with its heat velocity, v ($V_1 << v$). Then the motion of the atom can be described quasi-classically. Since we are interested in the space-averaged values, the rotating wave approximation can be used as before. Let us introduce the convenient notations

$$R = \frac{1}{\sqrt{2}}(\rho_{22} - \rho_{11})$$

$$\rho_{12} = R_{12}^+ \exp[-i(\Omega_1 t - ky)] + R_{12}^- \exp[-i(\Omega_p t - ky)]$$

$$\rho_{21} = R_{21}^+ \exp[-i(\Omega_1 t - ky)] + R_{21}^- \exp[-i(\Omega_p t - ky)]$$

and use the following variables

$$2^{1/2}\alpha_1 t = \tau; \frac{\gamma}{\alpha_1} = 2^{1/2}\varepsilon; \frac{\gamma_{12}}{\alpha_1} = 2^{1/2}\Gamma\varepsilon; \frac{k}{\alpha_1} = 2^{1/2}\kappa; \frac{\Omega_p - \omega}{\alpha_1} = 2^{1/2}\Delta_p;$$

$$\frac{\Omega_1 - \omega}{\alpha_1} = 2^{1/2}\delta; \frac{\alpha_p}{\alpha_1} = a_p\varepsilon; V_1 = v_1\varepsilon, [v_1 = O(1))];$$

$$Dt = \delta_D\tau, (\delta_D = \frac{D}{\alpha_1\sqrt{2}}); \frac{\Lambda}{\alpha_1\sqrt{2}} = \lambda$$

They have similar meanings with those used in the previous section. Now $v_1 = O(1)$ is given here in such a way that a small perturbation of the atomic velocity caused by the mechanical vibrations corresponds to the term comparable to parameter ε which characterizes the strength of the field. Substituting the variables, we come to the following system of equations

$$\frac{\partial}{\partial \tau} R = -\varepsilon R + i[R_{21}^+ - R_{12}^+ - a_p(R_{21}^- - R_{12}^-)] + \lambda \qquad (A.33)$$

$$\frac{\partial}{\partial \tau} R_{12}^+ = -iR + [-\Gamma \varepsilon + i(\delta - \kappa V)]R_{12}^+$$

$$\frac{\partial}{\partial \tau} R_{21}^+ = iR + [-\Gamma \varepsilon - i(\delta - \kappa V)]R_{21}^+$$

$$\frac{\partial}{\partial \tau} R_{12}^- = -ia_p R + [-\Gamma \varepsilon + i(\Delta_p - \kappa V)]R_{12}^-$$

and $R_{21}^- = \overline{R_{12}^-}$. The probe wave absorption coefficient can be found as

$$\chi(\Omega_p) = \int \exp(-\frac{V^2}{u_T^2}) Im R_{21}^-(V) dV \qquad (A.34)$$

where u_T is the atomic heat velocity. So, in order to find the absorption coefficient, we have to solve the last equation of the system eq. (A.33) for the case when $\varepsilon << 1, \Gamma = O(1)$.

It seems natural to assume that the amplitude, a_p of the probe wave is so small that it doesn't affect the dynamics of the atom. Then the last equation of the system eq. (A.33) separates and can be solved independently from the others. Then the expression for R which is present in this last equation can be obtained from the solution of the first three equations of the system eq. (A.33). Thus, we first have to solve the system

$$\frac{\partial}{\partial \tau} \mathbf{W} = [\mathbf{Q}_0 + \varepsilon \mathbf{Q}_1(\tau)]\mathbf{W} + \mathbf{C}(\tau) \qquad (A.35)$$

where

$$\mathbf{W} = \begin{pmatrix} R \\ R_{12}^+ \\ R_{21}^+ \end{pmatrix}; \mathbf{Q}_0 = i \begin{pmatrix} 0 & -1 & 1 \\ -1 & \sigma & 0 \\ 1 & 0 & -\sigma \end{pmatrix}; \mathbf{C} = \begin{pmatrix} \lambda \\ 0 \\ 0 \end{pmatrix};$$

$$\mathbf{Q}_1 = \begin{pmatrix} -1 & 0 & 0 \\ 0 & -\Gamma - i\kappa v_1 \cos(\delta_D \tau) & 0 \\ 0 & 0 & -\Gamma + i\kappa v_1 \cos(\delta_D \tau) \end{pmatrix}$$

and $\sigma = \delta - \kappa v$. Our goal is to find the main term of the asymptotic expansion of the stationary solution of system eq. (A.35) in the powers of ε. The mathematical details are similar to those used above, and again it is important to underline that the condition of the parametric resonance is essential for the solution.

Let us first consider the homogeneous equation corresponding to the system eq. (A.35) as before

$$\frac{\partial}{\partial \tau}\mathbf{N}(\tau) = (\mathbf{Q}_0 + \varepsilon \mathbf{Q}_1(\tau))\mathbf{N}(\tau); \mathbf{N}(0) = \mathbf{I} \qquad (A.36)$$

where $\mathbf{I}$ is a unity matrix. Now we construct the matrices $\mathbf{U}$ and $\mathbf{E} = diag\{0, iJ, -iJ\} = \mathbf{U}^{-1}\mathbf{Q}_0\mathbf{U}$ with the help of eigenvalues, $\pm J = \pm(\sigma^2 + 2)^{1/2}$ of matrix $\mathbf{Q}_0$

$$\mathbf{U} = \frac{1}{J} \begin{pmatrix} \sigma & 1 & 1 \\ 1 & -\frac{1}{J-\sigma} & \frac{1}{J+\sigma} \\ -1 & \frac{1}{J+\sigma} & -\frac{1}{J-\sigma} \end{pmatrix}$$

Introducing $\mathbf{L}(\tau) = \exp(-\mathbf{E}\tau)\mathbf{U}^{-1}\mathbf{N}(\tau)$, we transform eq. (A.35) into

$$\frac{\partial}{\partial \tau}\mathbf{L} = \varepsilon \mathbf{G}(\tau)\mathbf{L} \qquad (A.37)$$

where

$$\mathbf{G}(\tau) = \exp(-\mathbf{E}\tau)\mathbf{U}^{-1}\mathbf{Q}_1(\tau)\mathbf{U}\exp(\mathbf{E}\tau)$$

In order to solve the last equation, we use the many scales method. Introducing the slow time, $\tau_1 = \varepsilon\tau$, we look for the solution of equation

$$(\frac{\partial}{\partial \tau} + \varepsilon\frac{\partial}{\partial \tau_1})\mathbf{L}(\tau, \tau_1) = \varepsilon \mathbf{G}(\tau)\mathbf{L}(\tau, \tau_1)$$

in the form

$$\mathbf{L}(\tau, \tau_1) = \mathbf{L}_0(\tau, \tau_1) + \varepsilon\mathbf{L}_1(\tau, \tau_1) + \dots$$

The direct substitution gives $\mathbf{L}_0(\tau, \tau_1) = \mathbf{L}_0(\tau_1)$, and we get the equation for $\mathbf{L}_0(\tau_1)$ demanding that on the right-hand side of eq. (A.37) matrix $\mathbf{G}$ has no terms with frequencies of the order of $O(1)$. Then we get for $\mathbf{L}_0(\tau_1)$

$$\frac{\partial}{\partial \tau_1}\mathbf{L}_0(\tau_1) = <\mathbf{G}(\tau)> \mathbf{L}_0(\tau_1) \tag{A.38}$$

where $< ... >$ means the absence of all the high frequency terms. The parametric resonance takes place when the following condition is sufficed

$$J = \delta_D + \varepsilon v; v = O(1) \tag{A.39}$$

Substituting eq. (A.39) into eq. (A.38), we get

$$\frac{\partial}{\partial \tau}\mathbf{L}_0(\tau_1) = \langle \mathbf{G}_1(\tau) \rangle \mathbf{L}_0(\tau_1) \tag{A.40}$$

where

$$\langle \mathbf{G}_1(\tau) \rangle = -\frac{1}{J^2} \begin{pmatrix} \sigma^2 2\Gamma & -i\kappa v_1 \exp(iv\tau_1) & i\kappa v_1 \exp(-iv\tau_1) \\ -i\kappa v_1 \exp(-iv\tau_1) & 1+\Gamma(1+\sigma^2) & 0 \\ i\kappa v_1 \exp(iv\tau_1) & 0 & 1+\Gamma(1+\sigma^2) \end{pmatrix}$$

The system eq. (A.40) can be transformed into the system with constant coefficients. Introduce $\mathbf{Z}(\tau_1) = diag\{1, \exp(-iv\tau_1), \exp(iv\tau_1)\}$ and denote $\mathbf{L}_0(\tau_1) = \mathbf{Z}(\tau_1)\mathbf{S}(\tau_1)$. Then eq. (A.40) takes the form

$$\frac{\partial}{\partial \tau_1}\mathbf{S}(\tau_1) = \mathbf{K}\mathbf{S}(\tau_1) \tag{A.41}$$

where

$$\mathbf{K} = -\frac{1}{J^2} \begin{pmatrix} \sigma^2 + 2\Gamma & -i\kappa v_1 & i\kappa v_1 \\ -i\kappa v_1 & 1+\Gamma(1+\sigma^2) - iv & 0 \\ i\kappa v_1 & 0 & 1+\Gamma(1+\sigma^2) + iv \end{pmatrix}$$

Solving eq. (A.41), we get

$$\mathbf{L}_0(\tau, \tau_1) = \mathbf{Z}(\tau_1) \exp(\mathbf{K}\tau_1)$$

and the solution of the homogeneous equation is

$$\mathbf{N}(\tau) = \mathbf{U} \exp(\mathbf{E}\tau)\widetilde{Z}(\tau) \exp(\varepsilon \mathbf{K}\tau)\mathbf{U}^{-1}$$

where $\widetilde{Z}(\tau) = diag\{1, \exp(-i\varepsilon v\tau), \exp(i\varepsilon v\tau)\}$. We see that $\exp(\mathbf{E}\tau)\widetilde{Z}(\tau) = \exp(id\tau)$ where $\mathbf{d} = diag\{0, \delta, -\delta\}$, and the leading term of the asymptotic expansion of the solution of eq. (A.36) in the powers of ε is

$$\mathbf{N}(\tau) = \mathbf{U}\exp(id\tau)\exp(\varepsilon\mathbf{K}\tau)\mathbf{U}^{-1} \tag{A.42}$$

The solution of the non-homogeneous equation eq. (A.35) can be found with the help of eq. (A.42) as

$$\mathbf{W}(\tau) = \mathbf{N}(\tau)\mathbf{W}(0) + \mathbf{N}(\tau)\int_0^\tau [\mathbf{N}(x)]^{-1}\mathbf{C}dx$$

Since the real parts of the matrix $\mathbf{K}$ eigenvalues are strictly negative, we cam omit the first term on the right-hand side of the last equation and study the solution on the upper limit of the integral

$$\mathbf{W}(\tau) = \mathbf{U}\exp(id\tau)\exp(\varepsilon\mathbf{K}\tau)\int_0^\tau \exp(-\varepsilon\mathbf{K}x)\exp(-idx)\mathbf{U}^{-1}\mathbf{C}dx \tag{A.43}$$

Performing the integration and neglecting terms of the order of ε, we obtain the main term of the asymptotic for R, i.e. the first component of vector $\mathbf{W}$

$$R = T(v) + S(v) \tag{A.44}$$

$$T(v) = \frac{\lambda\sigma^2}{\varepsilon}\frac{J^5(B^2 + v^2)}{(\sigma^2 + 2\Gamma)(B^2 + v^2) + 2\kappa^2 v_1^2 B}$$

$$S(v) = -\frac{2\lambda\sigma\kappa v_1}{\varepsilon}\frac{J^5(\cos\delta_D\tau + B\sin\delta_D\tau)}{(\sigma^2 + 2\Gamma)(B^2 + v^2) + 2\kappa^2 v_1^2 B}$$

$$B = 1 + \Gamma(1 + \sigma^2)$$

The obtained result shows that if the atom is not subjected to the low-frequency action ($v_1 = 0$), then the main term of function R expansion in eq. (A.44) does not contain the oscillating component, and its form coincides with the known one with account to the notations. In order to find the probe wave absorption coefficient, let us substitute eq. (A.44) into the fourth equation of system eq. (A.33) with regard to $R_{21}^- = \overline{R_{12}^-}$. Then the expression for ImR_{21}^- takes the form

$$ImR_{21}^- = -\frac{a_p\lambda}{\det(\mathbf{K})J^6}\Big\{\frac{\Gamma\sigma^2(B^2 + v^2)}{\Gamma^2\varepsilon^2 + \sigma_p^2} \tag{A.45}$$

$$+\frac{\kappa v_1\sigma}{2\varepsilon}\Big[\Big(\frac{-\Gamma\varepsilon v + B(\sigma_p + \delta_D)}{\Gamma^2\varepsilon^2 + (\sigma_p + \delta_D)^2} + \frac{-\Gamma\varepsilon v - B(\sigma_p - \delta_D)}{\Gamma^2\varepsilon^2 + (\sigma_p - \delta_D)^2}\Big)\cos\delta_D\tau$$

$$+\Big(\frac{-\Gamma\varepsilon B - v(\sigma_p + \delta_D)}{\Gamma^2\varepsilon^2 + (\sigma_p + \delta_D)^2} + \frac{-\Gamma\varepsilon B + v(\sigma_p - \delta_D)}{\Gamma^2\varepsilon^2 + (\sigma_p - \delta_D)^2}\Big)\sin\delta_D\tau\Big]\Big\}$$

where

$$\sigma_p = \Delta_p + \kappa v$$

$$\det(\mathbf{K}) = -\frac{1}{J^6}\left[(\sigma^2 + 2\Gamma)(B^2 + \nu^2) + 2\kappa^2 v_1^2 B\right]$$

Substituting eq. (A.45) into eq. (A.34), we finally get the expression for the dependence of the probe wave absorption coefficient on frequency

$$\chi(\Omega_p) = \int \exp(-\frac{V^2}{u_T^2})\frac{a_p \lambda}{\det(K)J^6}\left[\frac{\Gamma\sigma^2(B^2 + \nu^2)}{\Gamma^2\varepsilon^2 + \sigma_p^2} + \frac{A\sigma\kappa v_1}{2\varepsilon}\cos(\delta_D \tau - \phi)\right]dV \tag{A.46}$$

where

$$A = 2\sqrt{B^2 + \nu^2}\frac{\sqrt{\sigma_p^4\delta_D^2 + \sigma_p^2(2\Gamma^4\varepsilon^4 + \Gamma^2\varepsilon^2 - 2\delta_D^4) + (\Gamma^2\varepsilon^2 + \delta_D^2)^3}}{\sigma_p^4 + 2\sigma_p^2(\Gamma^2\varepsilon^2 - \delta_D^2) + (\Gamma^2\varepsilon^2 + \delta_D^2)^2}$$

$$\phi = \arctan\frac{\sigma_p^2(\Gamma\varepsilon B - \nu\delta_D) + (\Gamma^2\varepsilon^2 + \delta_D^2)(\Gamma\varepsilon B + \nu\delta_D)}{\sigma_p^2(\Gamma\varepsilon\nu - B\delta_D) + (\Gamma^2\varepsilon^2 + \delta_D^2)(\Gamma\varepsilon\nu - B\delta_D)}$$

Let us discuss the origin and meaning of the terms in eq. (A.46). The first term in square brackets corresponds to Lorentz contour in the vicinity of the atomic transition frequency. The expression for it coincides with the known one and describes the scanning of an atom in the strong resonant field by the probe wave. The second term in square brackets is an oscillating one. One can see that it corresponds to the previous (larger) order in the asymptotic expansion due to ε in the denominator. This term appears only, if the conditions of the optic-mechanical parametric resonance are fulfilled. Notice the difference with the result of the previous section: there the principal term of the asymptotic expansion of the solution was proportional to ε^0 while now it is proportional to ε^{-1}. It doesn't mean a discrepancy, because then we concerned the force acting on an atom in the field of two waves, and now we discuss the influence of the parametric resonance on the probe wave scanning the system, so, the physical situations and target objects are different. This kind of parametric resonance is natural to call the inverse optic-mechanical parametric resonance. The part of the signal which oscillates with time will be called the non-stationary component (here: of the probe wave absorption spectrum).

The analysis of eq. (A.46) can be illustrated by the following figures. A small central peak drawn on

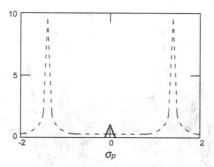

Fig. A.1   Absorption coefficient spectrum.

Figure A.1 by solid line presents the stationary component of the absorption coefficient spectrum. It corresponds to the first term in brackets in the expression for $\chi(\Omega_p)$ and has the known Lorenzian form. The dashed line on Fig. A1 presents the dependence of the amplitude of the non-stationary component that appeared because of the parametric resonance on the detuning from the atomic frequency transition. The heights of the sideband peaks are essentially larger than the height of the central peak corresponding to the stationary component. Mathematically, this is due to the presence of ε in the denominator of the second term in eq. (A.46).

Figure A.2 gives the 3D-plot of the probe wave absorption coefficient dependence on time and on the detuning calculated with the help of formula eq. (A.46).

When the measurement of the absorption spectrum are performed in the usual way and the stationary component is registered, the high sideband peaks cannot be observed because the amplification during the one half of the period (negative $\chi(\Omega_p)$) is compensated by the absorption during the other half of the period (positive $\chi(\Omega_p)$), and the observable effect cancels out with averaging. But with the help of some simple additional instrumentation like gate detector, the effect can be observed. Its frequency region must be such that the selection of the atomic mechanical vibrations frequency is possible. The frequency of the non-stationary component is equal to the frequency of the mechanical vibrations of an atom which is related to the intensity of the excitation field and its Rabi frequency. Choosing a phase, one can find either a dip or a spike on the absorption spectrum. The other way is to discover the parametric resonance signal is to use the special processing of the obtained signal in order to detect the alternating

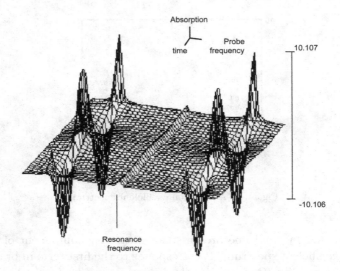

Fig. A.2 Time-dependence of the absorption-amplification spectrum.

component.

One can see that the phase of the non-stationary component of the absorption coefficient depends on the atomic velocity in a complicated way, (recollect that $B = 1 + \Gamma(1 + \sigma^2); \sigma = \delta - \kappa v; \sigma_p = \Delta_p + \kappa v$). Therefore, one should make sure that it wouldn't essentially affect the effect. Figure A.3 illustrates the dependence of the phase of the non-stationary component of the absorption coefficient on the atomic velocity at room temperature for the arbitrary chosen set of parameters.

As can be seen, there are two values of velocities in whose vicinities the phase differs from zero. Integrating in velocities, that is taking into account the scattering of velocities of many atoms that have the velocities providing the OMPR, one can see that the effect is not destroyed, and the influence of various phase shifts of the separate atoms on the picture as a whole is negligible.

The additional condition of the appearance of the specific feature related to the parametric resonance, is the presence of the non-zero detuning of the falling radiation frequency, Ω_1 from the atomic transition frequency, ω, i.e. $\sigma \neq 0$. If the radiation frequency is equal to the resonance frequency, the parametric resonance effect disappears. Thus, we see that the presence of the parametric resonance reveals through the appearance of the oscillating

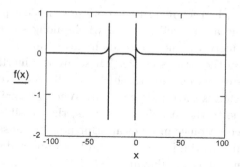

Fig. A.3   Phase dependence on velocity. Here $\Gamma = 1.2; \Delta = 0.9; \Delta_p = 1.1; \epsilon = 0.01; \delta = 1.3; \kappa = 0.1; \nu = 0.8.$

terms in the expression for R that result in the additional high peaks on the plot for $\chi(\Omega_p)$ which can be drawn with the help of eqs. (A.34, A.45). As can be seen from eq. (A.45), these peaks have the amplitudes that are much larger than the height of the stationary peak corresponding to the resonance frequency, ω, they are located symmetrically around ω and the shifts are equal to the frequency of atomic vibrations, D.

Though the obtained results have the quantum mechanical origin some heuristic reasoning might be helpful.

When we consider the dynamic Stark effect, the low-frequency external electric field produces the additional dipole moment of the neutral atom, which is placed into the strong optical field resonant to the atomic transition. Note, that when the atomic transition takes place, the effective radius of the atom changes (actually, what is changing is the coordinate of the maximum of the charge distribution function). When the frequency of these radius pulsations coincides with the frequency of the external field producing the additional dipole moment, the quasi-energies overlap and the parametric resonance occurs. Therefore, the important feature which has to be considered while regarding a quasi-classical model of the atom is its space characteristics – the radius. The displacements of the electrically positive and negative parts of the atom due to a low-frequency external field take place in the opposite directions, and their magnitudes provide almost stationary position of the center of inertia, the force acting on the atom as a whole being negligibly small.

On the contrary, when we regard the optic-mechanical parametric resonance and make the atom oscillate as a whole along some axis, we act upon the center of inertia. The atom resembles a rattle with the heavy nucleus suspended on the strings inside. These are the inertial forces that produce the relative displacements of the electrically positive and negative parts of the atom – that is the dipole moment. When we again consider the radius pulsations due to the atomic transitions, the possible origin of the parametric resonance becomes more clear. But now its cause is different.

Another possibility to think about the problem under discussion is the TLA vector model introduced in [Feynman *et al.* (1957)] . Since the problem discussed in the paper treats essentially a case of frequency modulation, the obtained results can be regarded from the point of view of the trajectories on the Bloch sphere in the same way as in [Eberly and Popov (1988)] or [Wu *et al.* (1994)]. Here the role of the initial phase angle was not regarded, though judging by the known theoretical [Eberly and Popov (1988)] and experimental [Wu *et al.* (1994)] results, this angle can affect the spectrum. As it is mentioned in [Eberly and Popov (1988)], the absence of the line-shape on the probe wave absorption spectrum does not mean that there is no interaction, but means that there is no time-independent interaction. Here the time-dependent term is found explicitly.

A.3.2   *Assumptions and demands*

Let us discuss the meaning of the assumptions that were used while deriving the results and recognize what demands do they infer on the experimental situation.

1) $\alpha_p << \alpha_1$. This condition corresponds to the low intensity of the probe wave in the experiment and is easily achieved.

2) $\frac{\gamma}{\alpha_1\sqrt{2}} = \varepsilon << 1$. This is a condition on the laser intensity. The decay rate must be sufficiently smaller than Rabi parameter α_1. This demand can also be sufficed.

3) $V_1 << v$ and $V_1 = v_1\varepsilon; v_1 = O(1)$. This is an amplitude condition of the optic-mechanical parametric resonance. It means that the change of the atomic velocity for the one quarter of a period must be small in comparison with the heat velocity of atom ($v \sim 10^3$ m/s at room temperatures). Thus, the amplitude of the low-frequency action on an atom or an ion must not be large. If an ion is made vibrating by the low-frequency electric field, the stress of the field must be rather small. If a TLA is an admixture to the gas, the intensity of the acoustic field must be rather small. In

both cases condition 3) is easily controlled. We considered V_1 and ε to be the values of the same order, and this is important. If $V_1 << \varepsilon$, then matrix $\mathbf{Q}_1$ in eq. (A.35) degenerates, and the oscillating component in the expression for the absorption coefficient of the probe wave vanishes. In this case the atomic vibrations are to the effect of a little increase of the absorption coefficient due to the increase of the number of atoms inside Doppler contour because of the oscillations of their velocities. But as we will see in the next item, 4), the scattering of velocities of the atoms taking part in the parametric resonance is proportional to $\varepsilon^{1/2}$, and the vibrational atomic motions with amplitudes $V_1 << \varepsilon$ are negligible. Thus, condition 3) is essential and should be considered every time with regard to the system under investigation.

4) $J = \delta_D + \varepsilon\nu$ ($\nu = O(1)$). This is the frequency condition of the parametric resonance, i.e. the combination of parameters that causes the appearance of the specific feature (non-stationary component) on the absorption spectrum of the probe wave. Since J is defined as $J = \sqrt{\sigma^2 + 2}$, we get

$$\delta - \kappa v = \pm[(\delta_D + \varepsilon\nu)^2 - 2]^{1/2}$$

Substituting the initial variables as before, we see that the parametric resonance is possible only if the threshold condition, $D \geq \alpha_1\sqrt{2}(1 - \frac{\varepsilon\nu}{\sqrt{2}})$ is fulfilled. Then if $\alpha_1\sqrt{2} - |D| = \varepsilon\eta$ (η can be negative too), the velocities of the atoms taking part in the parametric resonance appear to be

$$v = \frac{\Omega_1 - \omega}{k} \pm 2^{3/4}\varepsilon^{1/2}(\nu - \eta)^{1/2}\kappa^{-1}$$

The velocity scattering is proportional to $\varepsilon^{1/2}$ in the units of κ^{-1} where $\kappa^{-1} = \frac{\alpha_1 c\sqrt{2}}{\omega}$ and c is the light speed. So, this condition means that to achieve a parametric resonance, there must be a certain relation between the frequency of the external action and Rabi frequency of the field.

Thus, we have shown that if a TLA or a two-level ion in the strong resonant field is subjected to the action of a low frequency external force of an arbitrary nature, and this force makes an atom or an ion mechanically vibrate in such a way that the conditions of the optic-mechanical parametric resonance are fulfilled, then the absorption coefficient of the probe wave obtains the oscillating non-stationary component whose amplitude is much larger than the value of the stationary (regular) component of the absorption spectrum of the probe wave on the resonant frequency. The locations of the additional peaks on the absorption spectrum depend on the frequency of the perturbing force. In order to provide the maximal sensitivity in experiment, one could change the location of the additional peaks, changing simultaneously the intensity of the laser field and the frequency of external force.

A.4 Fluorescence

The phenomenon of resonance fluorescence constitutes one of the most interesting problems in quantum optics and has been explored theoretically and experimentally over the years. In [Weisskopf (1931)] a quantum perturbation approach was used to describe the weak-field resonance fluorescence. Later it was shown [Burshtein (1962); Mollow (1969)] that the resonance fluorescence spectrum from a strongly driven atom had a three-peaked structure which was subsequently observed in experiments [Schuda *et al.* (1974); Grove *et al.* (1977)].

A two-level atom or ion again presents a simple but very useful model which possesses principal features of some real atoms observed in an experiment. When we place such an object into the field of electromagnetic radiation we transform a TLA into a "TLA+field" system or as they say sometimes, we regard a dressed atom. An obvious generalization in the investigation of the fluorescence is to use a polychromatic field (bichromatic in the simplest case) of optical frequency or drive a TLA by some other additional field.

The idea of the optic-metrical parametric resonance is to make the atoms (or ions) subjected to the resonant electromagnetic radiation oscillate at the Rabi frequency. There are several ways to perform this. The simplest way is to place a two-level ion into a low-frequency electric field. The oscillations will result in the reconstruction of the fluorescence spectrum. This resembles the dynamic Stark effect measurements. But now the electric field does not create the dipole moment of the neutral atom without affecting the atom's center of inertia. The charged ion starts to move – to vibrate – and the inertial forces contribute to the effect. Another way is to place the TLAs into the buffer gas and use the acoustic wave, though the atomic collisions will make the picture rather complicated. In order to exclude Doppler effect (and, thus, to assume $v = 0$) the registration of the fluorescence radiation should be made from the direction orthogonal to the oscillations. With the help of the mathematical technique already used above, we will obtain the pump terms for the equations describing the resonance fluorescence when the optic-mechanical parametric resonance takes place and find the fluorescence spectrum. It turns out that the sidebands substantially more intensive than the central peak appear on the plot.

The commonly used [Stenholm (1984)] Bloch's equations for the density matrix of the TLA in the resonant electromagnetic field are

$$i\frac{d}{dt}\rho_{22} = -i\gamma\rho_{22} - \alpha(\tilde{\rho}_{12} - \tilde{\rho}_{21}) \tag{A.47}$$

$$i\frac{d}{dt}\tilde{\rho}_{21} = (\Delta - i\frac{\gamma}{2})\tilde{\rho}_{21} - \alpha(\rho_{11} - \rho_{22})$$

$$\rho_{11} + \rho_{22} = 1$$

$$\tilde{\rho}_{21} = \overline{\tilde{\rho}_{12}}$$

Here ρ_{ik} are the elements of the density matrix; $\Delta = \omega - \Omega$ is the detuning of the laser angular frequency Ω from the angular frequency of the atomic transition ω; γ is the excited state decay rate; $\alpha = E\mu/2\hbar$ is the strong field coupling constant (E is the amplitude of the electric field of the laser radiation; μ is the dipole moment; $\hbar$ is the Planck's constant; $1/2$ factor is now present in the definition of Rabi parameter which often takes place in literature on fluorescence, this factor wandering is irritating but helps to keep in touch with both streams); "tildes" suggest that the rotating wave approximation has already been used. Usually the steady state is regarded and the left-hand parts of the differential equations in the system (A.47) are considered to be zero. But now we demand that the atoms oscillate at frequency D, which is low in comparison with Ω, and thus we cannot neglect the time derivatives. The amplitude V_1 of the additional velocity obtained by the atom in these oscillations is considered to be small. Thus, we change the full time derivatives for the following

$$\frac{d}{dt} = \frac{\partial}{\partial t} + kV_1\cos Dt$$

To start the perturbation approach we have to choose the small parameter. Let the decay rate γ be much less than the Rabi frequency α of the resonant field $\varepsilon = \gamma/\alpha$. Besides, let $kV_1/\alpha = \varepsilon v_\kappa$, where v_κ is a dimensionless constant, $v_\kappa = O(1)$. Then the system (A.47) becomes

$$\frac{\partial}{\partial \tau}\rho_{22} = -\varepsilon \rho_{22} - i\tilde{\rho}_{21} + i\tilde{\rho}_{12}$$

$$\frac{\partial}{\partial \tau}\tilde{\rho}_{21} = -2i\rho_{22} + [-i\delta + \varepsilon(-1/2 - iv_\kappa \cos \delta_D \tau)]\tilde{\rho}_{21} + i \quad \text{(A.48)}$$

$$\frac{\partial}{\partial \tau}\tilde{\rho}_{12} = 2i\rho_{22} + [i\delta + \varepsilon(-1/2 + iv_\kappa \cos \delta_D \tau)]\tilde{\rho}_{12} - i$$

where the following designations were used: $\delta = \Delta/\alpha$, $\tau = \alpha t$, $\delta_D = D/\alpha$, c is the light velocity and k stands for the wave vector of the laser field which is parallel to the direction of the mechanical oscillations. In this section we omit the factor $\sqrt{2}$ which was used before for technical reasons. In the matrix notation the system (A.48) transforms into

$$\frac{\partial}{\partial \tau}\mathbf{W} = (\mathbf{Q_0} + \varepsilon \mathbf{Q_1})\mathbf{W} + \mathbf{C} \quad \text{(A.49)}$$

$$\mathbf{W} = \begin{pmatrix} \rho_{22} \\ \tilde{\rho}_{21} \\ \tilde{\rho}_{12} \end{pmatrix}, \mathbf{Q_0} = i\begin{pmatrix} 0 & -1 & 1 \\ -2 & -\delta & 0 \\ 2 & 0 & \delta \end{pmatrix}, \mathbf{Q_1} = \begin{pmatrix} -1 & 0 & 0 \\ 0 & \Phi & 0 \\ 0 & 0 & \overline{\Phi} \end{pmatrix},$$

$$\mathbf{C} = i\begin{pmatrix} 0 \\ 1 \\ -1 \end{pmatrix}, \ \Phi = -1/2 - iv_\kappa \cos \delta_D \tau$$

Again our goal is to find the stable solution of the system eq. (A.49) and the leading term in the asymptotic expansion in the powers of ε of the components of the vector $\mathbf{W}$. Following the scheme that we have already mentioned, we can get

$$\mathbf{W}_{\text{hom}}(\tau) = \mathbf{U}\exp(i\mathbf{d}\tau)\exp(\varepsilon \mathbf{K}\tau)\mathbf{U}^{-1}.$$

Here $\mathbf{U}$ is the matrix of matrix's $\mathbf{Q_0}$ eigenvectors and $\mathbf{K}$ is the matrix obtained from the matrix $\mathbf{G} = \exp(-\mathbf{E}\tau)\mathbf{U}^{-1}\mathbf{Q_1}\mathbf{U}\exp(\mathbf{E}\tau)$ with excluded secular components: $\mathbf{K} = \langle \mathbf{G} \rangle - diag\{0, i\eta, -i\eta\}$,

$$\mathbf{U} = \mathbf{U}^{-1} = \frac{1}{J}\begin{pmatrix} -\delta & 1 & 1 \\ 2 & \frac{-2}{\delta+J} & \frac{-2}{\delta-J} \\ 2 & \frac{-2}{\delta-J} & \frac{-2}{\delta+J} \end{pmatrix},$$

$$\mathbf{K} = \frac{1}{J^2}\begin{pmatrix} -\delta^2 - 2 & \frac{1}{2}iJv_\kappa & -\frac{1}{2}iJv_\kappa \\ iJv_\kappa & -B - i\nu & 0 \\ -iJv_\kappa & 0 & -B + i\nu \end{pmatrix}$$

$$J = \sqrt{\delta^2 + 4}, B = 3 + \frac{1}{2}\delta^2, \mathbf{E} = diag\{0, iJ, -iJ\}, \mathbf{d} = diag\{0, \delta_D, -\delta_D\}$$

Parameter $\nu = O(1)$ characterizes the parametric resonance condition $J = \delta_D + \varepsilon\nu$ as before. Since all the real parts of the matrix $\mathbf{K}$ eigenvalues are negative, we are going to obtain the steady (time independent) solution of the non-homogeneous equation (A.49) in the form of

$$\mathbf{W}(\tau) = \mathbf{W}_{\text{hom}}(\tau) \int_0^\tau [\mathbf{W}_{\text{hom}}(x)]^{-1} \mathbf{C} dx$$

where we regard only the upper limit of the integral. Fulfilling the integration and neglecting the terms of the order of ε we obtain [Siparov (1998)] the leading terms of the expansion in the powers of ε for the first two components of vector $\mathbf{W}$

$$\rho_{22} = T + S \tag{A.50}$$
$$T = \frac{2}{\delta^2} - \frac{\delta^2}{\delta_D^2 \Theta}(B^2 + \nu^2 - \frac{1}{2}B\,\delta_D^2 v_\kappa^2),$$
$$S = \frac{\delta}{\delta_D \Theta} 2v_\kappa(2 + \frac{1}{2}\delta^2)\sqrt{B^2 + \nu^2}\cos(\delta_D\tau + \varphi),$$

$$\tilde{\rho}_{21} = H + P \tag{A.51}$$
$$H = \frac{\delta}{\delta_D^2 \Theta}[1 + 2(B^2 + \nu^2 - \frac{1}{2}B\delta_D^2 v_\kappa^2)],$$
$$P = \frac{\delta}{\delta_D \Theta} v_\kappa(2 + \frac{1}{2}\delta^2)\sqrt{B^2 + \nu^2}[\delta\cos(\delta_D\tau + \varphi) - i\delta_D\cos(\delta_D\tau + \psi)],$$

$$B = 3 + \frac{\delta^2}{2}, \ \Theta = (\delta^2 + 2)(B^2 + \nu^2) + B\delta_D^2 v_\kappa^2, \ \varphi = \arctan\frac{B}{\nu}, \ \psi = -\arctan\frac{\nu}{B}$$

Both equations (A.50) and (A.51) contain constant (T, H) and oscillating (S, P) terms, their form being determined by the optic-mechanical parametric resonance. From the mathematical point of view it means that the eigenvalues of the matrix which transforms $\mathbf{Q_0}$ into the diagonal matrix coincide with the external frequency δ_D within the accuracy of ε

$$\sqrt{\delta^2 + 4} = \delta_D + \varepsilon\nu, \ \nu = O(1) \tag{A.52}$$

The parametric resonance here means that the frequency of the time dependent periodical coefficient (the atom's velocity) in the dynamics equation coincides with the frequency of the level population variance (Rabi frequency). Physically this means that we produce a kind of Doppler modulation upon a wave falling on an atom.

The obtained expressions eq. (A.50) and eq. (A.51) give the components of the density matrix and describe the dynamics of the TLA in the strong field in the conditions of optical-mechanical parametric resonance. Besides, eq. (A.51) contains not only the imaginary part of $\tilde{\rho}_{21}$ which was needed earlier when we calculated the absorption coefficient of the probe wave and obtained (in a slightly different notation) the expression eq. (A.45). Now we will use eq. (A.50) and eq. (A.51) to calculate the fluorescence. Obviously, the inverse character of this parametric resonance, i.e. the influence of the mechanical motion of the atom on the properties of the radiation interacting with it, could cause certain modification of the result.

A.4.1 *Calculation of the fluorescence spectrum*

The experimentally observable value in the resonant fluorescence measurements is the photon number rate, which can be calculated [Stenholm (1984)] as follows:

$$N_{ph} = -2\lambda(q)Re\langle q, 1 \mid \rho_q \mid 0, 2\rangle \qquad (A.53)$$

where $\mid q\rangle$ stands for the excited photon state after the decay, $\lambda(q) = (e\mu/\hbar)(\hbar\Omega_q/2\epsilon_0 V_c)^{1/2}$ is the coupling constant (e is the electron charge, Ω_q is the frequency of the emitted photon, ϵ_0 is the absolute dielectric permittivity, V_c is the empty cavity volume), ρ_q is the so-called one-photon density matrix of the TLA in the strong field of laser radiation, and it is used to describe the photon emission. The components of this matrix can be calculated from the following system of equations:

$$\frac{d}{d\tau}\tilde{\rho}_{21q} = (\Delta_q + \Delta - i\frac{1}{2}\gamma)\tilde{\rho}_{21q} - \alpha\tilde{\rho}_{11q} + \alpha\tilde{\rho}_{22q} - i\lambda(q)\rho_{22} \quad (A.54)$$

$$\frac{d}{d\tau}\tilde{\rho}_{11q} = -\alpha\tilde{\rho}_{21q} + \Delta_q\tilde{\rho}_{11q} + i\gamma\tilde{\rho}_{22q} + \alpha\tilde{\rho}_{12q} - i\lambda(q)\tilde{\rho}_{12}$$

$$\frac{d}{d\tau}\tilde{\rho}_{22q} = \alpha\tilde{\rho}_{21q} + (\Delta_q - i\gamma)\tilde{\rho}_{22q} - \alpha\tilde{\rho}_{12q}$$

$$\frac{d}{d\tau}\tilde{\rho}_{12q} = \alpha\tilde{\rho}_{11q} - \alpha\tilde{\rho}_{22q} + (\Delta_q - \Delta - i\frac{1}{2}\gamma)\tilde{\rho}_{12q}$$

where $\Delta_q = \Omega - \Omega_q$ is the difference between the laser frequency and the frequency of the emitted photon. In vector form, we obtain the following system

$$\frac{d}{d\tau}\mathbf{W_q} = (\mathbf{Q_{q0}} + \varepsilon\mathbf{Q_{q1}})\mathbf{W_q} + \mathbf{C_q}$$

$$\mathbf{W_q} = \begin{pmatrix} \tilde{\rho}_{21q} \\ \tilde{\rho}_{11q} \\ \tilde{\rho}_{22q} \\ \tilde{\rho}_{12q} \end{pmatrix}, \mathbf{Q_{q0}} = i\begin{pmatrix} -\Delta_q - \sigma & 1 & -1 & 0 \\ 1 & -\Delta_q & 0 & -1 \\ -1 & 0 & -\Delta_q & 1 \\ 0 & -1 & 1 & -\Delta_q + \sigma \end{pmatrix},$$

$$\mathbf{Q_{q1}} = \begin{pmatrix} -1/2 & 0 & 0 & 0 \\ 0 & 0 & 1 & 0 \\ 0 & 0 & 1 & 0 \\ 0 & 0 & 0 & 1/2 \end{pmatrix}, \mathbf{C_q} = -i\lambda(q)\begin{pmatrix} \rho_{22} \\ \tilde{\rho}_{12} \\ 0 \\ 0 \end{pmatrix}$$

In the regular case (no optic-mechanical parametric resonance) in the right-hand side of the first two equations of this system there are constant values for ρ_{22} and $\tilde{\rho}_{12}$ ($\rho_{22} = 1/2$, $\tilde{\rho}_{12} = 0$ for the strong field), and if we regard the stationary solution, the derivatives can be neglected, and we get the known symmetrical three-peaked structure. But now we have the expressions eq. (A.50) and eq. (A.51) for the specific pumping terms, that contain both constant and oscillating parts that appeared because of the atomic mechanical vibrations with the specially chosen frequency. Let us regard constant and oscillating parts separately, and then discuss the final result as their sum. The constant parts T and H described by eq. (A.50) and eq. (A.51) can be substituted into the system of equations (A.54) without the derivatives as in the regular case, and we obtain

$$(\Delta_q + \Delta - \frac{1}{2}\gamma)\tilde{\rho}_{21q} - \alpha\tilde{\rho}_{11q} + \alpha\tilde{\rho}_{22q} = i\lambda(q)T \qquad (A.55)$$
$$-\alpha\tilde{\rho}_{21q} + \Delta_q\tilde{\rho}_{11q} + i\gamma\tilde{\rho}_{22q} + \alpha\tilde{\rho}_{12q} = i\lambda(q)H$$
$$\alpha\tilde{\rho}_{21q} + (\Delta_q - i\gamma)\tilde{\rho}_{22q} - \alpha\tilde{\rho}_{12q} = 0$$
$$\alpha\tilde{\rho}_{11q} - \alpha\tilde{\rho}_{22q} + (\Delta_q - \Delta - i\frac{1}{2}\gamma)\tilde{\rho}_{12q} = 0$$

This is an algebraic system of equations. Solving (A.55), we get the constant part of $\tilde{\rho}_{21q}$ which is due to the optic-mechanical parametric resonance, substitute it into eq. (A.53) (notice, that $\tilde{\rho}_{21q} = \langle q, 1 \mid \rho_q \mid 0, 2 \rangle$) and obtain the corresponding time-independent part of the spectrum that could be directly compared with the measurements. The determinant Det for the system (A.55) itself and the determinant Det_q for the calculation of $\tilde{\rho}_{21q}$ have the forms

$$Det = \Delta_q^2(\Delta_q^2 - \frac{5}{4}\gamma^2 - 4\alpha^2 - \Delta^2) + i\Delta_q\gamma(-2\Delta_q^2 + \Delta^2 + \frac{1}{4}\Gamma\gamma^2 + 2\alpha^2) \quad (A.56)$$

$$Det_q = T\Delta_q(\Delta_q^2 - \Delta_q\Delta - 2\alpha^2 - \frac{1}{2}\gamma^2) + H\alpha(\Delta_q^2 - \Delta_q\Delta - \frac{1}{2}\gamma^2) \quad (A.57)$$

$$+i[T\Delta_q\gamma(\Delta - \frac{3}{2}\delta\Delta_q) + H\alpha\gamma(\Delta - \frac{3}{2}\Delta_q)]$$

Substituting (A.56) and (A.57) into

$$\tilde{\rho}_{21q} = \frac{Det_q}{Det} \quad (A.58)$$

and then into eq. (A.53), we obtain the expression for the photon number rate which characterizes the fluorescence spectrum in the case of the optical-mechanical parametric resonance:

$$N_{ph} = -2\lambda^2(q)\frac{Re(Det_q)Im(Det) - Im(Det_q)Re(Det)}{(Re(Det))^2 + (Im(Det))^2} \quad (A.59)$$

The results of the calculation are shown in Fig. A.4, which illustrates the time-independent part of the emitted radiation spectrum which corresponds to the constant components that appear when the optic-mechanical parametric resonance takes place.

Let's compare the obtained result with the known one. For the usual case of the TLA in the strong field when there is no parametric resonance of any kind, the upper level population ρ_{22} at zero detuning is equal to $1/2$, while $\tilde{\rho}_{21} = O(1/\alpha)$. It means that the three-peaked structure of the usual resonance spectrum must originate from the particular features of the determinant Det of the matrix of coefficients on the left-hand part of the system (A.55). From the expressions for T and H in the equations (A.50), (A.51), it follows that if we regard the optic-mechanical parametric resonance for zero detuning, i.e. $\Delta = 0$, $\delta_D = 2$, $(\Omega = \omega, D = 2\alpha)$, then $\rho_{22} = 1/2$, $\tilde{\rho}_{21} = 0$, and we get the same three-peaked structure as usual. This is relevant to the additional condition mentioned in the previous section, when we found out that the effect of the parametric resonance disappears for the zero detuning of the driving radiation. Figure A.5 presents the theoretical plot resulted from the calculation of the photon number rate with the help

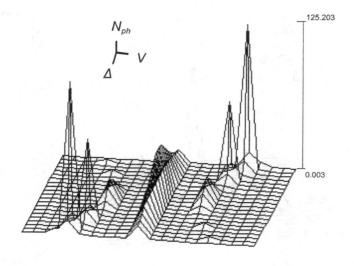

Fig. A.4   Fluorescence spectrum (OMPR driving).

of the corresponding expression in [Stenholm (1984)] for the usual case of $\rho_{22} = 1/2$, $\tilde{\rho}_{21} = 0$ for various Δ.

The particular deformation of the theoretical spectrum presented in Fig. A.4 predicts the new features that could be found in experiment. Notice that the heights of the sidebands in Fig. A.4 are much larger than the height of the central peak.

The existence of the oscillating terms S and J in (3.5) and (A.51) means that we cannot neglect the time derivatives when calculating the oscillating components of the one-photon density matrix. Moreover, if any of the $\mathbf{Q_{q0}}$ eigenvalues (which are $\lambda_{1,2} = -\Delta_q$, $\lambda_{3,4} = -\Delta_q \pm \sqrt{\delta^2 + 4}$) coincide with the pumping frequency δ_D, the solution of the system will be governed not by the parametric resonance but by the resonance, i.e. the corresponding terms in the expansion for the $\tilde{\rho}_{21q}$ will be of the order of ε^{-1}. Thus, the optical-mechanical parametric resonance provides the additional low-frequency non-stationary components with great amplitudes which contribute to the emitted radiation spectrum, their locations being at the frequencies $\Omega_q = \Omega + \delta_D$, $\Omega_q = \Omega + \delta_D \pm \sqrt{\delta^2 + 4}$.

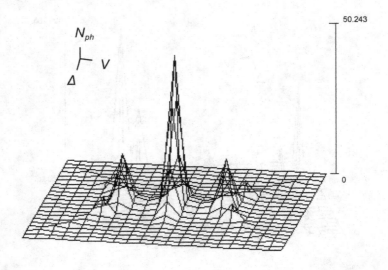

Fig. A.5   Regular fluorescence spectrum (no parametric resonances).

A.4.2   *Driving the TLAs by the bichromatic radiation*

Two apparent possibilities to make TLAs vibrate are the low-frequency electric field for the two-level ions and the acoustic field for the admixtures. But the effect of the direct optic-mechanical parametric resonance discussed above produces the same result. We saw that the bichromatic resonant radiation with one strong and one weak mode (the distuning is equal to the doubled Rabi frequency) acts on a TLA with the force which has not only the well-known constant component proportional to the excited state decay rate, but the oscillating component too. Moreover, the amplitude of this oscillating component appeared to be of the larger asymptotic order than the value of the constant component, and it is proportional to the Rabi parameter of the strong component of the bichromatic field. Thus, such bichromatic radiation produces the oscillating force and can also be the source of the mechanical vibrations needed for the realization of the inverse optical-mechanical parametric resonance. So, we have to solve the corresponding equations for the density matrix again, because the bichromatic field can well lead to the levels reconstruction. Various aspects of this problem were discussed in [Stenholm (1978b); Smyth and Swain (1996)].

Strictly speaking, when we get the result which will be the modification of the fluorescence spectrum when there are two specifically chosen electromagnetic waves acting on atoms, we won't have any need to involve mechanical vibrations into the speculations. Nevertheless, in the previous subsection it was shown that such vibrations with the specifically chosen frequencies and amplitudes do cause the modifications of the fluorescence spectrum. If these two modifications are of the same type, it will mean that the approaches are equivalent and could be regarded in the same context when needed.

Not to repeat all the reasoning and calculations which are also based on the Bloch equations in [Kazakov and Siparov (1997)] and given above, let us only present the result which was obtained in a similar way. In case of the TLA in the bichromatic field and the parametric resonance due to the coincidence of the Rabi frequency and the distuning of the modes of the bichromatic field with one strong and one weak component, the elements of the density matrix have the following form

$$\rho_{22b} = T_b + S_b \tag{A.60}$$

$$T_b = \frac{\sqrt{2}}{D^2}[1 + \sigma^2 \frac{a^2 D^2 B - (1-\gamma)(B^2 + \eta^2)(D-\sigma)^2}{2a^2 D^2 B + (\sigma^2 + 2\gamma)(B^2 + \eta^2)(D-\sigma)^2}]$$

$$S_b = \frac{\sqrt{2}a\sigma(\sigma^2 + 2)(D-\sigma)\sqrt{B^2 + \eta^2}}{D[2a^2 D^2 B + (\sigma^2 + 2\gamma)(B^2 + \eta^2)(D-\sigma)^2]} \cos(D\tau + \varphi)$$

$$\tilde{\rho}_{21b} = H_b + J_b \tag{A.61}$$

$$H_b = \frac{\sigma}{D^2\sqrt{2}}[1 + 2\frac{a^2 D^2 B - (1-\gamma)(B^2 + \eta^2)(D-\sigma)^2}{2a^2 D^2 B + (\sigma^2 + 2\gamma)(B^2 + \eta^2)(D-\sigma)^2}]$$

$$J_b = \frac{a\sigma(\sigma^2 + 2)(D-\sigma)\sqrt{B^2 + \eta^2}}{\sqrt{2}[2a^2 D^2 B + (\sigma^2 + 2\gamma)(B^2 + \eta^2)(D-\sigma)^2]}[\sigma \cos(D\tau + \varphi)$$

$$+iD\cos(D\tau + \psi)] \tag{A.62}$$

In these expressions the notations are again the same as in Section 2 of the Appendix: α_1 is Rabi frequency of the strong mode; $\delta_d = \frac{\Omega_2 - \Omega_1}{\alpha_1\sqrt{2}}$ is the dimensionless distuning, and here it coincides with the frequency, D of the atomic vibrations, $\delta_d = D$; $\sigma = \delta + \kappa v$; $\delta = \frac{\omega - \Omega_1}{\alpha_1\sqrt{2}}$; $\kappa = \frac{k}{\alpha_1\sqrt{2}}$; v is the heat velocity (or maybe the atoms beam propagation velocity) of the TLA; $\varepsilon = \frac{\gamma}{\alpha_1\sqrt{2}}$ is a small parameter; γ is the excited state decay rate; $\varepsilon a = \frac{\alpha_2}{\alpha_1}$, $a = O(1)$; $B = 1 + \Gamma_{12}(1 + \sigma^2)$; $\varepsilon\Gamma_{12} = \frac{\gamma_{12}}{\alpha_1\sqrt{2}}$, γ_{12} is the transversal decay rate; $\tau = t\alpha_1\sqrt{2}$; $\varphi = -\arctan\frac{B}{\nu}$; $\psi = \arctan\frac{\nu}{B}$; $\nu = O(1)$ is a tuning parameter; $\eta = O(1)$. The condition for the parametric resonance in this problem was

$$\sqrt{\sigma^2 + 2} = D + \varepsilon\eta$$

Substituting eqs. (A.60) and (A.61) into eq. (A.54), we get the fluorescence spectrum for the case of the TLA in the bichromatic field. As before, we use the constant parts of the density matrix components T_b and H_b to calculate the spectrum with the help of (A.59). Figure A.6 presents the result of this calculation. We can see that the qualitative character appears to be the same to what it was for the case of the external force of the arbitrary nature.

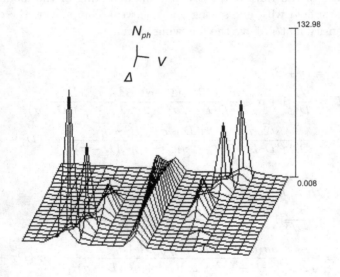

Fig. A.6 Fluorescence spectrum (bichromatic driving).

A.4.3 *Assumptions, demands and possible applications*

Let us again discuss the meaning of the assumptions that were used while deriving the results and recognize what demands do they infer on the experimental situation.

1) $\gamma \ll \alpha$ ($\gamma = \varepsilon\alpha$). This is a strong field condition, and it can be easily sufficed.

2) $kV/\alpha = \varepsilon v$, $v = O(1)$, where k is the wave vector of the resonant laser field, V is the additional velocity obtained by the atom through oscillations, v is the dimensionless constant. Here we should bear in mind that the effect can reveal itself only when the $\mathbf{Q_1}$ matrix is not degenerated, that is the combination $kV/\alpha = \frac{\Omega}{\alpha}\frac{V}{c}$ has to be not just small but be strictly of the order of ε. For example, when we take $v = 1$ and choose $\varepsilon = 0.01$ it means that we can use $\Omega \sim 10^{15}\text{s}^{-1}$, $\alpha \sim 10^{8}\text{s}^{-1}$, while the needed value of the additional velocity V due to the oscillations appears to be 0.1m/s (the velocity of an atom at room temperatures is proportional to 10^{3}m/s).

3) Let us evaluate the amplitude E_{ex} of the low-frequency electric field which makes a singly charged two-level ion vibrate at frequency $D \sim 10^{8}\text{s}^{-1}$. The amplitude A of these vibrations in the case corresponding to the example in item 2) is proportional to $VD^{-1} \sim 10^{-9}$m. The work of the electric field is equal to the kinetic energy of an ion $eE_{ex}A = mV^2/2$, which gives $E_{ex} \sim 10^{-1}$V/m for the e.g. hydrogen atom with mass $m \sim 10^{-27}$kg.

The experimental setup for this type of the vibrations stimulation will slightly differ from the usual setup for the resonance fluorescence measurements which provides the possibility to observe the three-peaked structure on the spectrum. Now the particle beam has to pass between the two plates of the condenser in which there are holes for the laser beam. The registration can be performed from the third orthogonal direction, that is the fluorescence is observed through the gap between the plates. The presented calculation suggests that when the optic-mechanical parametric resonance takes place, the sideband peak substantially higher than the central peak appears on the spectrum of the fluorescence radiation at non-zero detuning. Note, that the sign of the detuning is essential when observing and affects the sign of the shift of the sideband. The exact position of this feature depends on the Rabi frequency, i.e. on the intensity of the laser beam. This means that varying the laser intensity (and simultaneously the frequency of the external field created by the condenser), we can locate the feature onto the position on the spectrum which is the most convenient for the registration.

The results for the TLAs driven by the bichromatic radiation analogous to those obtained above are relevant to the dependence of the probe transmission spectrum on the probe and pump detunings from the resonance measured in [Gruneisen *et al.* (1988)] for the sodium vapor. In the experimental setup described in [Wu *et al.* (1994)] or [Bai *et al.* (1985)], we can expect that the results obtained above will look like those obtained in the mentioned papers.

The method based on the described effect of the inverse optic-mechanical parametric resonance can be applied to the investigation of the atomic levels structure and the atom's dynamics alongside with the methods based on Raman scattering and stimulated Brillouin scattering [Postan *et al.* (1992)]. In order to obtain a more realistic description of a physical situation of possible interest, we can complicate the model of an atom and regard a three-level atom [Siparov (2001b)] which has additional parameters that will provide additional possibilities. Another potentially interesting field is the amplification of light in absorbing media containing TLAs which was predicted in [Rautian and Sobelman (1961)] and subsequently observed in experiment in [Wu *et al.* (1977)]. The nonlinear interaction between the atom and radiation, which is the cause of such amplification is now influenced by the mechanical driving (see also [Brown *et al.* (1997)]). The corresponding calculations were performed in [Siparov (2002)].

The new result and the main point is the possibility to modulate the frequency in the resonance fluorescence investigations by means of modulating the mechanical motion of an atom in a parametric resonance way. The effect is not just a Doppler type frequency shift, but the appearance of high features, some of them have non-stationary character.

A.4.4 *Conclusion*

Concluding the discussion of the influence of the kinematics of atom motion on the properties of the electromagnetic radiation with which this atom interacts, we would underline the following. It turns out that if certain conditions including frequency part and amplitude part are fulfilled, the effect of optic-mechanical parametric resonance takes place and the radiation obtains a specific feature which can be observed on the spectrum. 'Can be observed' here means that there is a need for a modification either of the observation itself or of the signal processing. The thing is that the specific feature characterizing the effect presents a periodical change of the amplification to absorption at a certain frequency close to the frequency of the atomic transition. Observing the signal of probe wave absorption or of resonance fluorescence in a usual way, one wouldn't notice this feature because of averaging with time, the frequency of absorption/amplification change for standard laboratory laser experiments falls within radio region.

Therefore, in order to observe it, one has either to use a gated detector or process the signal in such a way as to find in it the periodic component with the frequency defined by the experimental conditions - laser intensity and atomic vibration frequency. These two conditions reflect the frequency condition of the optic-mechanical parametric resonance. It is important to notice that the amplitude conditions that must be fulfilled to provide the effect have specific character - the amplitude of the vibrations has to be neither large enough nor small enough but exactly of the specified order of magnitude. This means that the effect is of the zeroth order and is not directly proportional to the amplitude of its cause.

From the point of view of physics, it has the same origin as several already observed phenomena that can be found in the given references and does not seem something exotic. Really, we see that the due choice of components in the poly-chromatic (particularly, bichromatic) falling radiation spectrum leads to the modulation of the motion of atom, the due choice of atomic motion parameters leads to the modulation of the absorbed and emitted radiation, there are conditions when all these phenomena take place in a sense most effectively. The essential thing for the next step is the kinematical character of the optic-mechanical parametric resonance effect: we are not concerned about the physical reasons of the periodical change of the atom velocity with regard to the receiver of the radiation. It means the possibility to involve both the inertial and the gravitational forces acting on the atoms into the practice of usual spectroscopic experiments and, thus, could provide new perspectives in the experimental investigations of relativistic effects.

Bibliography

Aguirre, A., Burgess, C., Friedland, A. and Nolte, D. (2001). *CQG* **18**, p. R223.

Allen, B. (1989). *Phys. Rev. Lett.* **63**, pp. 2017–2020.

Amaldi, E. and Pizzella, G. (1979). *In Astrofisika e Cosmologia Gravitazione Quanti e Relativita* (Mir, Moscow), p. 241, (rus. trans.).

Anderson, A. (1987). *Proc. Int. Assoc. Geod. Symp.* **1**, pp. 83–90.

Anderson, J., Laing, P., Lau, E., Liu, A., Nieto, M. and Turyshev, S. (1998). *Phys. Rev. Lett.* **81**, p. 2858.

Andersson, N., Kokkotas, K. and Schutz, F. (1999) **510**, p. 846.

Argon, A., Greenhill, L. and Moran, J. (1994). *ApJ* **422**, p. 586A.

Arnold, V. (1974). *Mathematical methods of classical mechanics* (Nauka, Moscow), (rus).

Ashkin, A. (1970). *Phys. Rev. Lett.* **24**, 4, pp. 156–159.

Autler, S. and Townes, C. (1955). *Phys. Rev.* **100**, p. 703.

Bacry, H. and Levy-Leblond, J.-M. (1968). *J. Math. Phys.* **9**, p. 1605.

Bai, Y., Yodh, A. and Mossberg, T. (1985). *Phys. Rev. Lett.* **55**, p. 1277.

Balan, V. and Brinzei, N. (2005). *HNGP* **4**, p. 114.

Bao, D., Chern, S. and Shen, Z. (2000). *An Introduction to Riemann-Finsler Geometry* (Springer Verlag).

Begeman, K., Broeils, A. and Sanders, R. (1991). *MNRAS* **249**, p. 523.

Bekenstein, J. (2007). ArXiv:0701848v2.

Bekenstein, J. and Milgrom, M. (1984). *Ap. J.* **286**, p. 7.

Berman, P. (1999). *Phys. Rev. A* **59**, pp. 585–596.

Berman, P., Dubetsky, B. and Guo, J. (1995). *Phys. Rev. A* **51**, pp. 3947–3958.

Bertotti, B. (1983). Proc. XIIIth Int. Conf. on Gen. Rel. pp. 255–261.

Berulis, I., Lekht, E. and Munitsyn, V. (1998). *Astron. Zh.* **75**, p. 394.

Besse, A. (1978). *Manifolds all of whose geodesics are closed* (Springer, Berlin).

Bigelow, N. and Prentiss, M. (1990). *Phys. Rev. Lett.* **65**, p. 555.

Blioh, P. and Minakov, A. (1989). *Gravitational lenses* (Naukova dumka, Kiiv).

Bochkarev, N. (1992). *Foundations of the interstellar medium physics* (MGU, Moscow).

Bogoslovsky, G. (1973). *DAN SSSR* **213**, p. 1055.

Bogoslovsky, G. and Goenner, H. (1998). *Phys. Lett. A* **244**, p. 222.

Bohm, D. (1965). *The special theory of relativity* (W.A.Benjamin, Inc, New York, Amsterdam).

Bonch-Bruevich, A., Vartanian, T. and Chigir', N. (1979). *Zh. Eksp. i Teor. Fiz.* **77**, p. 1899, (rus).

Bonifacio, R., DeSalvo, L. and Robb, G. (1997a). *Opt. Comm* **137**, pp. 276–280.

Bonifacio, R., Robb, G. and McNeil, B. (1997b). *Phys. Rev. A* **56**, pp. 912–924.

Born, M. (1962). *Einstein's Theory of Relativity* (Dover Publications, New York).

Boyd, R. and Sargent III, M. (1988). *JOSA* **B5**, p. 99.

Braginsky, V. and Grischuk, L. (1985). *JETP* **89**, pp. 744–750, (rus).

Braginsky, V. and Thorne, K. (1987). *Nature* **327**, pp. 123–128.

Brans, C. and Dicke, R. (1961). *Phys. Rev.* **124**, p. 925.

Brecher, K. (1977). *Phys. Rev. Lett.* **39**, p. 1051.

Bridges, T. (1997). *Mon. Not. R. Astron. Soc.* **284**, p. 376.

Brinzei, N. and Siparov, S. (2007). *HNGP* **8**, p. 41.

Brown, W., Gardner, J., Gauthier, D. and Vilaseca, R. (1997). *Phys. Rev. A* **55**, p. R1601.

Burshtein, A. (1962). *Zh. Eksp. i Teor. Fiz.* **49**, p. 1362, (rus).

Callagari, G. (1991). *Acta Phys. Hung.* **70**, pp. 29–33.

Carroll, S. (1997). ArXiv [gr-qc]:9712019v1.

Chwolson, O. (1924). *Astronomische Nachrichten* **221**, p. 329.

Clifford, W. (1979). *Albert Einstein and the theory of gravitation* (Mir, Moscow).

Clowe, D. (2006). *Astrophys. J.* **648**, pp. L109–L113.

Cook, A. (1977). *Celestial masers* (Cambridge University Press, London).

Cropper, M. (1998). *MNRAS* **293**, p. L57.

Danileiko, M. (1984). *Lett. JETP* **39**, pp. 428–430, (rus).

Dauphole, B. and Colin, J. (1995). *Astron. Astrophys.* **300**, p. 117.

De Sitter, W. (ed.) (1913). *Proceedings of the Section of Sciences*, Vol. 15 (Koninkijke Academie van Wetenschappen, te Amsterdam).

De Sitter, W. (1917). *MNRAS* **78**, p. 3.

Dehlmelt, H. (1976). *Nature* **262**, 5571.

Denisov, V. (1989). *In Experimental tests of the relativity theory* (Moscow).

DeWitt, B. (1964). *Relativity, groups and topology* (Gordon and Breach, New York).

Drever, R. (1983). 10 Int. Conf. Padua, 10, pp. 397–412.

Dvali, G., Gabadadze, G. and Shifman, M. (2001). ArXiv [astro-ph]:0102422v2.

Eberly, J. and Popov, V. (1988). *Phys. Rev. A* **37**, p. 2012.

Einstein, A. (1905a). *Ann. Phys.* **17**, p. 891.

Einstein, A. (1905b). *Ann. Phys.* **18**, p. 639.

Einstein, A. (1915). *Sitzungsber. d. Berl. Akad.* **2**, p. 831.

Einstein, A. (1916a). *Ann. Phys.* **49**, p. 769.

Einstein, A. (1916b). *Sitzungsber. Preuss. Akad. Wiss.* **2**, p. 1111.

Einstein, A. (1916c). *Sitzungsber. Preuss. Akad. Wiss.* **1**, p. 688.

Einstein, A. (1917). *Sitzungsber. d. Berl. Akad.* **1**, p. 142.

Einstein, A. (1920). *Verlag von Julius Springer* .

Einstein, A. (1931). *Sitzungsber. Preuss. Akad. Wiss.* , p. 235.

Einstein, A. (1936). *Science* **84**, pp. 506–507.

Einstein, A. and Grossmann, M. (1914). *Zeitschrift fur Mathematik und Physik* **62**, pp. 225–261.

Einstein, A., Infeld, I. and Hoffmann, B. (1938). *Ann. Math.* **39**, p. 65.

Elitzur, M. (1992). *Astronomical masers* (Kluwer Acad. Publ., Boston).

Fakir, R. (1993). *Astophys. J.* **418**, pp. 202–207.

Feynman, R., Vernon, F. and Hellwarth, R. (1957). *J. Appl. Phys.* **28**, p. 49.

Ficek, Z., Freedhoff, H. and Rudolph, T. (1999). *Opt. Spectr.* **87**, pp. 670–675.

Fischer, V. (1994). *Class. Quantum Grav.* **11**, pp. 463–474.

Fizeau, H. (1850). *Ann. Phys.* **79**, p. 167.

Fock, V. (1939). *JETP* **9**, p. 375.

Fock, V. (1964). *The theory of space, time and gravitation* (Pergamon Press, Oxford, London, New York, Paris).

Fradkin, E. (1983). *JETP* **84**, p. 1654, (rus).

Freedhoff, H. and Ficek, Z. (1997). *Phys. Rev. A* **55**, pp. 1234–1238.

Friedmann, A. (1922). *Z. Phys.* **10**, p. 377.

Frisch, D. and Smith, J. (1993). Netpbm, `http://webpages.uah.edu/~bonamem/AST371_Fall09/frisch1963.pdf`.

Fux, R. (1999). ArXiv [astro-ph]:9903154v1.

Gertsenshtein, M. and Pustovoit, V. (1962). *JETP* **43**, pp. 605–627.

Gladyshev, V. (2000). *Irreversible electromagnetic processes in the astrophysical problems* (Bauman MSTU Publ., Moscow), (rus).

Godbillon, C. (1969). *Geometrie differentielle et mecanique analytique* (Hermann, Paris).

Goldhaber, G. (2001). *Ap. J.* **558**, p. 359.

Gromov, N. A. (1990). *Contractions and analytical continuations of classical groups. Unified approach* (Syktyvkar), (rus).

Grove, R., Wu, F. and Ezekiel, S. (1977). *Phys. Rev. A* **15**, p. 227.

Gruneisen, M., MacDonald, K. and Boyd, R. (1988). *JOSA* **B5**, p. 123.

Guo, H., Huang, C., Wu, H. and Zhou, B. (2008). ArXiv:0812.0871.

Guo, H., Wu, H. and Zhou, B. (2009). *Phys. Lett. B.* **670**, p. 437.

Helling, R. (1983). Proc. XIIIth Int. Conf. on Gen. Rel. pp. 485–493.

Hilbert, D. (1915). *Nachrichten K. Gesselschaft Wiss. Gottingen, Math. - Phys. Klasse* **3**, p. 395.

Hipparcos (1997). Hipparcos Variability Annex: Tables. Part 2: Unsolved Variables, 11th, No. 52308.

Hizhniakov, V. (1988). *Izv. AS ESSR, Phys. and Math.* **37**, pp. 241–243.

Hobson, M., Efstathion, G. and Lasenby, A. (2009). *General relativity* (Cambridge).

Huang, C. (2009). ArXiv:0909.2773v1.

Huang, C., Guo, H., Tian, Y., Xu, Z. and Zhou, B. (2007). ArXiv:0403013v3.

Hubble, E. (1929). *Proc. Nat. Acad. Sci.* **15**.

Huei, P. and Bo, P. (1990). *Gen. Relat. Gravit.* **22**, pp. 45–52.

Hulse, R. and Taylor, J. (1975). *Ap. J.* **195**, pp. L–51.

Iacopini, E. (1979). *Phys. Lett.* **A73**, pp. 140–142.

Iorio, R. (2008). ArXiv [gr-qc]:0806.3011v3.

Isaacson, R. (1968). *Phys. Rev.* **166**, p. 1263.

Israel, G. (2002). *A&A* **386**, p. L13.

Jafry, Y., Cornelisse, J. and Reinhard, R. (1994). *ESA J.* **18**, pp. 219–228.

Javanainen, J. (1980). *Appl. Phys.* **23**, 2.

Javanainen, J. (1981). *J. Phys.* **B14**, 21.

Jeeva, A. (1995). *Phys. Lett.* **A110**, pp. 446–450.

Johansson, S. and Letokhov, V. (2002). *Pis'ma v ZhETF* **75**, p. 591.

Kaluza, T. (1921). *Sitzungsber. d. Berl. Akad.* **2**, p. 966.

Kaplan, S. and Pikel'ner, S. (1979). *Fizika mejzvezdnoy sredy* (Nauka, Moscow).

Kauffman, L. (2004). *New J. Phys.* **6**, p. 173.

Kazakov, A. (1990). *Opt. i Spectrosk.* **69**, p. 244, (rus).

Kazakov, A. (1992). *JETP* **102**, p. 1484, (rus).

Kazakov, A. (1993a). *JETP* **103**, p. 1548, (rus).

Kazakov, A. (1993b). *Opt. i Spectrosk.* **75**, p. 1109, (rus).

Kazakov, A. and Siparov, S. (1997). *Opt. i Spectrosk.* **83**, p. 961, (rus).

Kazantsev, A., Surdutovich, G. and Yakovlev, V. (1991). *Mechanical action of light on atoms* (Nauka, Moscow).

Kerner, E. (1976). Proc. Natl. Acad. Sci. (USA), p. 1418.

Kerr, R. (1963). *Phys. Rev. Lett.* **11**, p. 237.

Kinney, W. and Brisudova, M. (2000). ArXiv [astro-ph]:0006453.

Kokarev, S. (2010). *Introduction to the general relativity theory* (Yaroslavl').

Konacki, M. and Wolszezan, A. (2003). *ApJ* **591**, p. L147.

Kulagin, V., Polnarev, A. and Rudenko, V. (1986). *JETP* **91**, pp. 1553–1564, (rus).

Kursunoglu, B. (1991). *Jour. Phys. Essays* **4**, p. 439.

Kyrola, E. and Stenholm, S. (1977). *Opt. Commun.* **22**, p. 123.

Landau, L. and Lifshits, E. (1963). *Quantum mechanics* (Nauka, Moscow).

Landau, L. and Lifshits, E. (1967). *Electrodynamics of the continuous media* (Nauka, Moscow), (rus).

Landau, L. and Lifshitz, E. (1967). *The theory of field* (Nauka, Moscow).

Leen, T., Parker, L. and Pimentel, L. (1983). *Gen. Relat. Gravit.* **15**, pp. 761–776.

Lense, J. and Thirring, H. (1918). *Phys. Z.* **19**, p. 156.

Letohov, V. and Chebotaev, V. (1990). *Non-linear laser spectroscopy of super-high resolution* (Nauka, Moscow).

Letokhov, V. and Pavlik, B. (1976). *Appl. Phys.* **9**, 3, pp. 229–237.

Liaptsev, A. (1994). *Opt. Spektrosk.* **77**, p. 705.

Lin, C. and Shu, F. (1964). *Ap. J.* **140**, p. 646.

Liu, F. and Tian, Y. (2008). ArXiv [hep-th]:0806.1310v2.

Lobachevsky, N. (1979). *Albert Einstein and the theory of gravitation* (Mir, Moscow).

Lorentz, H. (1892). *Zittingsverlag Akad. v. Wet.* **1**, p. 74.

Lounis, B., Jelezko, F. and Orrit, M. (1997). *Phys. Rev. Lett.* **78**, p. 3673.

Luminet, J.-P. (2003). *Nature* **425**, p. 593.

Mach, E. (1911). *History and root of the principle of the conservation of energy*, 2nd edn. (Open Court, Chicago).

MacMillan, D. (2003). ArXiv [astro-ph]:0309826.

Majorana, Q. (1918). *Phys. Rev.* **11**, p. 411.

Manida, S. (1999). ArXiv [gr-qc]:9905046v1.

Mannheim, P. (1997). *Astrophys. J.* **479**, p. 659.

Mannheim, P. (2006). *Prog. Part. Nucl. Phys.* **56**, p. 340.

Mannheim, P. and Kazanas, D. (1994). *Gen. Relat. Gravit.* **26**, p. 337.

Markwardt, C. (2002). *ApJ* **576**, p. L137.

Markwardt, C., Juda, M. and Swank, J. (2003). *IAU Circ.* **2**, p. 8095.

Marshakov, A. (2002). *Uspekhi Fiz.* **172**, p. 977.

Mashhoon, B. (2001). *Lect. Notes Phys.* **562**, p. 83.

Michelson, A. and Morley, E. (1887). *Am. J. Sci. Ser. III* **34**, p. 333.

Milgrom, M. (1983). *Ap. J.* **270**, p. 384.

Miller, M. and Hamilton, D. (2001). *ApJ* **550**, p. 863.

Minier, V. (2002). Proc. 6-th VLBI Network Sypm. 6 (Bonn), p. 205.

Minkowski, H. (1909). *Z. Phys.* **10**, p. 104.

Minogin, V. and Letohov, V. (1986). *Laser radiation pressure on atoms* (Nauka, Moscow).

Minogin, V., Letokhov, V. and Zueva, T. (1981). *Opt. Comm.* **38**, pp. 225–229.

Miron, R. and Anastasiei, M. (1987). *Vector bundles. Lagrange Spaces. Applications to the theory of relativity* (Ed. Acad. Române, Bucuresti).

Miron, R. and Anastasiei, M. (1994). *The geometry of Lagrange spaces: Theory and applications* (Kluwer Acad. Publ.).

Misner, C., Thorne, K. and Wheeler, J. (1973). *Gravitation* (W.H. Freeman).

Moffat, J. (1995). *Phys. Lett. B* **355**, p. 447.

Moffat, J. (2006). *JCAP* **3**, p. 004.

Molaro, P. (2002). *A&A* **381**, p. 64L.

Mollow, B. (1969). *Phys. Rev.* **188**, p. 1969.

Moore, M. and Meystre, P. (1998). *Phys. Rev. A* **58**, pp. 3248–3258.

Mornev, O. (2001). Proc. VIII All-Russia Congress on Theor. and Appl. Mechanics, (Perm'), (rus).

Muller, H. (2010). *Nature* **18**.

Myers, S. (1999). Astronomy 12, http://www.aoc.nrao.edu.

Naife, A. (1984). *Introduction to the perturbation theory methods* (Mir, Moscow), (rus. trans.).

NASA/ESA (1990). Hubble telescope image, http://hubblesite.org.

Nelemans, G. (2003). ArXiv [astro-ph]:0310800v1.

Newton, I. (1687). *Philosophiae naturalis principia mathematica* (Joseph Streater for the Royal Society, London).

NOAO (1999). http://www.noao.edu/image_gallery/html/im0553.html.

Palagi+ (1993). http://vizier.u-strasbg.fr/cgi-bin/Vizier?-source=3DJ/A+AS/101/153.

Papadopoulos, D. (2002a). *A&A* **396**, p. 1045.

Papadopoulos, D. (2002b). *Class. Quantum Grav.* **19**, p. 2939.

Pavlov, D. and Garas'ko, G. (2007). *HNGP* **7**, p. 52.

Pimenov, R. (1987). *Anisotropic Finsler extensions of relativity theory as a structure of order* (Syktyvkar).

Poincaré, H. (1904). The Foundations of Science (The Value of Science) (Science Press, New York), pp. 297–320.

Postan, A., Rai, J. and Bowden, C. (1992). *Phys. Rev. A* **45**, p. 3294.

Pound, R. and Rebka, J. (1959). *Phys. Rev. Lett.* **3(9)**, pp. 439–441.

Raigorodski, L., Stavrinos, P. and Balan, V. (1999). *Introduction to the physical principles of differential geometry* (Univ. of Athens).

Randall, L. and Sundrum, R. (1999). *Phys. Rev. Lett.* **83**, p. 4690.

Rashevsky, P. K. (2003). *The geometrical theory of partial differential equations*, 2nd edn. (Editorial USSR, Moscow), (rus).

Rautian, S. and Sobelman, I. (1961). *Zh. Eksp. i Teor. Fiz.* **41**, p. 456, (rus).

Reiss, A. (1998). *Ap. J.* **116**, p. 1009.

Reiss, A. (2001). *Ap. J.* **560**, p. 49.

Reiss, A. (2004). *Ap. J.* **607**, p. 665.

Reshetnikov, V. (2003). *Priroda* **2**, (rus).

Richards, A. (2005). *Astron. Zh. Astrophys. Space Sci.* **295**, p. 19.

Roberts, M. and Whitehurst, R. (1975). *Ap. J.* **201**, p. 327.

Rowland, P. R. and Cohen, R. J. (1986). *MNRAS* **220**, p. 233R.

Rudolph, T., Freedhoff, H. and Ficek, Z. (1998). *Phys. Rev. A* **58**, pp. 1296–1309.

Ruggiero, M. and Tartaglia, A. (2002). ArXiv [gr-qc]:0207065v2.

Rund, H. (1981). *Differential geometry of Finsler spaces* (Nauka, Moscow), (rus).

Samodurov, V. (2006). *AAT* **25**, 5, p. 396.

Samodurov, V. and Logvinenko, S. (2001). *Astron. Zh.* **78**, p. 396.

Sandage, A. and Lubin, L. (2001). *Ap. J.* **121**.

Sardanashvili, G. A. (1996). *Modern methods of field theory. v.1, geometry and classical fields* (URSS, Moscow), (rus).

Sazhin, M. (1982). *Vestnik MGU, Phys. Astron.* **23**, pp. 45–48, (rus).

Schuda, F., Stroud, C. and Hercher, M. (1974). *J. Phys. B* **7**, p. L198.

Schwarzschild, K. (1916). *Sitzungsber. d. Berl. Akad.* **1**, p. 189.

Silagadze, Z. (2008). *Acta Phys. Pol.* **39**, p. 812.

Siparov, S. (1997). *Phys. Rev. A* **55**, p. 3704.

Siparov, S. (1998). *J. Phys. B* **31**, p. 415.

Siparov, S. (2001a). *Space-time and Substance* **2**, p. 44.

Siparov, S. (2001b). *J. Phys. B* **34**, p. 2881.

Siparov, S. (2002). *J. Tech. Phys.* **72**, pp. 125–128, (rus).

Siparov, S. (2004). *A&A* **416**, p. 815.

Siparov, S. (2005). Proc. PIRT-05, (Moscow), p. 349.

Siparov, S. (2006a). *HNGP* **6**, p. 155.

Siparov, S. (2006b). Proc. Int. Conf. PIRT-2006, (London), p. 155.

Siparov, S. (2007). *In Space-time structure, algebra and geometry* (Moscow).

Siparov, S. (2008a). *HNGP* **10**, p. 64, arXiv [gr-qc]:0809.1817v3.

Siparov, S. (2008b). *AMAPN* **24**, p. 135.

Siparov, S. (2009). *HNGP* **12**, p. 144, arXiv [gr-qc]:0910.3408.

Siparov, S. (2010a). BGS Proc. 17, p. 205.

Siparov, S. (2010b). Int. Conf. FERT-2010, (Moscow), http://www.polynumbers. ru/section.php?lang=ru\&genre=75.

Siparov, S. and Brinzei, N. (2008a). ArXiv [gr-qc]:0806.3066v1.

Siparov, S. and Brinzei, N. (2008b). *HNGP* **10**, p. 56.

Siparov, S. and Samodurov, V. (2009a). *Computer Optics* **33**, p. 79, (rus).

Siparov, S. and Samodurov, V. (2009b). ArXiv [astro-ph]:0904.1875.

Slysh, V. (2003). ASP Conf. Ser. p. 239.

Smith, S. (1987). *Phys. Rev. D* **36**, pp. 2901–2904.

Smyth, W. and Swain, S. (1996). *Phys. Rev. A* **53**, p. 2846.

Sofue, Y. and Rubin, V. (2001). *Ann. Rev. Astron. Astrophys.* **31**, p. 137.

Srevin, M., Brodin, G. and Marklund, M. (2001). *Phys. Rev. D* **64**, p. 024013.

Stenholm, S. (1978a). *Appl. Phys.* **16**, 2, pp. 159–166.

Stenholm, S. (1978b). *Phys. Rep.* **43**, p. 151.

Stenholm, S. (1984). *Foundations of laser spectroscopy* (Wiley, N.Y.).

Tamello, P. (1987). preprint NF-38, Phys. and Astron. Dep. of Estonian Academy of science.

Taylor, E. and Wheeler, J. (1992). *Space-time physics* (W.H. Freeman and Company, New York).

Taylor, J. and McCulloch, P. (1980). *Annals of the New York Academy of Sciences* **336**, p. 442.

Teissier, C. (1985). *Ann. Phys.* **10**, pp. 263–286.

The ATNF Pulsar Database (2005). http://www.atnf.csiro.au/research/pulsar/psrcat.

Thorne, K. (1987). *In Three hundred years of gravitation* (Cambridge University Press).

Tian, Y., Guo, H., Huang, C., Xu, Z. and Zhou, B. (2005). ArXiv [hep-th]:0411004v2.

Tomaschek, R. (1924). *Ann. Phys.* **73**, p. 125.

Trujillo, I., Carretero, C. and Patiri, S. (2006). *Ap. J.* **640**, p. L111.

Varshalovich, D. (1982). *In Astrophysics and cosmic physics* (Nauka, Moscow).

Verbunt, F. (1997). *Class. Quantum Grav.* **14**, p. 1417.

Verbunt, F. and Nelemans, G. (2001). *Class. Quantum Grav.* **18**, p. 4005.

Vershik, A. and Faddeev, L. (1975). *In Problems of theoretical physics* (LGU, Leningrad).

Voicu, N. (2010). AIP Conf. Proc. 1283 (Melville, New York), p. 249.

Voicu, N. and Siparov, S. (2008). *HNGP* **10**, p. 44.

Voicu, N. and Siparov, S. (2010). BGS Proc. 17, p. 250.

Wagoner, R., Will, C. and Paik, H. (1979). *Phys. Rev* **D19**, pp. 2325–2329.

Warner, B. and Woudt, P. (2002). *PASP* **114**, p. 129.

Weber, J. (1960). *Phys. Rev.* **117**, p. 306.

Weber, J. (1961). *General relativity and gravitational waves* (Wiley, New York).

Weber, J. (1994). *Nuov. Cim.* **109**, pp. 855–862.

Weisskopf, V. (1931). *Ann. Phys.(N.Y.)* **9**, p. 23.

Weyl, H. (1918). *Sitzungsber. Preuss. Akad. Wiss.* **465**.

Wineland, D. and Dehlmelt, H. (1975). *Bull. Am. Phys. Soc.* **20**, p. 637.

Wu, F., Ezekiel, S. and Ducloy, M. (1977). *Phys. Rev. Lett.* **38**, p. 1077.

Wu, Q., Gauthier, D. and Mossberg, T. (1994). *Phys. Rev. A* **49**, p. R1519.

Zasov, A. and Postnov, K. (2006). *General astrophysics* (Vek 2, Friazino).

Zel'dovich, Y. and Novikov, I. (1988). *Physics of the interstellar medium* (Nauka, Moscow).

Zhu, Y., Lezama, A., Wu, Q. and Mossberg, T. (1989). Coherence and quantum optics VI, (Plenum Press, New York), p. 1297.

Zimmermann, M. (1978). *Nature* **271**, p. 524.

Zwicky, F. (1937). *Ap. J.* **86**, p. 217.

Index

anisotropic space, 115, 116, 125, 127, 168
  current, 161, 169
  Lorentz force, 164
  proper potential, 122
  proper tensor, 123
asymptotic expansion, 212, 224

cosmological constant, 72, 73, 95, 99, 112, 155
curvature, 4, 41, 55, 61, 73

dark energy, 75, 83, 91, 155
dark matter, 68, 75, 81, 83, 97, 102, 103, 108, 147, 155
data processing, 213, 230
detector of gravitation waves
  local, 193
  remote, 194

geodesic, 104
geodesics, 49, 56, 70, 86, 106, 116, 117, 132
geometrodynamics, 7, 48, 178
geometry, 37, 80, 98, 109, 116, 118, 171, 182
  Euclid, 3, 37, 161, 182
  Finsler, 183
  Lobachavsky, 86
  Minkowski, 43, 161
  non-Euclidean, 3
  Riemann, 5, 43, 56, 80, 86, 115, 161

OMPR conditions, 212, 214, 227, 231, 238, 245, 247

perturbation theory, 198, 210

Rabi frequency, 189, 191, 206, 210, 217, 221

space maser, 195

SERIES ON KNOTS AND EVERYTHING

Editor-in-charge: Louis H. Kauffman *(Univ. of Illinois, Chicago)*

The Series on Knots and Everything: is a book series polarized around the theory of knots. Volume 1 in the series is Louis H Kauffman's Knots and Physics.

One purpose of this series is to continue the exploration of many of the themes indicated in Volume 1. These themes reach out beyond knot theory into physics, mathematics, logic, linguistics, philosophy, biology and practical experience. All of these outreaches have relations with knot theory when knot theory is regarded as a pivot or meeting place for apparently separate ideas. Knots act as such a pivotal place. We do not fully understand why this is so. The series represents stages in the exploration of this nexus.

Details of the titles in this series to date give a picture of the enterprise.

Published*:

Vol. 1: Knots and Physics (3rd Edition)
by L. H. Kauffman

Vol. 2: How Surfaces Intersect in Space — An Introduction to Topology (2nd Edition)
by J. S. Carter

Vol. 3: Quantum Topology
edited by L. H. Kauffman & R. A. Baadhio

Vol. 4: Gauge Fields, Knots and Gravity
by J. Baez & J. P. Muniain

Vol. 5: Gems, Computers and Attractors for 3-Manifolds
by S. Lins

Vol. 6: Knots and Applications
edited by L. H. Kauffman

Vol. 7: Random Knotting and Linking
edited by K. C. Millett & D. W. Sumners

Vol. 8: Symmetric Bends: How to Join Two Lengths of Cord
by R. E. Miles

Vol. 9: Combinatorial Physics
by T. Bastin & C. W. Kilmister

Vol. 10: Nonstandard Logics and Nonstandard Metrics in Physics
by W. M. Honig

Vol. 11: History and Science of Knots
edited by J. C. Turner & P. van de Griend

*The complete list of the published volumes in the series, can also be found at
http://www.worldscibooks.com/series/skae_series.shtml

Vol. 12: Relativistic Reality: A Modern View
edited by J. D. Edmonds, Jr.

Vol. 13: Entropic Spacetime Theory
by J. Armel

Vol. 14: Diamond — A Paradox Logic
by N. S. Hellerstein

Vol. 15: Lectures at KNOTS '96
by S. Suzuki

Vol. 16: Delta — A Paradox Logic
by N. S. Hellerstein

Vol. 17: Hypercomplex Iterations — Distance Estimation and Higher Dimensional Fractals
by Y. Dang, L. H. Kauffman & D. Sandin

Vol. 18: The Self-Evolving Cosmos: A Phenomenological Approach to Nature's
Unity-in-Diversity
by S. M. Rosen

Vol. 19: Ideal Knots
by A. Stasiak, V. Katritch & L. H. Kauffman

Vol. 20: The Mystery of Knots — Computer Programming for Knot Tabulation
by C. N. Aneziris

Vol. 21: LINKNOT: Knot Theory by Computer
by S. Jablan & R. Sazdanovic

Vol. 22: The Mathematics of Harmony — From Euclid to Contemporary Mathematics and
Computer Science
by A. Stakhov (assisted by S. Olsen)

Vol. 23: Diamond: A Paradox Logic (2nd Edition)
by N. S. Hellerstein

Vol. 24: Knots in HELLAS '98 — Proceedings of the International Conference on Knot
Theory and Its Ramifications
*edited by C. McA Gordon, V. F. R. Jones, L. Kauffman, S. Lambropoulou &
J. H. Przytycki*

Vol. 25: Connections — The Geometric Bridge between Art and Science (2nd Edition)
by J. Kappraff

Vol. 26: Functorial Knot Theory — Categories of Tangles, Coherence, Categorical
Deformations, and Topological Invariants
by David N. Yetter

Vol. 27: Bit-String Physics: A Finite and Discrete Approach to Natural Philosophy
by H. Pierre Noyes; edited by J. C. van den Berg

Vol. 28: Beyond Measure: A Guided Tour Through Nature, Myth, and Number
by J. Kappraff

Vol. 29: Quantum Invariants — A Study of Knots, 3-Manifolds, and Their Sets
by T. Ohtsuki

Vol. 30: Symmetry, Ornament and Modularity
by S. V. Jablan

Vol. 31: Mindsteps to the Cosmos
by G. S. Hawkins

Vol. 32: Algebraic Invariants of Links
by J. A. Hillman

Vol. 33: Energy of Knots and Conformal Geometry
by J. O'Hara

Vol. 34: Woods Hole Mathematics — Perspectives in Mathematics and Physics
edited by N. Tongring & R. C. Penner

Vol. 35: BIOS — A Study of Creation
by H. Sabelli

Vol. 36: Physical and Numerical Models in Knot Theory
edited by J. A. Calvo et al.

Vol. 37: Geometry, Language, and Strategy
by G. H. Thomas

Vol. 38: Current Developments in Mathematical Biology
edited by K. Mahdavi, R. Culshaw & J. Boucher

Vol. 39: Topological Library
Part 1: Cobordisms and Their Applications
edited by S. P. Novikov & I. A. Taimanov

Vol. 40: Intelligence of Low Dimensional Topology 2006
edited by J. Scott Carter et al.

Vol. 41: Zero to Infinity: The Fountations of Physics
by P. Rowlands

Vol. 42: The Origin of Discrete Particles
by T. Bastin & C. Kilmister

Vol. 43: The Holographic Anthropic Multiverse
by R. L. Amoroso & E. A. Ranscher

Vol. 44: Topological Library
Part 2: Characteristic Classes and Smooth Structures on Manifolds
edited by S. P. Novikov & I. A. Taimanov

Vol. 45: Orbiting the Moons of Pluto
Complex Solutions to the Einstein, Maxwell, Schrödinger and Dirac Equations
by E. A. Rauscher & R. L. Amoroso

Vol. 46: Introductory Lectures on Knot Theory
edited by L. H. Kauffman, S. Lambropoulou, S. Jablan & J. H. Przytycki

Vol. 47: Introduction to the Anisotropic Geometrodynamics
by S. Siparov